Carbon Based Electronic Devices

Carbon Based Electronic Devices

Special Issue Editors

Alberto Tagliaferro
Costas Charitidis

MDPI • Basel • Beijing • Wuhan • Barcelona • Belgrade

Special Issue Editors

Alberto Tagliaferro
Department of Applied Science and Technology
Italy

Costas Charitidis
Research Lab of Advanced, Composite,
Nanomaterials and Nanotechnology Materials
Science and Engineering Department School of
Chemical Engineering National Technical University
of Athens
Greece

Editorial Office
MDPI
St. Alban-Anlage 66
4052 Basel, Switzerland

This is a reprint of articles from the Special Issue published online in the open access journal *Micromachines* (ISSN 2072-666X) from 2018 to 2019 (available at: https://www.mdpi.com/journal/micromachines/special_issues/Carbon_based_Electronic_Devices).

For citation purposes, cite each article independently as indicated on the article page online and as indicated below:

LastName, A.A.; LastName, B.B.; LastName, C.C. Article Title. *Journal Name* **Year**, *Article Number*, Page Range.

ISBN 978-3-03928-232-6 (Pbk)
ISBN 978-3-03928-233-3 (PDF)

Contents

About the Special Issue Editors

Alberto Tagliaferro is an associate professor of Solid State Physics at Turin Politecnico (Italy). He is member of Editorial Boards of International Journals as well as Chair of the Education Committee of IUVSTA. He has published more than 180 papers on International Journals. His research interest has been mainly focused on carbon materials, both in thin films and nanostructured forms. He has developed a strong expertise on the Raman characterization of such materials as well as investigated several type of applications, ranging from composites to sensors. In most recent years he has focused on biomaterials as a feedstock for the production of carbons. The investigation of feedstock such as pruned tree branches and wasted coffee grounds has led to interesting results both for composites and sensors and paved the way for new interesting results in the framework of carbon sequestration.

Costas Charitidis is Professor in the School of Chemical Engineering of the National Technical University of Athens and Director of the Laboratory of Advanced, Composite, Nanomaterials & Nanotechnology. Since 2018 he is President of the General Assembly of the Hellenic Foundation for Research and Innovation. He has been elected in the Deanship of the School of Chemical Engineering since 2017, while from 2011 is Director of the Interdisciplinary Postgraduate (MSc) Program: Materials Science & Technology. He has more than 25 years of experience in Materials Science & Nanotechnology, Carbon-based materials and their Safety. He has extensive R&D experience through international collaborations since he has participated in more than 60 European and National funded projects; in many of them as Scientific Coordinator. He is the author of several scientific books and chapters and has more than 250 scientific publications in peer reviewed international journals and conference proceedings (h-index 34, citations 4000).

Editorial

Editorial for the Special Issue on Carbon Based Electronic Devices

Alberto Tagliaferro [1,*] and Costas Charitidis [2,*]

[1] Department of Applied Science and Technology, Politecnico Torino, Corso Ducadegli Abruzzi, 24, 10129 Torino TO, Italy
[2] Research Lab of Advanced, Composite, Nanomaterials and Nanotechnology, School of Chemical Engineering, National Technical University of Athens, 9 Heroon Polytechniou str., Zographou, Athens GR-15780, Greece
* Correspondence: alberto.tagliaferro@polito.it (A.T.); charitidis@chemeng.ntua.gr (C.C.)

Received: 27 November 2019; Accepted: 2 December 2019; Published: 6 December 2019

For more than 50 years, silicon has dominated the electronics industry. However, due to resources limitations, viable alternatives are considered and investigated. Among all alternative elements, carbon is the predominant element for a number of reasons; last but not least the fact that it can be obtained from waste. Whereas the physical properties of graphite and diamond have been investigated for many years, the potential for electronic applications of other allotropes of carbon (fullerenes, carbon nanotubes, carbon nanofibres, carbon films, carbon balls and beads, carbon fibres, etc), has only been appreciated relatively recently. Carbon-based materials offer a number of exciting possibilities for new applications of electronic devices, due to their unique thermal and electrical properties. However, the success of carbon-based electronics depends on the rapid progress of the fabrication, doping and manipulation techniques.

The present Special issue has a twofold structure: on one side review papers dealing with the most developed fields; on the other innovative research papers that report new exciting results.

A wide spectra of carbon materials and a wide range of applications are described in the present issue. As per material type, papers deal with graphene and graphene-oxide [1–4], carbon nanotubes [2,5,6] and with other forms of carbon, such as porous carbon [7] and nanofibers [8,9]. A plethora of devices are witnessing the versatility of carbon materials: supercapacitors [1,9], non-volatile memories [8], pressure sensors [2,7], field-effect transistors [10], white-light photosensors [3], cold cathode electron emitters [5], gas and humidity detectors [6,11], MEMS and NEMS [12], carbon based inks for 3D microfluidic MEMS [13], transparent conductive electrodes [4].

Dywily et al. [1] describe the production of nanometal decorated graphene oxide anchored on PANI and its performance in supercapacitors, achieving specific capacitance values up to 227.2 F/g; a value that favorably compares with other literature data involving graphene based systems. Bondavalli et al. [8] focus on the fabrication of Resistive Random Access Memory (ReRAM) on flexible substrates based on oxidized carbon nanofibres (CNFs) showing that two different resistance states (ON, OFF) reversibly switchable can be obtained. Caradonna et al. [2] discuss the use of various carbon nanofillers to promote piezoresistivity in polymers by means of laser scribing treatment able to produce conductive tracks in an otherwise low conductive material. Porous carbon electrodes and their interesting piezoresistive properties are discussed by Dai et al. [7]. An innovative (faster and cheaper) method to produce liquid-metal electrodes for graphene field-effect transistors is discussed by Melcher et al. [10]. The use of a new technique (active-screen plasma) to functionalize and decorate carbon nanofibres with metals for supercapacitor applications is presented by Li et al. [9]. A method aimed to overcome the limitation of standard approaches in preparing graphene-based photosensors is discussed and detailed by Tu et al. [3]. Kim et al. [5] focused their work on the fabrication of stable CNT cold emitter using an aging technique. The last research paper is presented by Song et al. [6] and focuses on the ability of a two-nanotube sensor to selectively detect NO and NO_2.

The remaining papers of the issue are reviews aimed to provide to the reader an overview of several fields of interest, ranging from NEMS and MEMS [12] to functionalized carbon materials for electronic devices [14], from carbon-based humidity sensors [11] to graphene-based transparent conductive electrodes [4]. Finally, O'Mahony et al. [13] reviewed the rheological issues to be tackled in addictive manufacturing when using carbon-based inks for lab-on-a-chip applications.

Conflicts of Interest: The authors declare no conflict of interest.

References

1. Dywili, N.R.; Ntziouni, A.; Ikpo, C.; Ndipingwi, M.; Hlongwa, N.W.; Yonkeu, A.L.D.; Masikini, M.; Kordatos, K.; Iwuoha, E.I. Graphene Oxide Decorated Nanometal-Poly(Anilino-Dodecylbenzene Sulfonic Acid) for Application in High Performance Supercapacitors. *Micromachines* **2019**, *10*, 115. [CrossRef] [PubMed]
2. Caradonna, A.; Badini, C.; Padovano, E.; Veca, A.; De Meo, E.; Pietroluongo, M. Laser Treatments for Improving Electrical Conductivity and Piezoresistive Behavior of Polymer–Carbon Nanofiller Composites. *Micromachines* **2019**, *10*, 63. [CrossRef] [PubMed]
3. Tu, W.-C.; Liu, X.-S.; Chen, S.-L.; Lin, M.-Y.; Uen, W.-Y.; Chen, Y.-C.; Chao, Y.-C. White-Light Photosensors Based on Ag Nanoparticle-Reduced Graphene Oxide Hybrid Materials. *Micromachines* **2018**, *9*, 655. [CrossRef] [PubMed]
4. Woo, Y.S. Transparent Conductive Electrodes Based on Graphene-Related Materials. *Micromachines* **2019**, *10*, 13. [CrossRef] [PubMed]
5. Kim, J.H.; Kang, J.S.; Park, K.C. Fabrication of Stable Carbon Nanotube Cold Cathode Electron Emitters with Post-Growth Electrical Aging. *Micromachines* **2018**, *9*, 648. [CrossRef] [PubMed]
6. Song, H.; Li, K.; Wang, C. Selective Detection of NO and NO_2 with CNTs-Based Ionization Sensor Array. *Micromachines* **2018**, *9*, 354. [CrossRef] [PubMed]
7. Dai, K.; Wang, X.; You, Z.; Zhang, H. Pressure Sensitivity Enhancement of Porous Carbon Electrode and Its Application in Self-Powered Mechanical Sensors. *Micromachines* **2019**, *10*, 58. [CrossRef] [PubMed]
8. Bondavalli, P.; Martin, M.B.; Hamidouche, L.; Montanaro, A.; Trompeta, A.-F.; Charitidis, C.A. Nano-Graphitic based Non-Volatile Memories Fabricated by the Dynamic Spray-Gun Deposition Method. *Micromachines* **2019**, *10*, 95. [CrossRef] [PubMed]
9. Li, Z.; Qi, S.; Liang, Y.; Zhang, Z.; Li, X.; Dong, H. Plasma Surface Functionalization of Carbon Nanofibres with Silver, Palladium and Platinum Nanoparticles for Cost-Effective and High-Performance Supercapacitors. *Micromachines* **2019**, *10*, 2. [CrossRef] [PubMed]
10. Melcher, J.L.; Elassy, K.S.; Ordonez, R.C.; Hayashi, C.; Ohta, A.T.; Garmire, D. Spray-On Liquid-Metal Electrodes for Graphene Field-Effect Transistors. *Micromachines* **2019**, *10*, 54. [CrossRef] [PubMed]
11. Tulliani, J.-M.; Inserra, B.; Ziegler, D. Carbon-Based Materials for Humidity Sensing: A Short Review. *Micromachines* **2019**, *10*, 232. [CrossRef] [PubMed]
12. Neuville, S. Selective Carbon Material Engineering for Improved MEMS and NEMS. *Micromachines* **2019**, *10*, 539. [CrossRef] [PubMed]
13. O' Mahony, C.; Haq, E.U.; Silien, C.; Tofail, S.A.M. Rheological Issues in Carbon-Based Inks for Additive Manufacturing. *Micromachines* **2019**, *10*, 99. [CrossRef] [PubMed]
14. Kamran, U.; Heo, Y.-J.; Lee, J.W.; Park, S.-J. Functionalized Carbon Materials for Electronic Devices: A Review. *Micromachines* **2019**, *10*, 234. [CrossRef] [PubMed]

Review

Selective Carbon Material Engineering for Improved MEMS and NEMS

Stephane Neuville

Independent Consultant, F-77165 Cuisy, France; Steph.neuville709@orange.fr

Received: 3 June 2019; Accepted: 6 August 2019; Published: 16 August 2019

Abstract: The development of micro and nano electromechanical systems and achievement of higher performances with increased quality and life time is confronted to searching and mastering of material with superior properties and quality. Those can affect many aspects of the MEMS, NEMS and MOMS design including geometric tolerances and reproducibility of many specific solid-state structures and properties. Among those: Mechanical, adhesion, thermal and chemical stability, electrical and heat conductance, optical, optoelectronic and semiconducting properties, porosity, bulk and surface properties. They can be affected by different kinds of phase transformations and degrading, which greatly depends on the conditions of use and the way the materials have been selected, elaborated, modified and assembled. Distribution of these properties cover several orders of magnitude and depend on the design, actually achieved structure, type and number of defects. It is then essential to be well aware about all these, and to distinguish and characterize all features that are able to affect the results. For this achievement, we point out and discuss the necessity to take into account several recently revisited fundamentals on carbon atomic rearrangement and revised carbon Raman spectroscopy characterizing in addition to several other aspects we will briefly describe. Correctly selected and implemented, these carbon materials can then open new routes for many new and more performing microsystems including improved energy generation, storage and conversion, 2D superconductivity, light switches, light pipes and quantum devices and with new improved sensor and mechanical functions and biomedical applications.

Keywords: carbon-based material; carbon structure differentiation; NEMS quality; higher performances; revised Raman characterization; quantum electronic activation; carbon phase transition

---

1. Introduction

Although, micro, nano electromechanical and optomechanical systems are still often confronted to the lack of quality and longer life time and to the search of extended higher performances [1], huge progress has been recently achieved in MEMS and NEMS technology in using more performing carbon-based materials, which are presenting a large panel of various superior properties concerning their mechanical properties, such as young modulus, Poisson's ratio, fracture strength of nanocrystalline diamond for instance [2], tribological, electric, semicon, piezoelectric, heat conducting and optical/optoelectrical properties [3,4], diamond micro and nano resonators [5,6], piezo-resistivity obtained with carbon nanotubes [7], diamond-like carbon MEMS sensors [8] and many more which are making use of functionalized graphenic and related materials [9–20].

Key hurdles currently preventing the commercial application of many NEMS devices include low-yields and high device quality variability, concerning structure, physical and chemical stability, nucleation, adhesion, different kinds of internal and interface stress, tribology and wear rates, contamination diffusion barrier properties, stability and reproducibility of surface functionalization [16–19]. Before NEMS devices can actually be industrially implemented, reasonable integrations of carbon-based products must be created [21]. Next, the challenge to overcome is understanding carbon-based materials properties, with which efficient and durable NEMS with low failure rates can be achieved [22,23].

Basically, those can often be improved in using selectively different kinds of carbon materials including the different sorts of diamond, tetrahedral carbon, DLC, GLC, glassy carbon, nanowires and nanotubes and graphenic materials of specific properties [9] and we discuss that in more details in this study. Those have intrinsic properties, which are distributed over several order of magnitudes and have to be well distinguished from each other [2,3,12,13,21].

Limiting factors of their implementation are generally a consequence of degraded structure and defects and possible induced phase transformation which are depending on their elaborating device and application environment [23,24]. It then appears necessary to also have all sorts of appropriate characterizing devices, which have to be used for the different process optimization steps. Further on it is then also to be considered well understood the aspects of surface preparation and processing, nucleation and growth mechanisms and especially the phase transformation they can be subject and which have been generally neglected and omitted to be considered although appearing of greatest importance [12,13,25–27].

Necessary mastering of differentiate carbon material depositing and characterizing can be achieved with recently revisited fundamentals on carbon atomic rearrangement [25] and carbon Raman spectroscopy [26]. In addition to several other revisited subjects, we briefly recall concerning diamond-like carbon coatings [27], energy storage and conversion using different kinds of carbon-based materials [28], superconductivity (because of some analogy with electron ballistic properties in graphenic materials [29], for which carbon materials need to be correctly selected and associated. This appears all the more important to be achieved, considering those, can open new routes for improved microsystems and 2D devices concerning mechanical functions, light switches, light pipes and quantum calculation devices [30].

2. Brief Review on Main MEMS and NEMS Characteristics

2.1. Early Stage of Microelectromechanical Systems (MEMS)

MEMS is the technology of microscopic devices, particularly those with moving parts smaller than a human hair with outstanding accelerating practical application interest during the last decades and which have rapidly achieved actuator dimensions in the 100 nm range, before becoming much smaller with nano-technologic means. They are now used for many applications, and a future form of NEMS is expected to exceed the IC industry in both size and impact on society [1].

These devices replace bulky actuators and sensors with a micron scale equivalent that can be produced in large quantities by the fabrication process used in integrated circuits. They reduce cost, bulk, weight and power consumption while increasing performance, production volume and functionality by orders of magnitude [21,23].

Those are concerning multidisciplinary fields in the areas of engineering, chemistry, material science, physics, and any specialized field for applications in bioengineering or medicine. Their future holds revolutionary breakthroughs in a wide range of: Nanocircuits, actuators (piezo-electrostatic, shape memory, electromagnetic), sensors, radar, locators, materials, imaging, nanocontact and nano-relays [31–41], micromotors and micropumps [42–49], optical functions and optoelectronic devices, such as switches, integrated energy harvesting, wave length filtering, optical grating switch and optical sensors—these MEMS are called micro optical electromechanical systems (MOMS) [50–52], energy storage [28,53], nanocapacitors, data storage and nano-computers [1,30,54–56] including 2D superconducting devices to be used for quantum computers [29], and such as, bio-functionalizing, etc. [1,11,16,57,58].

2.2. MEMS Fab

MEMS became practical once they could be fabricated using modified semiconductor device fabrication technologies, normally used to make electronics. They usually consist of a central unit that processes data (the microprocessor) and several components that interact with the surroundings such as microsensors [1,21,30]. The fabrication of MEMS evolved from the process technology in semiconductor device fabrication, i.e., the basic techniques are the deposition of material layers, patterning by photolithography and etching to produce the required shapes and to which different types of bulk and surface micromachining of different materials can be associated [59–61].

The addition of specific material properties with the availability of inexpensive high-quality materials, and ability to incorporate electronic functionality make silicon attractive for a wide variety of MEMS applications. However, considering the large surface area to volume ratio of MEMS, and complex interface mechanical, chemical, electrical and electro-magnetic phenomena, their design requests particular attention and several other materials have been also considered for MEMS manufacturing, such as polymers, metals, and ceramics before considering advanced carbon materials [1,4–16] for many specific reasons on which we focus on in Sections 3–5.

2.3. NEMS

2.3.1. Definition and General Features

MEMS technology evolves into smaller nanoelectromechanical systems (NEMS) and nanotechnology [59–62]. These devices replace bulky actuators and sensors with a micron scale equivalent that can be produced in large quantities by the fabrication process used in integrated circuits in photolithography. They reduce cost, bulk, weight and power consumption while increasing efficiency, performance, production volume and functionality by orders of magnitude. They achieve a reduced size down to 10 nm range and less considering the size of some of their subsystems [63,64]. Miniaturization of MEMS fab could be achieved with two complementary approaches. (A) Top-down approach uses the traditional microfabrication methods. While being limited by the resolution of these methods, it allows a large degree of control over the resulting structures such as nanowires, nanorods, and patterned nanostructures are fabricated from metallic thin films or etched semiconductor layers [62]. (B) Bottom-up approaches, in contrast, use the chemical properties of single molecules to cause single-molecule components to self-organize or self-assemble into some useful conformation, or rely on positional assembly and allows fabrication of much smaller structures [63]. Those can be made at the VLSI scale, possibly co-integrated with CMOS well suited for autonomous, highly sensitive or dense sensors. They include complex gas portable recognition systems, mass spectrometry, or bio-sensors [11–14] and open several opportunities for integrated solutions in emerging domains as chemical analysis and life science [16,21–24,30,57,58].

2.3.2. Early NEMS Application Fundamentals

Nanoscale mechanical sensors offer a greatly enhanced performance that is unattainable with microscale devices [59–67]: Ultrasensitive sensors [68], high quality factor diamond resonators [69], high mass and spatial resolution basing on mechanical resonance and cantilever vibration up to high frequencies [70], which is enabling chemisorption measurements in air at room temperature, with very high mass resolution below 1 atto-gram (10–18 g) [71]. Further on, high-sensitive liquid/airflow meter could be produced [72], nano ph sensor [73] and nano-molecular machines [74–80].

A molecular machine is a group of molecular components that are able to produce quasi-mechanical movements when exposed to specific stimuli.

There are three categories of the molecular machines, namely natural or biological, synthetic, and natural-synthetic hybrid machines [81–83]. Synthetic molecular machine includes motors, propellers, switches, shuttles, tweezers, sensors and logic gates. Biological motors convert chemical energy into linear or rotary motion as well as controlling many biological functions. Examples of linear motions:

Proteins, muscle contraction, intracellular transport, signal transduction, ATP synthesis, membrane translocation proteins and the flagella motor. Natural-synthetic hybrid systems are mechanical motors inspired from DNA duplication and partition [83–88].

The NEMS technology is distinguished from molecular nanotechnology or molecular electronics in that the latter must also consider surface chemistry and solid-state phonon and electronic quantum mechanical aspects which can affect mechanical, electric and optoelectronic properties, friction and that can cause high signal/noise ratio ([89–92]. Mechanical deformation and electrical contact properties and adhesion between carbon nanotubes are important aspects of their quality and dynamic performances [93–98] and explaining why they must be selected and controlled upon their characteristics, size and defect content.

In spite of already numerous probated applications many potential others corresponding to bench lab prototypes could not yet be sufficiently mastered on their quality, reliability and life time in consequence of insufficiently understood fundamentals [21–23]. Among them, those concerning stability and material structure modification being induced by local quantum electronic effects which used to be ignored up to recent past [25,99] and another important quantum mechanical revisited aspect concerning the characterization of carbon material: The recently revised Raman fundamentals with which carbon material structure and their defects can be better and more correctly sorted out [26] and we recall and discuss this in more details next in Sections 3–5.

Considering selective electronic activation of semiconducting surface material caused by adsorption and chemisorption and with which transversal polarization effects can appear [100], much sensitive and selective physicochemical interactions can be considered between nanoparticles and the biologic material. This is illustrated with size dependence of Au and Ag particle on different biological metabolism [101,102]. Different corresponding nano-effects are considered for nanomedicine and dentistry applications [103,104] and are also a subject of toxicologic investigations [105–107].

3. Increased NEMS Performances with Advanced Carbon Material

3.1. Progress in NEMS Technologies

They are less mature than that of MEMS due to the difficulty to reliably couple the micro-actuators to the macroscopic world and to achieve requested quality and performances, specially concerning longer life time, higher strength, better adhesion and tribology and better mechanical and chemical stability and better reproducible size, bulk and surface optoelectronic effects and heat/electric conducting properties [21–23]. However, owing to the use of different carbon-based materials with corresponding superior solid-state bulk and surface properties and which have been synthetized with more or less empirical means [3,10], the fast-growing number of MEMS/NEMS applications could be achieved. For instance, the scanning tunneling microscopes (STMs), inertial, pressure, thermal, optical, flow, capacitive position sensors, biochips for detection of hazardous chemical and biological agents, high-throughput drug screening and selection, optical switches, valves, RF switches, micro-relays, electronic noses, etc. [1,44,45,61–65,74,108–110].

3.2. Diamond and Related Materials

3.2.1. Different Categories of Diamond and Diamond-Like Materials

Diamond materials offer great potential for electronic and biomedical application. With very high stiffness, high thermal conductivity, optical transparency range, chemical stability and wear resistance for the diamond-based materials extend their applicability for MEMS/NEMS [24]. Besides, these diamond materials which are nevertheless presenting a large panel of different structures and properties, and which have to be distinguished from each other's [3–68], many other different kinds of diamond-like materials have either to be considered and distinguished from each other's in so far they often present an underestimated wide range of specific combined properties. Those are depending on specific composite structure, defect, contamination and atomic disorder including physical and chemical, optical and optoelectronic properties, thermal, mechanical, chemical, tribological, and wear resistance, internal mechanical stress, electric properties and their possibility to be doped. Depending also, on their surface micro and nano rugosity, porosity and surface chemisorbing and adsorbing properties, and the way they can be produced [27,111–113]. Different categories of diamond and diamond-like carbon (DLC) materials have to be considered:

(a) *Polycrystalline diamond of different crystallite size*, including the hexagonal and epitaxial diamond. To be observed that the denser and smoother micro- and nano-crystalline diamond is almost containing a significant part of graphitic material where more or less ordered/disordered diamond crystallites are imbedded. However, besides interesting tribological properties, those have generally reduced others (optical, optoelectronic, chemical and mechanical) [114–119].
(b) *Amorphous diamond and degraded tetrahedral amorphous carbon*, ranging from materials with high internal stress such as the amorphous diamond and ta-C:H and others for which the internal stress could be reduced without graphitic degradation [120–126].
(c) *Polymeric carbon* [127,128], although not being really diamond-like, but which contain many sp3 carbon chains and have a similar optoelectronic gap.
(d) *Diamond-like a-C:H and composite material* such as diamond-like glassy-carbon CNx and sp3 rich carbon nanowire and fiber [24,27,129–133].

3.2.2. Upholding of Combined Properties

It must be emphasized that all these diamond and diamond-like materials have much different combined properties, concerning their Csp2 and Csp3 content, their thermal and chemical stability, and their simultaneous mechanical, tribological and optoelectronic properties. This is especially concerning the mechanical elastic/plastic/hardness properties which are obtained after an annealing process reducing the internal stress [3]. The resulting material will generally no longer combine higher diffusion barrier, optical and electric and/or dielectric properties. The harder homogeneous isotropic amorphous diamond combines many different superior interesting properties, except its electric conductivity and its internal stress which can only be annealed without degrading the initial properties with particular appropriate means (e.g., hard UV laser annealing) and when the temperature stays lower than the thermal graphitization temperature threshold [120–124].

When it is degraded to a more graphitic material, the amorphous diamond and harder ta-C will generally lose at least at nanoscale their amorphous homogeneity with the formed sp2 clusters (islands of sp2 material within the amorphous sp3 material). Then, they merely correspond to some nanocomposite material containing different adjacent phases and this is often leading to dramatic confusions between materials supposed to belong to the same category, whether being stress annealed or not and presenting important material structures and property differences [27,111–113]. Diamond-like carbon produced with non-optimized depositing equipment design and lower optimized depositing process or which is resulting from the thermal annealing of crystalline diamond and from highly stresses harder ta-C has been often considered as an amorphous diamond, although being merely less hard, less homogeneous less smooth, less dense composite material with reduced optical properties and reduced thermal stability (in comparison to the diamond and harder tetrahedral amorphous carbon) [114–125].

To be observed that the temperature induced exodiffusion of hydrogen and other gaseous species can induce tensile stress and cracks [27]. This effect can be amplified when a graphitic material is transformed into a denser diamond-like material. These sorts of materials can be obtained with other means than combined pressure/heat and thermal spikes [111–124]. In such a case, contradicting effects have to be considered (either favorizing graphitization, or favorizing diamond-like structures) and which can be in competition to each other [25]. Those are suggested as explaining, for instance, why glassy carbon is not always showing graphenic properties, but can be harder and more diamond-like [132,133]. Similar effects have also to be considered for the growth of carbon nano-fibers of irregular cylindrical structure [134–136] and which in contrast to CNT are filled tubes containing also significant amounts of Csp3. For their accurate characterization, anticipating on the next chapter IV, revision of different characterizing such as Raman spectroscopy and interpretation appeared to be necessary and could helpfully be achieved [26].

3.3. CNT and Graphene

3.3.1. Definition and Technologic Trends

CNT corresponds to scrolled graphene sheets which are composed of juxtaposed hexagonal cyclic sp2 hybridization carbon atoms [13]. Since discovery in 1991 [3,137], carbon nanotubes (CNTs) have aroused a high amount of interest in their use as building blocks for many new applications based on outstanding mechanical properties [12] and which can be combined with their other interesting chemical and physical solid-state properties especially for various electrical, and opto-electronical devices and future integrated circuits due to their outstanding electrical, mechanical, thermal, opto-electric and solid-state combined surface properties [138,139].

Improved performances achieved with several of their specific properties, are well illustrated with the development of carbon-based material field emissions beginning with a-C:H, improving with doped diamond and ta-C and achieving much more performing results with CNT, which are used for atomic force, scanning tunneling (STM) and magnetic force microscopy [140–145]. They could improve performances of catalytic nanomotors [146] and many new applications could be developed after controlling synthesis conditions, size (diameter/length) and structure (chirality, semi-conducting/metallic properties, single or multi-walled) [147,148]. However, many difficulties appeared for industrial application owing to contamination, defect and various still often little understood effects yet [63,64] (such as rippling, and phase transformation) for which we propose some clarification next.

3.3.2. Early Fundamentals on Graphene and CNT

Solid-state quantum mechanical calculations could predict many electronic level and phonon modes distribution for a well-defined carbon material bulk structure, which elementary optoelectronic properties and Raman frequency could be determined especially for diamond, graphite and graphenic materials [117,149–151]. However, more complex often not well understood Raman spectra have been experimentally observed for thin film carbon materials containing different phases (carbon composite materials), which have been mainly used as spectroscopic fingerprints of corresponding materials. Those could be first used for the empiric observation of their structure modification and for which refined aspects give more precise information [113,137–139].

Considering some quantum electronic confinement effects for nano particles, very fine opto-electronic structure modifications are observed (Figure 1). This is especially the case for one and two-dimensional materials, which will be very sensitive to physical adsorbtion and chemisorption and to any bulk and surface structure modification [152–158].

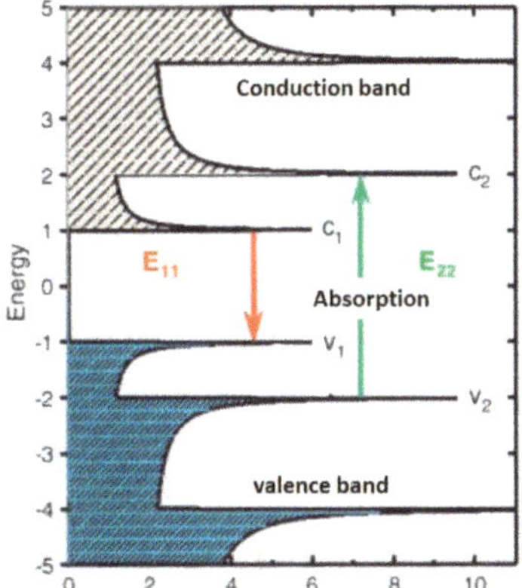

Figure 1. Scheme of optoelectronic band structure in single walled carbon nanotube by Bachilo et al. [158] reproduced with permission of the Journal of Science.

They can be functionalized and will offer, for instance, selective IR optical response to molecular adsorbtion [159,160], which can be used, for instance, for selective photosensitive intrinsic polarimetry [161]. Additionally, improved electric conductivity can be obtained [162–164] with incorporation of iodine or copper, for instance, and also providing possible conduction anisotropy of expected interest for electric energy storage and antistatic encapsulating [165,166]. Very low resistivity is achieved with the electron ballistic transport effect, similar to superconductivity [167,168] being actually well understood in using revisited description of super-conductivity with electron/phonon synchronic gating during which no electron/electron and electron/phonon scattering occur [29].

All these results will be also very sensitive to radiation damage [169] material contamination and defect content [149–151,170–172] (whenever with questionable conclusions we will discuss in next chapter) and which in addition, are also depending on specific material structure edge configuration and possible thermally induced modification [153–155] (as discussed in more details in Sections 5 and 6). However, interpretation and characterizing anomalies could be identified with the refined Raman theory [26], with which several observed contradicting aspects could be sorted out. Important new aspects of carbon phase transformation could especially be identified [25] with competing opposite effects that we believe can broadly explain the MEMS devices quality and performance distribution, which need to be comprehensively mastered and we briefly review next.

4. Brief Review of Quantum Activated Atomic Rearrangement

4.1. Atomic Rearrangement during Synthesis of Carbon Materials

4.1.1. Graphitic Thermal Degradation and Diamond-Like Material Reforming

Graphite being the ground state of carbon, any carbon material can be degraded to more graphitic material by thermal effects ruled by the Arrhenius law and with low energy thermal activation. This effect is observed during longer thermal annealing of diamond, DLC, a-C:H and glassy carbon [3,15,27,111–117,129–131,173] below their sublimation temperature (~3500 °K) [174] (corresponding to ~0.3 eV electronic activation).

However, it is also known, that higher quantum electronic activation produced by different means and especially with photonic activation which can strongly influence chemical synthesis [175]. This is also the case during the formation of diamond and diamond-like material, which are not only formed with temperature/pressure and ionic thermal spikes producing compressive stress [111,112,124] but also with many other effects and without energetic ions (flame depositing, hot filament, high flux low ion energy, plasma jet, UV, X rays, electric neutralization energy, catalytic effects) [25–27,119,122,176–180] and especially the chemical recombination energy release (CRER) of C–C (~7 eV) H_2 (~5 eV) and N_2 (~12 eV) [25,121,181], which are exciting electrons with relatively high energies (>1 eV up to more than 10 eV compared to thermal activation in the ~0.1 eV range) before producing phonons and heat, which can produce a noticeable higher quantity of rearranged sp3 material structure.

With Figure 2 we reproduce the Raman spectra from an N+ ion irradiated glassy carbon, which is transformed into a diamond like ta-C. To be observed that the annealing of CNx with a lower content of dissociated N will not produce the sp3 rearrangement effect because of insufficient N2 CRER activation events [182,183].

Updated Raman spectroscopy [26] shows that in many published thermal processing of carbon-based material supposed to produce graphitic material, in fact, resulting materials are also containing DLC, amorphous or nanocrystalline diamond [129–131,176,183,184] (what can be verified with measured sp2/sp3 obtained with electron energy loss spectroscopy and other more accurate, easy and fast XPS /Auger characterizing method, with which the measurement of the intensity of the Auger peaks of an XPS spectrum corresponding to Csp2 and Csp3 does not need any hazardous peak fitting assumptions and no deconvolution) [27,120–123]. In such a case appearance of isolated diamond crystallites can correspond to increased abrasive friction. The diamond like phase transformation is also well documented, with a brief high temperature annealing of polymeric material that can be transformed into a diamond-like glassy carbon [132,133], meanwhile a longer annealing at higher temperature will produce a graphenic glassy carbon [133].

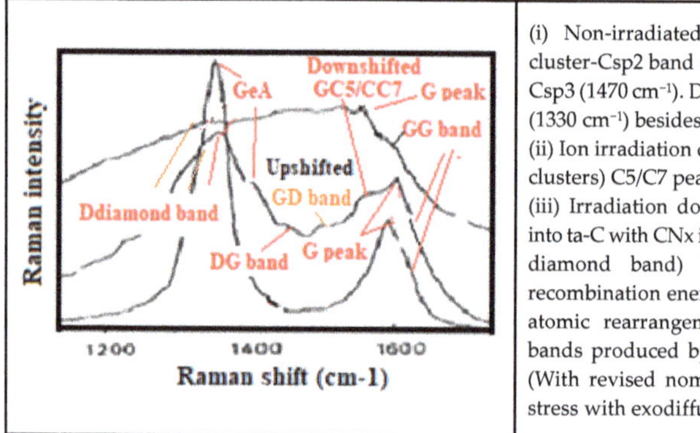

<table>
<tr><td>

(i) Non-irradiated glassy carbon. Aliphatic Csp2 cluster-Csp2 band (1620 cm^{-1}). Aliphatic Csp2 cluster-Csp3 (1470 cm^{-1}). Disordered diamond side band (Dd) (1330 cm^{-1}) besides the Aedge band (1350 cm^{-1}).
(ii) Ion irradiation dose 10^{15} (partial destruction of sp2 clusters) C5/C7 peaks band (GC5(1500 cm^{-1}).
(iii) Irradiation dose >10^{16} transforms glassy carbon into ta-C with CNx inclusions (increased Dd amorphous diamond band) by ion neutralization and N_2 recombination energy release are able to activate sp3 atomic rearrangements. Note: Existence of C5/C7 bands produced by the high energy ion irradiation. (With revised nomenclature of Table 1) and tensile stress with exodiffusion of N_2.

</td></tr>
</table>

Figure 2. Glassy carbon 25 KeV N+ ion irradiated with additional annotations from M. Iwaki et al. [182] with permission of the journal of J. Mater. Res.

4.1.2. Surface Polarization and Diamond-Like Atomic Rearrangement

It has been evidenced that adsorbtion energy can be directly transferred to the valence band electrons of the substrate, which after excitation are decaying with possible emission of corresponding photons of the same energy [185]. Thus, the adsorption of atoms and molecules on semiconducting material can produce electron/hole pairs and can produce a transient electric field (consequence of different mobility of electrons and holes being activated by released adsorbtion energy and which are then decaying [100]). In addition, it has been shown how an electric field transverse to a graphitic carbon thin film can enhance its sp3 content [186].

Above some electric polarization threshold energy, the carbon material can be transformed in a metastable material of higher density of cohesion energy. The hexagonal planar sp2 ring can be transformed in a buckled hexagonal chair plane (Figure 3) [25] (Similar to Silicene obtained with Si epitaxy on graphene [187], which is suggested to correspond to the caned structure of some carbon nanowires with irregular diameter [188].

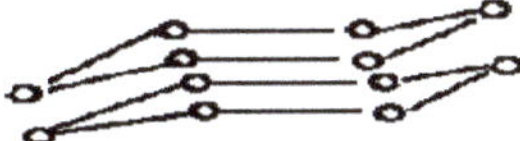

Figure 3. Transformation of graphene into H6 diamond with electric field polarization perpendicular to the graphene plane [25]. H6 diamond.: Stack of chair shape carbon hexagonal rings (sp3 bonded to four others). Raman frequency at~1325 cm^{-1} according to McNamara et al. [116] and Spear et al. [189] and Phelps et al. [190].

Most carbon material types have more or less a semiconducting electronic band structure and metallic CNT are in fact semimetal materials with valence band and conduction band [27,134,149–151,191]. Therefore, atomic rearrangement in consequence of more energetic electronic activation and formation of internal electric field with different decay time of activated electrons and holes (higher energy quantum electronic activation) can appear in many solid-state systems.

4.1.3. Criterion of Quantum Electronic Sp3 Activation

During an atomic rearrangement (Figure 4), the activated electron must always occupy authorized energy levels. Achievement of ordered atomic rearrangement can only happen with fulfilled conditions [25,26]:

(1) Valence band electrons must be excited up to the conduction band of the initial and final state.
(2) More electrons must be activated than atoms to be rearranged.
(3) Atomic rearrangement can only be achieved with its local (proximity) activated electrons.
(4) The kinetic and density of activation events must be compatible with the decay and diffusion kinetic.

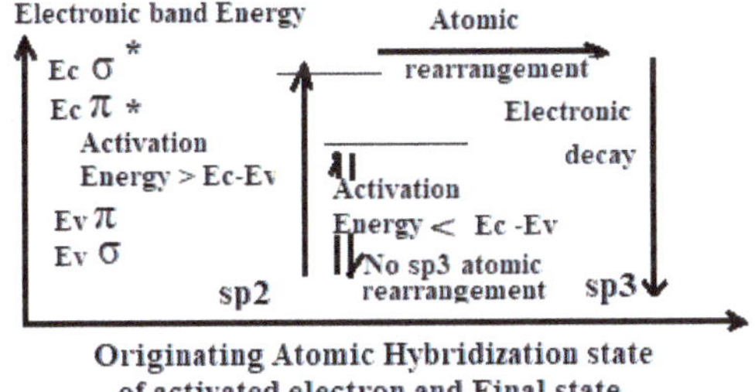

Figure 4. Principle of quantum electronic activated atomic rearrangement by S. Neuville [25].

4.1.4. Ordered Atomic Rearrangement and Stress Modification

Stress reduction is a major concern, in so far that internal stress can strongly affect adhesion and bulk diffusion [27]. Longer thermal annealing at a higher temperature produces graphitic degradation and stress relaxation, meanwhile a shorter thermal treatment of DLC, can reduce the stress without degrading the material as long as the limit of thermal stability has not been overpassed and the activation energy is not higher than the phase transition barrier potential [25,27,111,133]. This thermal stability depends on the structure of the carbon material (~900 °C) for polycrystalline diamond [117], ~600 °C/700 °C for hard ta-C [121–125] and less than 200 °C during the ta-C growth, considering that the flux of energetic ions bombarding the growing film surface during an arc deposition generates additional heating [111] and local transient higher temperature than on the substrate holder.

At a higher temperature than the thermal stability limit, the material will be degraded into more graphitic material, unless a higher electronic activation is able to counterbalance the graphitization process and which can then enhance the diamond-like character of the considered material. For instance, during hard UV irradiation [122,176] and with chemical recombination energy release (CRER) of H in a growing diamond film [25,192,193] and during glassy carbon synthesis, when the thermal process time did not exceed the time at which exo-diffusion of all hydrogen content has been achieved and the material has not yet be fully degraded to a more graphitic material [132,133].

With the exodiffusion of recombined H_2 (or other gaseous material), tensile stress can appear [27,130]. However, uniformly distributed sp3 atomic rearrangement is generally forming higher packing density and stress relaxation, when no other effects such as thermal dilatation and/or ion peening is hampering this stress reduction process. This is typically achieved with catalytic sp3 rearrangement [194,195]. Catalytic effects produced with boron atoms can reduce C–H binding energy and for which consequently the H_2 and C–C recombination energy release will be higher [25]. This explains why the added Boron in a high temperature diamond deposition process (where thermal dilatation can compensate the reduction of packing density) can produce better diamond quality in addition to the doping effect providing better electric conductivity [3,114,125,195].

4.1.5. Examples of Diamond Like Atomic Rearrangement of Graphenic Material

Several other illustrative examples can be cited for the diamond-like atomic rearrangement:

CNx Annealing with Formation of Diamond Crystallites

During the annealing of some CNx materials where larger amounts of N_2 CRER can exist, many nano-diamond crystallites appear [181]. This is only achieved when many dissociated nitrogen atoms exist, otherwise only few N_2 recombination events will be produced and at a higher annealing temperature the material will be degraded into some more graphitic material [183,184].

Graphene Transformation into a Dielectric Material

Adsorbed hydrogen on graphene has been reported to convert highly electric conducting graphene into a dielectric buckled 2D planar graphane [196,197]. However, the existence of graphane is questioned with different arguments. The so-called No-Go criteria of the Landau phase transition theory [198] predicts that graphane cannot exist for thermodynamic reasons. This effect appears to be very important to be accurately sorted out for many CNT/graphene NEMS applications.

On a graphene surface, C–H bonds are expected to be formed only on external graphene flake edges or on internal edges of voids and vacancies where dangling bonds exist, considering that the C–H bond energies (between 4 and 4.5 eV) is lower than the H_2 binding energy (~5 eV) and much lower than the graphene or diamond C–C bond energy (~7 eV). Whenever the existence and stability of graphane was predicted by theoretic calculation [199], the question will be how this material can be formed, considering the low adsorption energy of the H_2 molecule on a graphene surface and the low chemical reactivity of graphite and graphene for many chemical compounds (such as Cu etc.).

Considering the low adsorbtion energy on a defect free graphene surface, H_2 molecules have a high surface mobility, except on the graphene defects (vacancies, voids, Smith Wales defects –C5/C7 odd sp2 ring pairs [200] where adsorbed H_2 can be transformed in local C–H). Considering that H can easily recombine to H_2 (~5 eV), the formation of alternating the C–H bond (~4.3 eV) on each graphene flake face will be unlikely. However, with the high H_2 recombination energy release graphene can be transformed into a buckled H6 diamond (according to Figure 3) with enough available activation events and in agreement with observed formation of Csp3 [197] and looking in more details to the corresponding Raman spectra, with a Raman peak H6 structure at ~1325 cm^{-1} [199,200]. It can be observed that high pressure H_2 on a graphene surface, can hardly be dissociated except on defects where dangling bonds exist, explaining why only local diamond structure can be expected [201].

Transformation of Graphene into H6 Diamond with N and O

Another demonstrative experiment suggests that graphane cannot exist. This is given when a similar transformation of graphene into a dielectric material with other atoms than hydrogen, and also, which is observed with nitrogen and oxygen. Quantum electronic activation can be produced with the N_2 recombination energy release, similarly to what can be produced during some CNx annealing as precedingly reported [181] with the formation of diamond crystallites. The phase transformation can be enhanced in comparison to a similar experiment with H, because the $N\equiv N$ recombination is releasing higher energy (~12 eV) than the H_2 recombination and because the C–N bond energy (~3.16 eV) is weaker than C–H (~4 eV up to 4.5 eV) [25]. Binding energy differences of N_2, C–C and C–N also explain in contrast to the early quantum mechanical calculated prediction [202] why the c-C_3N_4 formation is unlikely (because the C–C and N_2 bond energy are much higher than C–N) and why this virtual material cannot be harder than diamond considering the equivalence of hardness with the density of cohesion energy [27].

However, graphenic G-C_3N_4 can be formed with the substitution N during a graphitization process and with the assembling of CN radicals. A material which is thought to be suitable for photo-catalytic H_2 production from water [203]. However, we suggest that similar conditions can be expected with the same optoelectronic gap with more stable material corresponding to doped ta-C-like material, and also, which can combine the appropriate gap, electric conductivity and enough thermal/chemical stability.

Oxygen Sensitivity of CNT Electric Conductivity

Strong sensitivity to the oxygen environment of CNT appears to be a major problem to be solved for many NEMS applications [204]. Anticipating on the next Sections 4 and 5 and in making usage of the revisited Raman spectroscopy, very demonstrative experiments with oxygen plasma etching of multilayer graphene are showing how on nearly defect free graphene sheets (no Raman C5/C7 band and no D peaks of any kinds) an important part of the graphene multilayer is transformed into an H6 diamond [205] and optoelectronic and electric conducting properties will be significantly modified. For such case the CNT cylindrical geometry is preserved because diamond structure islands are only formed on vacancies and around void edges.

4.2. *Importance of Diamond-Like Phase Transformation on Carbon Material Properties*

4.2.1. Phase Transformation and Internal Stress Formation

Rapid and strong phase transformation toward atomic denser packed material can induce tensile stress before forming cracks, all the more this effect can be enhanced with the exodiffusion of lose bonded gaseous material (H_2 or N_2 or CH_4 as observed for instance with the synthesis of porous glassy carbon). This is also observed during the annealing of some H-rich a-C:H or N-rich CNx materials [130,131,177] when some denser packed material can be formed containing diamond crystallites and for which higher friction can appear.

This sort of mechanism is suggested to explain how energetic UV will destroy graphene and CNT when these materials are subject to phase transformation toward isotropic denser packed sp3 with formation of cracks [206]. An effect which is also concerning metallic alloys, which can become brittle with cracks formation when incorporated hydrogen recombines to H_2 and this effect is the consequence of different structures with different atomic packing density (corresponding to different oxidation states), which are also used in aeronautical and nuclear metallurgy [207].

4.2.2. Incidence on Carbon Film Nucleation

Vacuum film nucleation is generally produced with disordered randomly distributed condensation, surface migration and cluster formation before being coalescent. If strong interlayer bonds are formed on the substrate surface or on the growing material surface, then surface migration is hampered and the structure of this first layer will be generally amorphous with reduced atomic packing density, especially when the growing film material is subject to ion bombardment which is heating the growing material and which will keep atomic disorder [208] except if the epitaxial order can be imposed and/or if the material can be activated towards better crystallinity. The first grown atomic layers is generally graphitic either amorphous or recrystallized in the form of Csp2 clusters depending on the achieved surface mobility of the condensing atoms on the substrate. This surface mobility can be very low in the case of strong bond formation corresponding either to diamond materials (~7.02 eV) or to graphenic materials (~7.03 eV).

However, nucleation of the diamond material appears where the chemical recombination energy release (CRER) activation is sufficiently available, which can be produced in larger quantities with different means: In the grooves of the substrate surface after scratching [25], which can be enhanced when instead of SiC bonds, higher energy C–C bonds can be formed and all the more, when clean contaminant free epitaxial conditions exist [111]. This is the case also, for heteroepitaxial conditions when additional activation energy can be delivered. For instance, with the diamond growth on an activated carbon pretreated crystalline molybdenum [119]. (This is no longer observed after decay of the surface activation) [111].

Graphene can be formed on a substrate with high surface mobility of condensed carbon atoms, which will have reduced adhesion on the substrate [13,172]. Meanwhile, single walled carbon nanotubes can nucleate on some catalytic substrate, with which some activation (much lower than necessary to produce diamond) will determine radius and chirality of the CNT [13,150].

Defects on the substrate, contamination and temperature can modify nucleation conditions and the adhesion of the growing material. The formation of interface bonds with higher binding energy on defect edges and causing electron excitation with released recombination energy, may produce some phase modification towards denser material structure, stress and additional defect formation [15,150–153]. Discontinuities, defects, vacancies and voids will be important to consider, when on their edges strong interface bonds can be formed, contrary to chemically neutral defect free graphenic surface.

4.2.3. Incidence on Carbon-Based NEMS Engineering

All precedingly mentioned effects concerning phase transformation, adhesion, heat and electric conductivity, and also, which are depending on the type, size and number of defects are expected to cause major problems in industrial implementations of NEMS [10,13–15,21–24,63,64,147,148]. They will affect the life time and many opto-electronic and sensor functions, friction, wear, geometric tolerances, adsorbtion and mechanical properties.

Therefore, the principle of these effects must be well understood and correctly characterized. In all cited cases, it appears important to know and to master the types and number of defects and for which many interpretations will have to be revisited with all newly discovered effects, which need to be considered and to be correctly characterized, and what we discuss next.

5. Brief Review of Revisited Carbon Raman Spectroscopy

5.1. General Aspects of Carbon Raman Spectroscopy

The Raman spectroscopic assignment and interpretation still present in the literature of many confusing and contradicting aspects need to be sorted out. Carbon materials are often more composite than originally assumed and for which persisting grave lacks on Raman fundamentals must be cured. A distinction must be made between the D diamond that peak normally at ~1330 cm^{-1} and the so-called D-disorder peak which is normally at ~1350 cm^{-1} when no stress-shift applies. It must be determined which stress is to be considered and to which substructure the observed Raman band corresponds (Figure 5) [208,209].

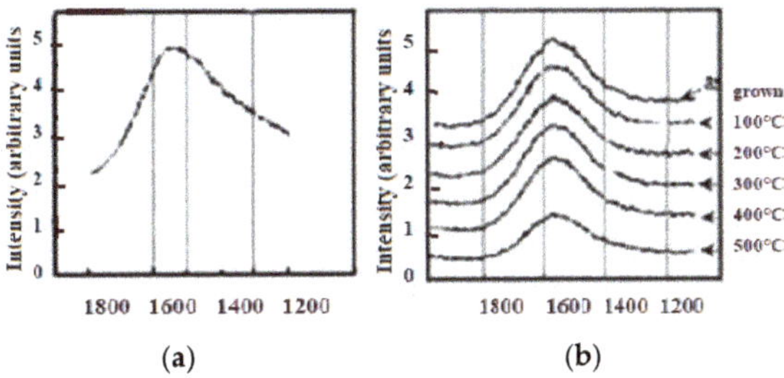

(a)   (b)

Figure 5. Raman spectra of stress shifted amorphous carbon graphitic carbon, ta-C and degraded ta-C (DLC) with no Csp2 clusters (no G sharp peak). Band max corresponding to (overlapping bands neighbor to the G peak at nominal 1580 cm^{-1}). (**a**) Tensile stress ~30 cm^{-1} downshifted spectrum of amorphous graphite by Rouzaud et al. [208] by permission of the journal of Thin Solid Films. Additional dangling Csp2-Csp2 and C5/C7 bands. (**b**) Compressive stress ~80 cm^{-1} upshifted spectra of ta-C thermally degraded to less hard ta-C with reduction of stress. By Anders et al. [209] by permission of the Journal of Thin Solid Films.

The analysis of the Raman spectra contrary to popular belief [210,211], cannot be limited to the sole measurement of the peak intensity and width of the respective so-called D and G bands [27,111–117,147–151,170,171,210,211], except when the material is either polycrystalline graphite [212] or diamond-like pure [113,173] (containing only sp3). This is well shown with Raman spectra, which have been recorded during a diamond annealing process, with which different peaks and bands appear, corresponding each of them to some specific substructure (Figure 6) [117]. Contrary to the elder persisting popular belief the sp2/sp3 ratio and the diamondlike character of carbon will generally not be able to be deducted from the ID/IG ratio [27].

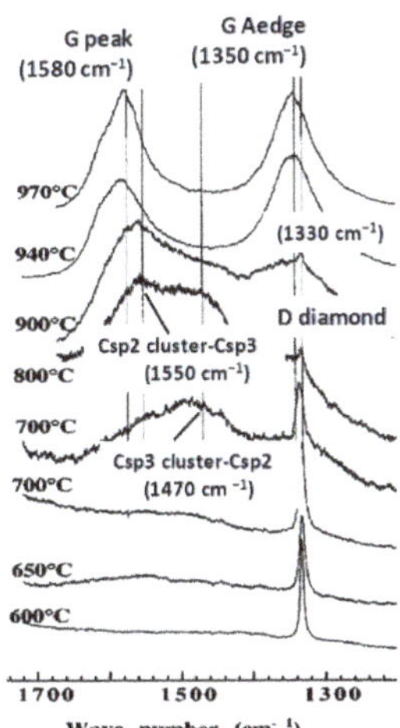

Figure 6. Diamond degradation with temperature by L. Fayette et al. [117] with permission of the Journal of Physical Review B. The D diamond peak decreases meanwhile Csp3-Csp2, G and G A edge peak are growing.

5.2. Carbon Structures to Be Considered with Raman Spectroscopy

5.2.1. Comparison of Raman Spectra from Different Carbon Materials

Different Raman peaks corresponding to different carbon material structures have been calculated with theory [189,190] and in agreement with experimental results from: (a) Micro-crystalline and amorphous diamond [116,173], (b) degraded variants and diamond-like carbon containing different kinds of subdomains [111,115,117,125,181]: Amorphous sp2/sp3 mixture, and amorphous/nanocrystalline sp3 parts) [130,173,209,210] (d) deformed sp2 clusters, graphene planes and graphenic fibers and tubes [134,150,151], (e) glassy carbon (porous mix of fullerenic GLC, graphenic material, diamond-like material and H6 hexagonal diamond) [111–113,132,133] and for which persisting confusion in literature needs to be sorted out.

The Raman spectra of carbon materials that are supposed to be different, can show a paradoxical similitude for instance between the ta-C and amorphous graphite (Figure 5) or between the micro-diamond [115] and glassy carbon [132,133] (as shown in Figure 7). However, the stress shift must be considered in order to correctly assign the considered peak (or band). In Figure 6, the band maximum will not correspond to a shifted G band (which is normally at ~1580 cm^{-1}), but to a band which normally would be observed at ~1500 cm^{-1} and which corresponds to a Csp3cluster-Csp2 band [25].

The D band designation had been extrapolated from the Raman peak of the single crystal diamond with a similar frequency (whenever different) and which contrary to elder belief is not always giving account for the sp3 concentration [26]. Proportionality to the sp3 concentration is only given when the carbon material is corresponding to totally mono- or poly-crystalline and to amorphous diamond materials [113,124,126,173]. For GLC (graphite like carbon) IG/ID has been shown (by Tuinstra and Koenig) to be proportional to the graphite crystallite size and surface/volume ratio [213] and which can be corroborated by the optoelectronic gap dependence on number of adjacent hexagonal cyclic sp2 rings [214].

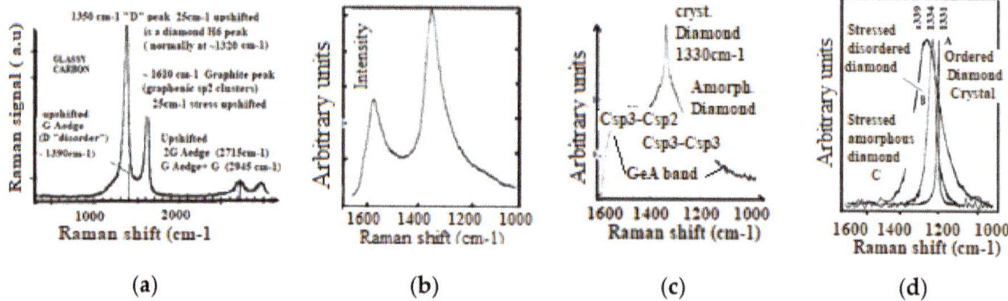

Figure 7. Comparison of Raman spectra of carbon materials that are supposed to be very different. (**a**) Glassy carbon (25 cm^{-1} stress up-shift) by Badzian et al. [132] with permission of Les Editions de la Physique. Glassy Carbon (GC) is not always graphitic. In the present case, ordered Diamond H6- (sharp up-shifted peak at ~1350 cm^{-1}). Upshifted G peak (~1610 cm^{-1}). Upshifted Ddisorder (GeA) (1370 cm^{-1}) and G + GeA (2945 cm^{-1}). Note similitude with b. (**b**) Diamond crystals imbedded in graphenic matrix by McNamara et al. [116] with permission of the Journal of Diamond and Related Materials. G peak, D diamond broad peak superimposed on GeA band (so-called D disorder band), (**c**) Superimposed D diamond peak with D diamond band and GeA band by Mc Namara et al. [116] showing existence of DG band (~1450 cm^{-1}) and DD band (~1140 cm^{-1}) with permission of the Journal of Diamond and Related Materials.. (Necessity of differentiating the D designation. (**d**) Comparison of (A) ordered diamond (B) disordered diamond (C) amorphous diamond broader peak (B) and (C) with stress shift by P.H. Huong et al. [173] with permission of the Journal of Diamond and Related Materials.

5.2.2. Raman Stress Shift and Atomic Disorder Band Broadening

It was generally admitted that a Raman D peak in graphenic materials containing no sp3 sites would correspond to the atomic disorder [112,170,208–210] and an assignment which can be questioned in taking into account following considerations

The disordered material has distributed an interatomic distance explaining the Raman peak broadening [26], considering interatomic energy potential $U = \alpha x^6$ and binding force $Bf = 2\alpha x^4$ and internal stress (which affects the interatomic energy potential) [215,216] it can be deduced:

$\delta\omega/\omega_0 = \eta$. $\delta x/x_0$ (η a proportionality factor, ω the Raman frequency and δx the interatomic distortion to the ordered material interatomic distance x_0 and with $\delta\omega/\omega_0 = 6(1 - \nu)/E_0$ (ν Poisson coefficient, E_0 the mean elastic and 6 the stress). Therefore:

(a)   Sharp Raman peak corresponds always to an ordered structure.
(b)   Atomic disorder corresponds to peak broadening
(c)   Raman shift is proportional to internal stress.

This suggests incoherent and confusing designation between the so-called D diamond and so-called "Ddisorder" band which appears necessary to be sorted out. It must first be taken into account that the stress shift of Raman frequencies is of greatest importance to be considered in order to avoid possible confusion between neighbor carbon Raman peaks.

A Raman peak/band might be erroneously assigned when the relevant local stress shift has been ignored. Additionally, awareness of possible atomic rearrangement may help to sort them out.

The relation between some so-called D and D′ disorder peaks and atomic disorder in a graphene plane has been proposed with a study on defect formation in a graphene plane [170] (Figure 8).

However, such a disorder is only appearing for larger ion doses for which the graphene material is amorphized and hexagonal cyclic rings specific to graphene will be almost destroyed, meanwhile is to be emphasized that the appearing sharp so-called "D disorder" peaks correspond to well defined frequencies and which can only be associated to well-ordered material structure. Therefore, suggesting that this "D disorder" peak is associated to void formation with Ar ion impact, which has perforated the graphene sheet and void internal edges have been formed.

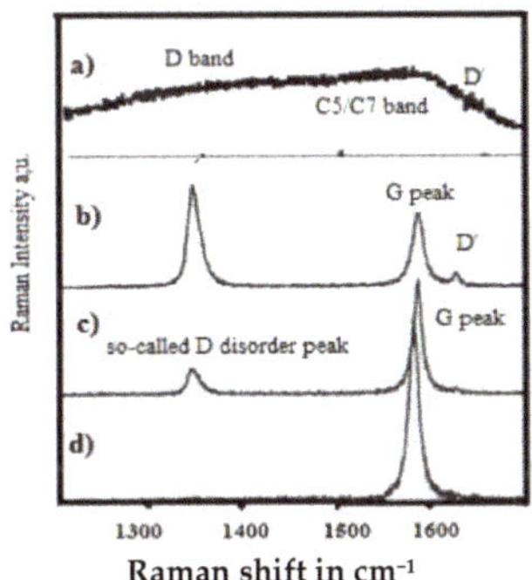

Figure 8. Relation between graphene void defects produced by Ar ion bombardment and so-called D disorder carbon Raman peak. By M.S. Dresselhaus et al. [170] with permission of Phil. Trans. R. Soc. A. (**a**) With Ion dose $>10^{15}$ destruction of hexagonal ring. Disorder band broadening, with ~30 cm⁻¹ stress shift. The broad "G" band corresponds to the upshifted band of C5/C7 odd rings. (**b**) D′ peak suggested to be assigned to dangling Csp2-Csp2 formed on internal void edges. (**c**) For ion dose 10^{11} the G peak is stress upshifted, without disorder band broadening, suggesting few incorporated vacancies. (**d**) Pristine graphene is an ordered structure (one sharp G peak).

5.2.3. Raman Peak Designation and Atomic Disorder

With the temperature annealing of some polycrystalline diamond film showing a single intense D diamond peak, a polycrystalline graphite structure appears with several peaks corresponding to specific subdomain structures for intermediate states of the graphitic degrading process [117].

Sp2 clusters and graphite crystallites are growing in number and size, meanwhile the diamond polycrystalline structure is reduced.

Csp3-Csp3 and Csp3-Csp2 dangling bonds appear on diamond crystallite edges and boundaries corresponding to the transition to graphitic clusters and graphite crystallites. Thus, also, Csp2-Csp3 and Csp2-Csp2 dangling bonds will appear on graphitic particle edges and boundaries (Figure 6) [25,117,210]. The D peak is growing, although the material is becoming graphitic and the sp3 content is decreasing, and the higher content of sp2 and graphenic particles have appeared. Obviously, this growing "D" peak is different from the D diamond peak and used to be called the "D disorder" peak, although no disorder peak broadening is to be considered.

With thermal annealing of some particular CNx and a-C:H [130,181] (Figure 9) for which graphite recrystallization would have been expected, it can be observed as a similar spectra than for the polycrystalline diamond [115,116] (Figure 7) (with a G peak for sp2 clusters inclusions) and similar, as well to some glassy carbon [132]. Here, a D peak/band (~1325 cm^{-1}) is superimposed on a D disordered/amorphous diamond band and on a broader "Ddisorder" side-band, before being converted into graphite (carbon ground state) [113,130,211]. With a 500 °C longer annealing (1 h range) any stress is reduced and all observed Raman peak frequencies correspond to their nominal assignment.

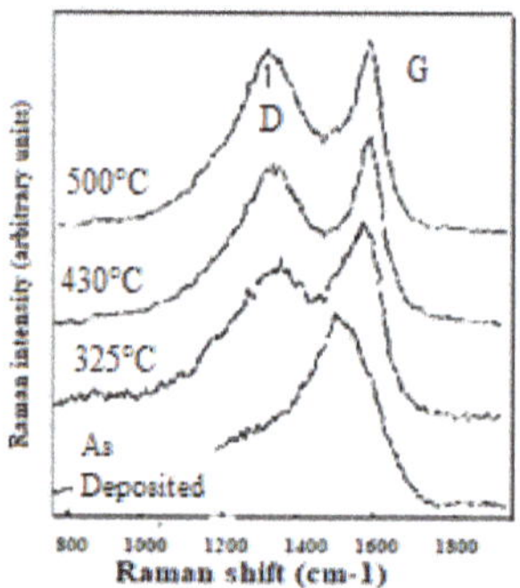

Figure 9. Annealing of H rich a-C:H. As grown, a-C:H ~70 cm^{-1} tensile down shift owing to H$_2$ exodiffusion. With annealing, stress is reduced and Csp2 clusters grow (G peak) A GeA band (Csp2 cluster edge) appears beside a growing diamond peak (~1325 cm^{-1}) (an H6 diamond). By Wagner et al. [130] with permission of Les Editions de la Physique.

5.3. Revision Necessity of Common Raman Scattering Description

Raman spectroscopy is widely used for graphenic material characterizing [217]. No so-called "D disorder" band is observed for the bulk of large graphite particles [117,212], in contrast to graphite dust [212] suggesting that the so-called "D disorder" peak corresponds to some edge effect.

This is clearly evidenced with the high-resolution micro-Raman [154,155] (Figure 10) showing that the phonon vibration mode of the so-called Raman "D disorder" peak (~1350 cm^{-1}) has a specific locality on the A edge, meanwhile this Raman peak does not appear on the ZZ edges. For some kinds of defect free bulk graphene, no D peak exist [170,218,219] meanwhile an intense 2D peak can be observed (Figure 11), depending on how the graphene was elaborated (many different elaboration processes exist today) [13,220,221].

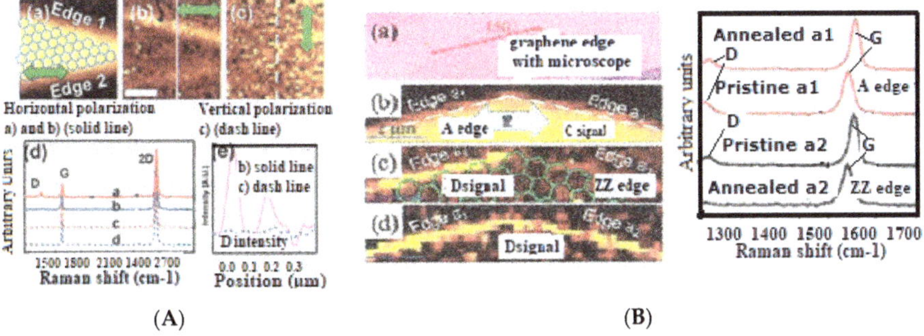

(A)   (B)

Figure 10. Micro-Raman showing locality on graphene A edges of the so-called D disorder Raman band. (**A**) Micro-Raman of graphene by Y.M. You et al. [154] with permission of Applied Physics Letters, Image of: (**a**) G peak, (**b**) "D disorder" peak (GeA peak) only on the A edge (**c**) GeA peak with vertical laser polarization. No D′ peak and C5/C7 band (no disorder). (**d**) Raman spectra on bulk. (**e**) Raman intensity geometric distribution of D signal (**B**) Micro Raman spectra showing graphene ZZ edge transformation into the Aedge with thermal annealing at ~300 °C adapted from Y.N. Xu [155] with permission of the ACS Nano. No C5/C7 band at ~1500/1530 cm^{-1}. Pristine non-annealed graphene shows tensile downshift of ~20 cm^{-1} indicating that it was prepared by CVD. Stress disappears with annealing temperature at ~300 °C. (**a**) graphene edge location, (**b**) Image of G peak on bulk, (**c**) Image of D peak only on A edge after annealing, (**d**) Image of D peak on pristine graphene flake.

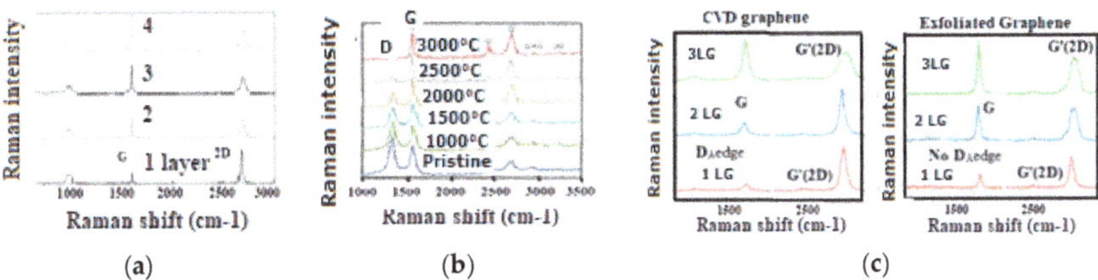

(a)   (b)   (c)

Figure 11. Raman spectroscopy of different quality of graphene. (**a**) High temperature longer time annealed graphene by Watanabe et al. [218] with permission of Diamond and Related Materials. No GeA peak (so-called "D disorder"). (**b**) Annealed graphene by J. Campos Delgado [172] with permission of Chemical Physics Letters suggesting many defects left, even after annealing. (c) Raman spectra by A. Reina et al. [219] with permission of Nano letters, showing no D disorder peak (GeA), sharp 2D and G on 1 L/2 L indicating reduced disorder for the exfoliated graphene and some disorder for the CVD graphene with broader G and G′ (2D) sheet without C5/C7 ring formation.

The Raman analysis and scheme of cut graphene (Figure 12) [222] can bring clarification on the corresponding structure of the so-called D′disorder band, considering that the Raman frequency for Csp2-Csp2 dangling bonds is the same than that for their IR stretching mode at ~1620 cm^{-1} and that such dangling bonds are likely existing after a graphene plane cut and noticing that the cutting process can involve compressive stress in the resulting edge structure [223].

However, in order to know more accurately the carbon structure of the material it is necessary to know to which structure the D peak corresponds and to consider the different types of subsystems including Csp3-Csp2 (~1520 cm^{-1} and ~1470 cm^{-1}) [26,117,182], Diamond edge Csp3-Csp3 bonds (often designated T band) ~1100 cm^{-1}, H6 sp3 hexagonal Diamond ~1325 cm^{-1}, D crystalline diamond/D amorphous diamond band ~1330 cm^{-1} [26,113,173]. Edge dangling bonds Csp2-Csp2 (so-called D′ peak) ~1620 cm^{-1} [170,222,223] (same IR and Raman frequency), C5/C7 odd rings ~1520/1550 cm^{-1} [224] (Table 1). For this achievement some revised aspects of the carbon Raman spectroscopy fundamentals are suggested to bring clarification and we recall next.

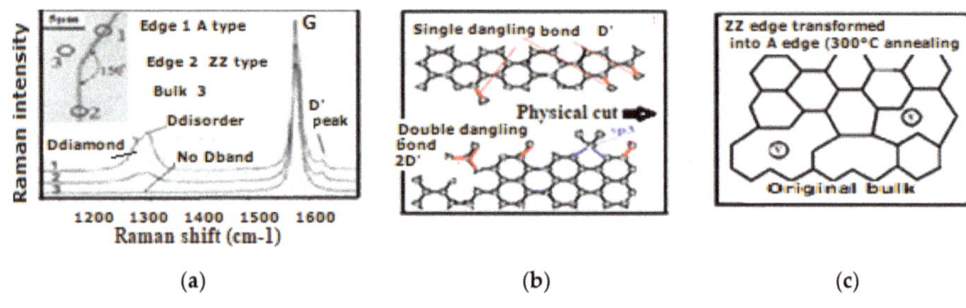

(a)   (b)   (c)

Figure 12. Graphene edge analysis of cut graphene plane and scheme of edge modification. (**a**) Raman spectra of cut graphene by Cançado et al. [222] by permission of the Physics Review Letters. D (1333 cm^{-1}), D disorder peaks: GeA (~1350 cm^{-1}) and D' disorder Csp2-Csp2 (~1620 cm^{-1}) are shown. Sharp G peak means that no disorder exists on hexagonal sp2 rings. (**b**) Our proposed scheme of graphene cut edge with the formation of sp3, single and double dangling Csp2-Csp2 corresponding to D diam, D' and 2D' peaks. (**c**) Our proposed scheme of voids formation associated to the thermally induced ZZ edge transformation into the A edge with internal ZZ and A edges on formed voids.

5.4. Refined Carbon Raman Spectroscopy

The classical Raman theory [225] has been refined with quantum mechanical aspects involving phonon modes distribution (Figure 13) and photon/electron/phonon scattering represented in the reciprocal space [150,151,226,227]. The Raman frequency corresponding to the homogeneous and ordered bulk structure could be satisfactorily calculated for diamond [189,190] and graphene [150,151]. However, this is not the case for the so-called "Ddisorder" peak as shown precedingly. It must be emphasized that among the different conditions which must be fulfilled, a Raman effect can only exist when the momentum and energy conservation laws are fulfilled [228]. The so-called "Ddisorder" peak was assigned to phonon backscattering on defects with the double resonance Raman effect. It is represented in the reciprocal space with the so-called "extra-valley" transitions between the Brilloin zones [150,151,170,226].

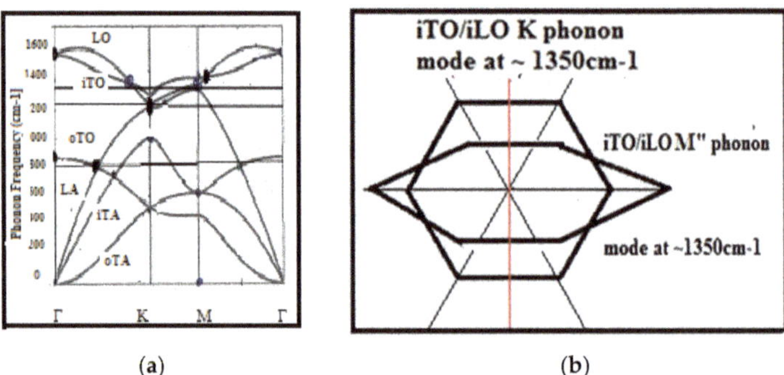

(a)   (b)

Figure 13. Phonon dispersion curves in graphene by Lazzari et al. [227] with permission of Physical Review B. (**a**) At 1350 cm^{-1} (the frequency of so-called Raman D "disorder" peak) two coupled phonon vibration modes corresponding to K and M modes and only the K mode is Raman active. (**b**) Representation of K and M coupled vibration mode for a hexagonal sp2 ring with perpendicular wave vectors. K mode corresponds to the vibration of A edge meanwhile the M mode is the vibration mode of the ZZ edge. iTO in-plane transverse optical mode, oTO out of plane transverse optical mode, LO Longitudinal optical mode, LA longitudinal acoustical mode, iTA in-plane acoustical mode, oTO out of plane acoustical mode

Table 1. Raman nomenclature for carbon material characterizing.

Raman (cm^{-1}) (Nominal)	Peak/Band	Type of Structure (Non-Stressed Structure)	Type of Bonds Energy in eV
~1330	D peak	Ordered Diamond cubic	Csp3-Csp3 ~7.02
~1325	DH6 peak	Ordered hexagonal diamond	~7.015 eV
~1200/1400	Dd band	Amorphous diamond, ta-C	Overlapping with DD band
~1150	DD	Edges of Diamond crystallites.	Aliphatic Csp3–Csp3
~1470	DG	Diamond and sp2 edges	Aliphatic Csp3–Csp2
~1580	G peak	In plane double degenerated Γ phonon mode stationary vibration of sp2 cyclic ring	Csp2-Csp2 ~7.03 eV collective bond vibration
~1560/1620	G band	Atomic disorder broadening	Superposition GG, DG, GC5/C7
~1620	GG	Csp2-Csp2 clusters edge in a-C and DLC	Csp2-Csp2 ~7.03 eV collective bond vibration
~1510	GD	Csp3-Csp2 cluster	
~1490	G$_{C5}$	C5 ring	Fullerene C5 (~1550 cm^{-1})
~1540	G$_{C7}$	C7 ring	(upshift by plane curvature)
~1350	GeA	A edge 0° CDR scattering Voids internal A edges 1st, 2d order disorder on edge	Free edges: not active with ⊥ vertical polarized laser light Bonded edges: all laser light
~1300/1400	GeA band	Broadening by edge disorder	polarization
~2690	G$_{2P}$	2 phonon CDR scattering Any polarized laser light	In plane 2K and 2M Preferential in-plane polarization
~150	RBM	Breathing mode of CNT (radius dependent)	Collective phasic stretching
~1600	G+	CNT in plane Longitudinal	Distorted Csp2-Csp2 by
~1560	G–	CNT in plane Transversal	sp2 plane curvature of CNT

However, the usual quantum mechanical Raman theory [150] presents some flaws [229] in agreement with the point that the quantum mechanical theory generally neglects the locality of energy states and that the law of energy conservation is not always respected [26,98]. We state that the description of the double resonance back scattering on defects and edges and which is supposed to give account for the so-called "Ddisorder" peak (at ~1350 cm^{-1}) does not fulfill the energy conservation law [26] (only the momentum conservation). Further on, the usual quantum mechanical description of the Raman effect is not considering the locality of the involved phonons. Therefore, the interference of the phonon being backscattered on a defect with the incident one (resulting into a stationary vibration modes) and the eventual coupling between phonon modes have not been considered either.

Considering that a wave vector in the reciprocal space corresponds to a wave propagation direction in the real space, we could refine this graphene Raman model with interferences of involved phonon modes and in taking into account the coupling between activated electrons and corresponding holes (the so-called Kohn effect) [230] and their different decay times in correspondence to each sort of scattering time [26]. The modified double resonance Raman effect corresponding to the backscattered phonon on edges and defects (we call "Coupled Double Resonance" CDR) can only exist when the backscattered phonon has the same direction with the original incident one (0° angle backscattering). Otherwise, the law of impulse conservation cannot be fulfilled. To be noted, that the erroneously so-called "D disorder" Raman peak (~1350 cm^{-1}) in graphenic structures was also assigned for the breathing mode of a hexagonal sp2 ring [112]. However, this mode will be evidently quenched with higher number of adjacent rings [25].

Considering that this CDR Raman peak (~1350 cm^{-1}) corresponds to an A-edge vibration mode as shown with micro-Raman [154,155] (Figure 10) and only to be considered on graphite hexagonal cyclic ring edges we have proposed the GeA designation for it (Table 1). It is here to be emphasized that the K mode vibration mode (of the A edges) is coupled to the M vibration mode (of ZZ edges).

The above described CDR model explains why the corresponding Raman effect is only appearing on the symmetric A edge on which a 0° angle backscattering is possible (Figure 14).

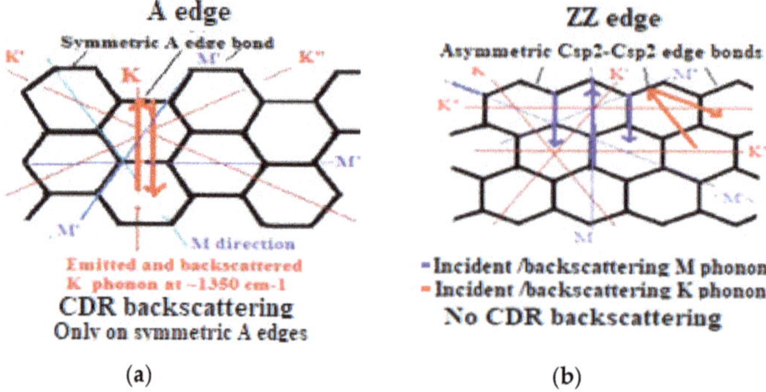

Figure 14. Phonon backscattering and Raman conditions at 1350 cm^{-1}. (**a**) CDR Raman scattering conditions are fulfilled on a symmetric A-edge and for which in-plane K phonon at ~1350 cm^{-1} is perpendicular to A-edge. No CDR Raman scattering on asymmetric graphene A edge of respective K′, K″ orientation. (**b**) No 0° angle backscattering on the ZZ edge for K and M phonon at ~1350 cm^{-1} where edge Csp2-Csp2 bonds are not symmetric to the K orientation. M phonon can be split in other phonon of lower energy and therefore, no CDR Raman at 1350 cm^{-1} can exist on ZZ edge.

This statement appears to be in agreement with the point that no equivalent Raman peak is observed either with non-symmetric A edges of the h-BN structure for instance [26,231]. This model will also explain why this so-called D "disorder" peak is only appearing on the smaller dust polycrystalline graphite particles and not for the defect free polycrystalline graphite bulk. This is suggested to be explained with signal/noise sensitivity and enhanced surface/volume distribution of relevant edge and bulk signals [26,212]. With this revised approach, different kinds of defect in graphenic materials can be characterized more accurately.

6. Defect Characterizing with Raman Spectroscopy

6.1. Phonon K Mode and M Mode Wave Scattering

From the phonon dispersion curves of graphene [150,151,226,227] (Figure 13) (corresponding also to some extent to sp2 clusters in DLC and to CNT, whenever related larger number of dispersion curves exists here [150], and considering coupled graphene sheet in-plane and out-of-plane vibration modes, it can be deducted that the G peak of graphite and graphenic material is only corresponding to the stationary vibrating mode specific to hexagonal cyclic rings (Γ mode) compatible with the vibration modes of adjacent material structures [26].

In the amorphous graphite (GAC) [208,228], no hexagonal cyclic ring clusters have been formed. Therefore, its broad so-called G band of the GAC material [208,228] cannot include the G peak corresponding to sp2 hexagonal cyclic ring clusters. It is the result of superimposed neighbor Raman peaks being broadened to a band by the atomic disorder.

This is likely the case for C5/C7 bands at ~1500 cm^{-1} (which is normally at ~1520/1530 cm^{-1}) and for the adjacent band corresponding to Csp2-Csp2 dangling bonds of sp2 clusters at ~1590 cm^{-1} (which is normally at ~1620 cm^{-1}) [26,223].

On such a Raman band, a stress shifted band corresponding to Csp2-Csp3 dangling bond of a Csp3 cluster is also observed at ~1430cm^{-1} (which is normally~1470 cm^{-1}). A stress shifted band corresponding to a Csp3-Csp2 dangling bond of a Csp2cluster structure which would have been expected at ~1520 cm^{-1} (which is normally ~1550 cm^{-1}) is unlike if no hexagonal cyclic ring can be considered, and when no G peak specific to them is observed.

For the so-called "Ddisorder" Raman peak frequency at ~1350 cm^{-1} two coupled vibration modes are identified on the phonon dispersion curves (Figure 13). The K mode (transverse to A-edge) and the M mode (diagonal direction in a hexagonal cyclic ring) (Figure 13b). However, according to theoretic predictions [150,151] only the K mode is subject to some Raman effect and can give account for the "Ddisorder" Raman signal with a double resonance backscattering, meanwhile the M mode will not (Figure 14). This explains why the so-called "Ddisorder" peak is not observed on the ZZ edges in agreement with the preceding discussion [154,155] (Figure 10).

An effect which is also confirmed with the Raman spectroscopy of the h-BN material for which no equivalent peak to the carbon D "disorder" peak exists [231], in contrast to the possible existence of an intense so-called 2D peak (2GeA) which is observed in some graphene Raman spectra which paradoxically are not showing any so-called D disorder peak (GeA) [218,219].

The addition of two M modes (complementary M′, M″) can form a new vibration mode with double frequency and the same direction than the K mode and for which a double resonance 0° angle backscattering can fulfill the impulse and energy conservation laws (thus, subject to a Raman effect).

The same is to be considered with the addition of two complementary K′ and K″ modes, which can form a new vibration mode of double frequency with a wave vector perpendicular to an A edge and corresponding to a so-called 2D Raman peak (2GeA).

Therefore, the observed 2D peak can result from the addition of some vibration modes, in which each of them separately considered cannot produce a so-called D disorder Raman peak (A edge K mode). Thus, explaining the possible much higher intensity of the so-called 2D peak compared to the so-called "D disorder" peak [218,219] (Figure 11) and considering that the overtones of the so-called D disorder peak used to be weaker than the basic corresponding vibration mode.

6.2. Defect Types to Be Considered for NEMS Engineering

Independent from those associated to doping (such as B and N) and contamination (such as H, O and N) and completing some description of defects in graphenic materials (multilayer stacking anomalies), several sorts of "intrinsic" defects have been identified which can appear in the form of network discontinuities in the graphenic bulk and on their external edges (Figures 12 and 15) and which have significant consequences on NEMS properties [21–23,150,151,169].

- Vacancies and voids with different "internal" edges. Single vacancy has only ZZ edges and can be identified with "2D" (2GeA) Raman peaks when no "Ddisorder" (GeA) peak is observed or when the "2D" peak is more intense than the "D disorder" (GeA) peak [171] (Figure 11).
- Interstitials which can be rearranged during annealing in forming odd ring C5/C7 ring (SW defects) [151,200] with possible additional vacancy formation (Figures 10b and 12c) [26].
- Edge discontinuities resulting for instance from graphene plane cut [222] can produce on the cutting edge single and double aliphatic Csp2-Csp2 edge dangling bonds at ~1620 cm^{-1} and 3240 cm^{-1}) and which are corresponding to so-called D′disorder peak (Figure 12a). Those can evidently modify chemical reactions and material rearrangement on graphenic external edges.
- Graphene rippling has a focused high interest strongly affecting their mechanical and optoelectronic properties and their electric and heat conduction [232–234].

On a single vacancy of graphenic material, no internal A edge is formed and only internal ZZ edges exist. This suggests that the single vacancy will not produce any GeA Raman peak, contrary to larger voids which have both internal A and ZZ edges. In order to reduce the confusion between the D diamond and so-called "D disorder" peaks and very high impact on carbon-based MEMS and NEMS quality and performances (opto-electronical, electric and thermal conductivity properties and density of cohesion energy), we have suggested to modify the designation nomenclature with Table 1 [26].

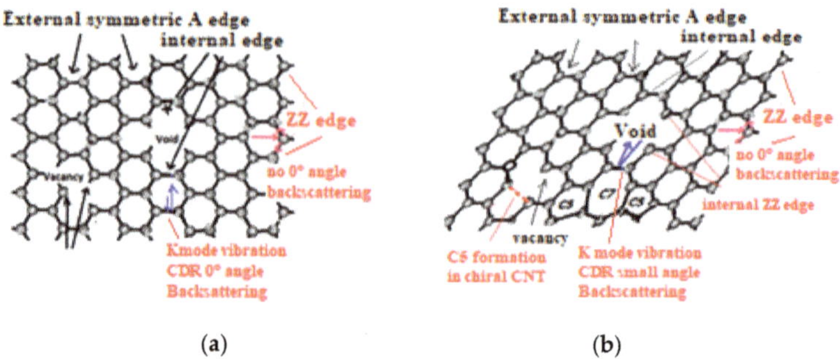

(**a**) (**b**)

Figure 15. Defect in graphene bulk (vacancies and voids). (**a**) Phonon backscattering on external and internal A edges and ZZ edges in a metallic graphene. No Raman double resonance is possible on the internal edges of vacancies (ZZ edges) in contrast to voids where symmetric A edges exist. (**b**) Formation of C5/C7 odd rings and some internal A edges on vacancies of chiral graphene in consequence of chiral distortion. The reduced A edge asymmetry allows local Raman double resonance (in consequence of Heisenberg incertitude criteria).

Considering edge dangling electrons, easy chemical reactions with recombination energy release can be considered. Atomic and molecular species can be chemisorbed on vacancies (via their internal edges) and with which some new electronic band configuration can be obtained and which can determine some specific graphene functionalizing [10,11,21–23].

However, those can also produce sp3 atomic rearrangement in the surroundings of the vacancy/void if the released energy is able to produce sufficient electron activation. In such a case a dielectric material can be formed which will strongly affect the originally expected high electric conductivity.

6.3. Local Atomic H6 Diamond Rearrangement

In addition to thermal transformation of the ZZ edge into more stable A edge [155] with the formation of additional vacancies (Figure 12c) [26], local diamond atomic rearrangement of graphenic materials can be produced with quantum electronic activation of higher energy, and with which graphene can be locally transformed into a dielectric H6 diamond structure [233] (Non shifted Raman peak at nominal ~1325 cm^{-1}) corresponding to a buckled hexagonal structure containing only sp3 (Figure 3).

We suggest these being at the origin of the graphene rippling and which is much affecting the electric conductivity of graphene [234] and the mechanical properties of graphenic material. This is suggested to explain the higher stiffness of nano poly-crystal diamond, where the crystallite boundary material is containing graphenic materials which have been partially transformed in sp3 substructures and the whole diamond material is interlinked by more isotropic diamond material [235].

Refined lecture of produced Raman spectra is clearly showing that an H6 diamond structure has been formed (~1328 cm^{-1}), meanwhile the G peak appears at ~1580 cm^{-1} (no longer compressive stress). Considering the reduction of graphene oxide, a process during which some O-X molecules are formed with corresponding higher energy release (X = H, O, etc.) [236] are able to produce some diamond like quantum electronic activation (Figure 16). This is also well shown with other experiments especially after oxygen etching of multilayer graphene [205] (Figure 17) and [237] (Figure 18) and with the modification of the sp2/sp3 content which can be evaluated from the plasmon spectra, and more easily with the sp2 and sp3 carbon AUGER peaks of XPS spectra [25].

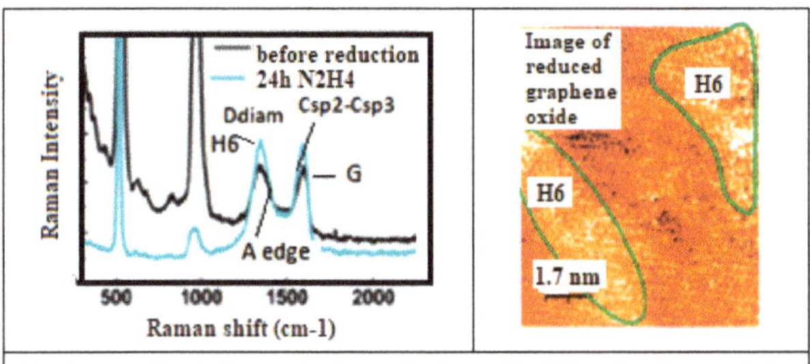

Graphene oxide before and after reduction with N2H4. The image of the material after reduction of the oxide shows juxtaposition of Diamond and Graphene. The Raman spectra show stress upshifted diamond structure H6 (~1330 cm⁻¹, Diamond (~1335 cm⁻¹), Csp2cluster-Csp3 (1580 cm⁻¹), G (~1590 cm⁻¹) and associated GeA edge (~1360 cm⁻¹).

Figure 16. Graphene rippling and local H6 structure formation. Note: Reduced GeA edge after reduction and increase of H6 diamond by C. Gómez-Navarro et al. [236] with permission of the New Journal of Physics.

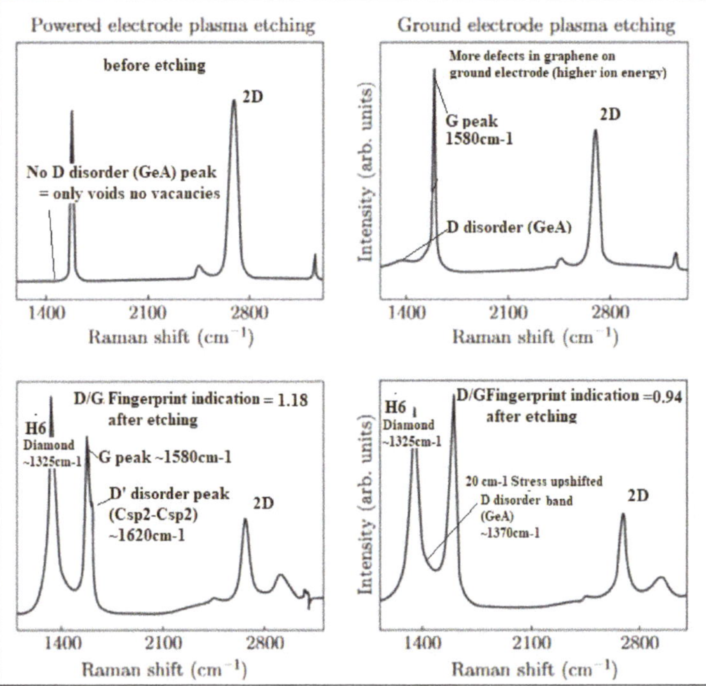

Refined Raman lecture shows that after etching the so-called D disorder (GeA) peak, is in fact a diamond H6 peak produced by neutralized/recombined adsorbed oxygen ions and atoms on a graphene flake surface inducing quantum electron activated sp3 structure atomic rearrangement [25]. The 20 cm⁻¹ compressive stress Raman upshift on ground electrode after etching is caused by ion peening (higher ion energy on ground electrode indicates that powered electrode is larger than the ground electrode Note significant reduction of 2D peak suggesting transformation of a part of the internal ZZ edges into a diamond like structure.

Figure 17. Raman spectroscopic results of oxygen plasma etched graphene by Al-Mumen et al. [205] with permission of Nano-Micro Lett.

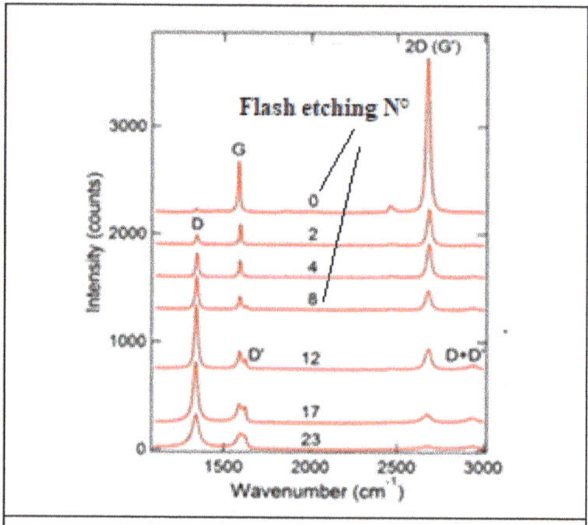

After 12 plasma flashes, a compressive stress upshift appears on the G peak which intensity is strongly reduced suggesting the transformation of the graphene into a diamond material. The so-called D peak is mainly an upshifted H6 diamond peak, beside a weak GeA band and a much reduced 2D (void defect) peak. The D' peak (dangling Csp2-Csp2) indicates that internal edges of voids are destroyed

Figure 18. Example of revised interpretation of Raman spectra of oxygen plasma etched graphene by Childres et al. [237] with permission of the New Journal of Physics.

In addition to defects which can affect the properties of graphenic material, the selective CNT adsorbtion efficiency (depending on adsorbtion energy) for molecular compounds of different shape must be considered, and which also depends on the CNT chirality and which are influencing the optoelectronic and mechanical properties and the volume and density of the formed CNT/adsorbate system. Recent high interest has been focused on the way to select the different types of SWCNT [238]. They can be separated with some selective buoyancy effects described elsewhere [239] and which appears of great importance for CNT based synthesis of complex helicoidal structures [11,81,82]. It is suggested to be at the origin of Precambrian abiotic synthesis of early NRA life molecules [240].

7. Application Developments

7.1. Role of Interface Structure on Composite Material Properties

7.1.1. Anchoring and Adhesion of Graphenic Materials with Counter-Facing Materials

Graphenic materials have been first used for more performing load-bearing applications [64,65,241]. CNT powders are mixed with metal alloys, polymers or precursor resins to increase stiffness, strength and toughness and also heat and electric conducting properties. For instance, bulk graphene (reduced graphene oxide)-reinforced Al matrix composites with significantly improved mechanical properties in comparison to Al based alloys, and which is reducing cost, brittleness and cracks formation [242]. MWNT-polymer composites reach conductivities as high as 10,000 s m⁻¹ at 10 wt % loading and is used for electrostatic assisted painting and which is thought to be used for anti-icing (with heat generated by electric resistivity) and microwave absorption [243,244].

However, these enhancements depend on nano particle size and shape, aspect ratio, alignment, dispersion and interfacial interaction. The last aspect will strongly depend on which type of chemical bonds can be formed on the internal edges of vacancies and void and on external edges of graphenic particles considering the discontinuities of the added graphenic material and on their number, size and density/unit area of vacancies and voids [200] and which need to be fully characterized.

7.1.2. Modification of Intrinsic Mechanical, Electric and Optoelectronic Properties

Graphenic particle properties can be functionalized for specific and improved characteristics corresponding to more or less combined intrinsic mechanical, electric, optoelectrical bulk and surface adsorbtion properties and reactivity. This is to be achieved upon which kind of doping and which sorts of adjacent materials is in contact with the graphenic particle and which can affect the energy distribution of their optoelectronic band structure [101,102,144,158,162,195].

For instance, a graphene coating makes carbon aerogel composite particular elastic and resistant to fatigue [245]. Mechanical properties of high-aspect-ratio CNT can be tailored upon added silicon carbide coating [246], better performing transparent conducting electrodes can be achieved with transparent CNT sheet [243,244] and in using graphene silica composite [247]. Electric conductance could be strongly increased with added iodine and especially with copper doping [162–164]. Noteworthy the achievement of very high frequency nano transistors in making use of very low resistivity and ballistic electron transport condition in graphene sheets [167,168]. This, in analogy to superconductivity and which can be comprehensively described with a new model which consider the synchronic electron-phonon gating effect and the reduced fermi surface atomic rugosity and the reduced amplitude of transverse phonon in a graphene plane [29].

Transparent electric conducting mechanical and chemical resistant epoxy could be produced with magnetic molecule functionalized CNT [165]. More performing anode [166] and optimization of MWCNT/LIFePO4 cathodes have been achieved for Li-Ion battery [248]. Those are expected to be further improved with better compromise between electron conductivity and proton diffusion barrier properties [28]. 3D macroscopic carbon material scaffold with combined electro-mechanical properties could be developed [249]. Making use of combined piezo-electric properties, electro-mechanical resonator [249] and improved micro-loudspeaker could be produced [250,251]. Arrays of vertically aligned helicoidal CNT could be produced for better mechanical deformation for high frequency electric contact [252], Superelastic CNT aerogel muscles for bio-application [253].

However, these enhanced properties are strongly dependent on defect type, size and number and which must be characterized in order to be able to achieve some desired compromise between higher solid-state properties and precedingly described anchoring effect which is influencing the adhesion, the mechanical and chemical stability and other interfacial electronic effects. Notwithstanding, that eventual phase transition must be comprehensively kept under control.

7.1.3. Local Activation of Phase Transitions on Edges and in Graphene Bulk

Incidence of high energy activation on phase transformation from graphenic material towards diamond material, could be many times unconsciously demonstrated for long. This can be stated in considering several corresponding Raman spectra features:

(a)   With the differentiation between so-called "Ddisorder" peaks (GeA) and the neighbor collective vibration modes of the D diamond peaks and band (~1330 cm^{-1})
(b)   The related Csp3cluster-Csp2 (~1470 cm^{-1}) and Csp3cluster-Csp3 structure (~1150 cm^{-1}).
(c)   In considering some possible stress up- and down-shift which can be checked on the G peak.
(d)   The differentiated analysis of the so-called G band which is not always containing a nominal G peak (corresponding to ordered graphitic and graphenic Csp2 material at ~1580 cm^{-1} when not stress shifted and which can be eventually broadened by disorder). They can correspond to the superimposition of neighbor shifted other peaks and bands [26,180–182,208,209].

Attention must be brought to the point that modification of mechanical properties of crumpled graphene sheet (graphene rippling) is not only the consequence of the new geometric non-flat surface shape [232,233], but almost basing on a phase transformation from a graphenic state toward dielectric

H6 and diamond state (as we observe on corresponding Raman spectra). This phase transformation can be activated by various means (UV, external polarization, chemical recombination and ion and electron electric neutralization energy release) [25,234,235]. However, this eventual phase transformation will not be always homogeneously distributed, as observed for instance with the reduction of graphene oxide by Gomez et al. [236]. This explains the observed increased ohmic resistance (by mix of juxtaposed dielectric and highly conducting materials). In addition, some associated material shrinking is to be considered, which can cause cracks and increased brittleness and reduction of heat transfer capacity.

7.2. Friction and Wear

Some NEMS system includes rotating parts for which reduced wear rates and friction can achieve longer life time and higher number of mechanical cycles [254]. Mechanical robustness of a polymer substrate can be increased with an adherent graphene coating [255]. Considering the reduced chemical reactivity of defect free graphene surface, stronger adhesion of graphene on a polymer substrate (or reverse situation) is only possible with the formation of chemical stronger interfacing bonds on graphene vacancies/voids internal edge. However, on these spots, we suggest that released energy from chemical recombination of H_2, C–C and new formed chemical bonds on the internal void edges can also transform the graphene material in the vicinity of these spots into a more wear resistant tribological diamond like material (which can be evidenced with corresponding Raman Ddiamond peak/band different from the so-called "Ddisorder" peak (GeA edge coupled double resonance vibration mode) [256].

These combined effects are suggested to explain why non-expected particularly strong reduction of wear and friction have been obtained on graphene coated metallic substrates sliding in dry nitrogen after the graphene coating begin to be corrugated with accumulation of sliding graphenic wear residues containing dissociated nitrogen [257]. This is to be considered with the very high N_2 chemical recombination energy release, which can enhance the diamond like atomic rearrangement process [26]. To be observed that epitaxial graphene grown on SiC, can also be transformed into an epitaxial diamond structure. However, because of the much smaller interatomic distance in graphene than in SiC, interfacial atomic mesh mismatch and induced tensile stress is expected to form many more vacancies and voids. Thus, creating new additional stronger C–C interface edge bonds (~7 eV) which can enhance precedingly described effect [258].

7.3. Yarn and Scaffolds

High performing fibers and scaffold could be manufactured with the spinning of longer CNT and wires in considering pressure induced interlinking of adjacent CNT [259–262]. Enhanced fiber strength is obtained when fiber manufacturing is associated to polymer between the tubes which can better interlink the CNT via void defect anchoring effects [263,264] and also in causing diamond like atomic rearrangement in the vicinity of the voids (considering that simple vacancies will not be large enough for receiving all reactants necessary to such atomic rearrangement and with which stronger interlinking can be produced). However, such a process involving some phase transformation from graphene to diamond H6 phase may reduce its electric conductivity.

We suggest that the higher electric conductivity of Cu doped metallic CNT [265] is obtained when vacancies and voids can be filled with some atomic species able to form an electric conducting continuous flat CNT surface (similar to superconductivity conditions, as previously discussed).

For such application semiconducting CNT must be avoided, correct orientation of metallic CNT particles must be secured. Diamond-like atomic rearrangement forming more dielectric material must be avoided (or reduced to thin interlayer material, through which the electric conductivity is obtained by electron tunneling, all the more that diamond and ta-C materials have low work function). Addition of copper can fill larger voids with reduced chemical recombination energy release (Cu-C binding energy is very low) and with electron enrichment of the electron conduction band edges and in agreement with achieved high electric conducting CNT/Cu composite materials [163,164]. It will be essential here to optimize size, density and number of voids in the precursor graphene material and its contamination being reduced, considering that atomic rearrangement towards more dielectric materials has to be minimized. Using biochar material, a graphitic electric conducting carbon dust [266], its 3D isotropic distributed orientation is contributing to easier reproducibility of carbon-based composited materials [267].

7.4. Micro and Nano Interconnecting and Thermal Management

With the possibility of increased strength, elasticity, heat and electric conductivity and reduced copper electromigration, low weight and low cost, carbon-based fibers appear to be interesting materials for micro and nano interconnection and thermal management of many electromechanical and electronic devices [268–274]. With bottom up technology vertical tubes and fibers can be catalytical nucleated with strong bonds on its specific substrate alloy materials [275]. However, great care must be brought to the achievable low ohmic, strong and stable electric contact to other added parts of the NEMS. Any possibility to have the contact wire/fiber material being transformed into a low graded electric conducting material must be avoided. It must be controlled and managed (a) the diamond atomic rearrangement forming dielectric material by higher chemical recombination energy release (CRER) and (b) the surface passivation with reactive contaminants which can affect both the adhesion and the mechanical strength of the link and the resistivity of the electric contact. Cleanliness and selection of fiber type on defect content, composition, mechanical and electric properties will be a major concern.

7.5. Electronical and Optoelectronic Functions and Field Emission Effects

Different nano electronic devices combine [134] chemically and electric field modifiable semiconducting properties, low work function and different mechanical properties, which can be associated to high thermal conductivity, high current density and very high mobility, and ballistic properties, low voltage field emission properties and fast switching.

This could be achieved and/or improved with associated efficient chips cooling being up to ten times better than with copper and with which electromigration of copper can be avoided [274], especially in making use of polymer/graphene and CNT composites nano thin films [243–246] and with which different devices could be elaborated. Among them, flexible electronic integrated circuits and transparent thin film could be produced on transparent polymer and glass substrates, thin film transistors (TFT) and Field effect transistors (FET) [166–168,276–281]. Those are thought to be used for low power/low energy consumption field emission display [140,141] with which organic light emitting diode (OLED) can be activated [142] or selective light emission can be produced on optoelectronic functionalized CNT field emitters [142,143]. Here, particular care has to be brought to reduced contamination and defect contents and to possible phase transitions especially induced by various chemical recombination energy release activation.

7.6. Solar Cells, Hydrogen and Energy Storage and Energy Conversion

7.6.1. Solar Cells and Energy Storage

Organic solar cells have focused much interest, since low cost flexible thin film photovoltaic system could be produced with them [282–284]. Graphene polymer composite could increase the conversion efficiency, in reducing electron/hole recombination, since electrons in the conduction band can be faster evacuated [285,286]. Such graphene polymer composite exhibit also transparent conductive properties which are thought to be used for displays and for solar cells, considering they are much cheaper than usual ITO (Indium Tin Oxide) and related TCO (Transparent Conductive Oxide) [243,244,247,282–284]. As a result of their extreme high electric conductivity, they can be used as transparent electrodes in very low film thickness. They can be deposited with spin deposition techniques without vacuum technologies. However, polymers are not best diffusion barrier and give little protection against humidity and low weight elements, and that thicker coatings can absorb a significant part of the solar light. Therefore, it will be useful to protect them with less permeating transparent ta-C encapsulating which can provide additional anti scratch resistance and resistance against hard UV [121] (Figure 19). Harder stress annealed transparent ta-C exhibiting very high antireflection in addition to other interesting properties, can generally enhance over the double of yearly solar light harvesting per unit surface, in collecting solar light with oblique incidence and especially azimuthal light during many days along the year where the sky presents some luminous coverage [287].

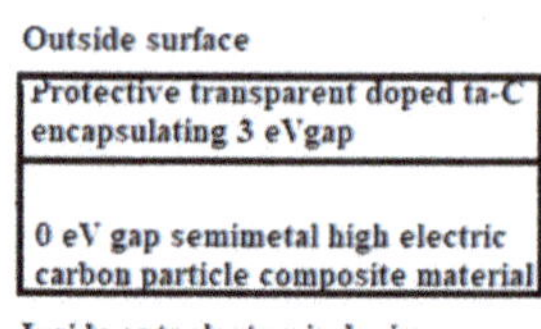

Figure 19. Scheme of conducting system with little conducting doped ta-C coating on highly electric conducting composite graphene.

Association of different carbon-based materials have been considered for improved electric energy storage and energy conversion (fuel cells) [28,288–290]. Several major aspects suggested to be considered are the extreme differences in work function (from lower than 1 eV up to over 5 eV) between different categories of carbon materials and other properties which must be correctly distinguished. On a low work function graphene surface, H+ can be neutralized and recombined to H_2 and induce phase transformation into more diamond like materials. Meanwhile, on a doped ta-C surface (highly diamond like) no chemo-structural change can be expected with such H_2 recombination energy release (CRER) activation mechanism.

Thorough distinction between different types of carbon materials must be achieved, considering their different porosity and surface rugosity, chemical inertness, electric conductivity, work function and diffusion barrier properties (for instance homogeneous, harder, dense packed ta-C have best diffusion barrier properties including for H^+ protons which other carbon materials will not have) [28]. Despite their poor electric conducting properties (whenever being doped), electric transport can be maintained through very thin ta-C layers by electron tunneling effect and can be used in multilayer nano materials with which low friction and high antiwear properties can be associated to sufficient electric conductivity for sliding contact brushes for instance [256].

7.6.2. Hydrogen Storage and Photocatalytic Production from Water

Considering hydrogen in modern clean energy management, much interest has been developed for its storage efficiency and low-cost production [291]. Hydrogen can be stored at ambient temperature with much less energy than necessary for cooling and gas compression in making use of temperature dependent and reversible adsorbtion capability on CNT external and internal wall surface [292,293]. Physical adsorbtion of H_2 on CNT is to be considered on its defects where dangling bond electron stay captive, meanwhile chemisorption corresponding to the formation of C–H bonds on the graphene surface and on its vacancies and voids internal edges (~4 eV) appears to be unlike, because the C–H bond energy is lower than the H–H bond (~5 eV) of a H_2 molecule. Therefore, hydrogen storage on graphenic material will be much dependent from contaminants which can screen the defects [294]. With the dissociation of adsorbed H_2 by catalytic and electrolytic effects [295], H atoms can be better chemically bonded to the graphene substrate defect edges. Consecutive temperature activated H_2 recombination can release the stored atomic hydrogen in form of molecular H_2. The desorbtion mechanisms is generally endothermic and is absorbing energy, meanwhile the H_2 recombination and process is releasing energy with which the CNT material can be locally converted into some denser H6 diamond like material and will enhance the formation of defect with tensile stress. Process optimization will depend on defect characteristics.

Hydrogen production with photocatalytic dissociation of water has been thought in using the CNx material which can be tailored on its optimized optoelectronic gap (~1.8 eV) and with sufficient electric conductivity and with which solar light can dissociate water molecules with H_2 formation [203,296]. Enhanced efficiency of similar process has been obtained with association of usual catalytic materials such as palladium/TiO_2 material [295]. However, these materials can be degraded by oxidative processes which will harm to their operational life time. We suggest that chemically more stable doped ta-C which can also be tailored at 1.8 eV gap and presenting some anti-soiling properties might give the route for better satisfactory solutions.

7.7. Sensors, Medical Applications and Miscellaneous

Functionalizing of graphenic materials, assembled to yarn could produce electrosensitive elastic fibers thought to be used for artificial muscles [253] and highly performing bio sensors [68]. Several Carbon based NEMS sensor principles can be distinguished: Selective mechanical properties (cantilever and membrane deformation, mass selective resonance) [35–38,69,71], selective molecular adsorbtion of functionalized graphenic materials [297–300] and selective optoelectronic modification [158,159]. A model for biologic and artificial olfaction had been proposed based on the appearance of transient polarization pulse which can be produced by selective adsorbtion on specific molecular compounds wrapping an electric conductor [100]. With the functionalized graphenic material, its optoelectronic band organization is modified and can present IR fluorescence being activated by specific adsorbtion energy release, and an effect which is used for instance for the monitoring of glucose in the blood [159,160].

A reverse effect is to be considered, when some functionalized graphenic material is adsorbed on biologic cells and tissues which can have toxic and antibacterial effects and being used for cancer therapy [101–107]. CNT scaffold composed by strongly interconnected fibers [249,301] can be used for bone tissue engineering [302]. Noteworthy, is the possibility to produce some helicoidal molecular structure with the adsorbtion of proteins [303] with are expected to have specific sensor properties. The same effect is suggested to be at the origin of first terrestrial abiotic RNA synthesis [240]. Many other applications have been developed, using graphenic composite scaffold for water nano-filtering and purification [304], as flame retardant material which combines heat conducting cooling effect with formation of diffusion barrier material [305], plasmon enhanced detection of microwave [306], photodetection with broadband polarimetry, with functionalized CNT upon corresponding photon energies and making use of their electric conductance anisotropy [161].

For all these systems, it appears the decisive role of defects of the graphenic sensor materials on which functionalizing molecules will be anchored with modified optoelectronics. Contaminants can screen and modify the defects and the optoelectronic organization of the functionalized graphenic materials and quantum electronic activation by specific chemical recombination energy release can transform a graphenic/polymeric material into diamond and diamond like dielectric material. The last effect which is thought to be at the origin of manufacturing of high performing diamond/diamond-like composite material for abrasive resistant and more performing cutting tools with 3D printing of liquid polymer/diamond crystallite mixture [307].

8. Conclusions

Carbon-based materials have extreme material properties of interest for more advanced NEMS applications. However, despite huge progress particularly with them, their implementation has to face several hindrances [1,10,23,65]. Changes to electronic and mechanical attributes of carbon-based materials must fully be explored before their implementation, especially because of high surface area which can easily react with environments. The NEMS design must receive particular attention to contamination and actual material structure and defects which can affect adhesion, surface rugosity (rippling corresponding to phase transformation towards H6 diamond and other diamond-like structures), friction (especially on local asperities), electric conductivity, inter connection, optoelectronic properties and surface functionalization. It is necessary to be fully aware about them, and all aspects must be correctly characterized and selected upon their manufacturing processes and their origin in order to be able to optimize and secure achieved results.

For this purpose, revised fundamentals have appeared as essential. This is concerning on one hand the characterizing achieved especially with Raman spectroscopy and on other hand some dramatically neglected effects inducing phase transformation and significant modification of material properties. For instance, with phase transformations of graphene and polymeric material toward diamond and diamond like material by quantum electronic activation and which can be particularly efficiently produced by high energy H_2, N_2, O_2 and C–C chemical recombination and electric neutralization energy release and which can produce diamond-like material glassy carbon and/or diamond particle inclusions and different kinds of amorphous diamond materials.

Diamond-like phase transformation appears to be very useful for many anti-wear and tribological application (effects being used for instance by Sandvik for the production of new diamond composite materials [307] or the contrary, when the main advantages of graphenic particles electric and opto-electronical properties have to be preserved and the appearance of defects and cracks by shrinking effects has to be avoided (similar effects exist also with different steel and aluminum based alloy when for instance with recombination of incorporated atomic hydrogen to H_2 [207]). For this purpose, the efficient diffusion barrier properties against atomic hydrogen diffusion of very thin harder ta-C coating appear to be of high interest [121]. When being stress annealed without graphitic thermal degradation it is suggested to be used for different kinds of transparent antireflecting, anti-erosion, anticorrosion and anti-soiling encapsulation in association with transparent highly electric conductive polymers for many applications [287].

Funding: This research received neither external nor internal funding.

Conflicts of Interest: I declare that up to my knowledge there is no conflict of interest.

References

1. Karumuri, S.R.; Srinivas, Y.; Sekhar, J.V.; Sravani, K.G. Review on Break through MEMS Technology. *Sch. Res. Libr. Arch. Phys. Res.* **2011**, *2*, 158–165.

2. Mohr, M.; Caron, A.; Herbeck-Engel, P.; Bennewitz, R.; Gluche, P.; Brühne, K.; Fecht, H.J. Young's modulus, fracture strength, and Poisson's ratio of nanocrystalline diamond films. *J. Appl. Phys.* **2014**, *116*. [CrossRef]

3. Vetter, J. 60 years of DLC coatings: Historical highlights and technical review of cathodic arc processes to synthesize various DLC types, and their evolution for industrial applications. *Surf. Coat. Technol.* **2014**, *257*, 213–240. [CrossRef]

4. Tibrewala, A.; Peiner, E.; Bandorf, R.; Biehl, S.; Lüthje, H. The piezoresistive effect in diamond-like carbon films. *J. Micromech. Microeng.* **2007**, *17*, S77–S82. [CrossRef]

5. Lu, J.X.; Cao, Z.; Aslam, D.M.; Sepulveda, N.; Sullivan, J.P. Diamond micro and nano-resonators using laser, capacitive or piezoresistive detection. In Proceedings of the 2008 3rd IEEE International Conference on Nano/Micro Engineered and Molecular Systems, Sanya, China, 6–9 January 2008. [CrossRef]

6. Kauth, C. Electronic Interfaces for Carbon Nanotube Electromechanical Oscillators and Sensors. Ph.D. Thesis, EPFL, Lausanne, Switzerland, 2014.

7. Mashman, M.R.; Ehlert, G.J.; Dickinson, B.T.; Phillips, D.M.; Ray, C.W.; Reich, G.W.; Baur, J.W. Bioinspired Carbon Nanotube Fuzzy Fiber Hair Sensor for Air-Flow Detection. *Adv. Mater.* **2014**, *26*, 3230–3234. [CrossRef] [PubMed]

8. Luo, J.K.; Fu, Y.Q.; Le, H.R.A.; Williams, J.; Spearing, S.M.I.; Milne, W. Diamond and diamond-like carbon MEMS. *J. Micromech. Microeng.* **2007**, *17*, S147–S163. [CrossRef]

9. Martin-Olmos, C.; Rasool, H.I.; Weiller, B.H.; Gimzewski, J.K. Graphene MEMS: AFM Probe Performance Improvement. *ACS Nano* **2013**, *7*, 4164–4170. [CrossRef]

10. Jiang, S.; Shi, T.; Zhan, X.; Xi, S.; Long, H.; Gong, B.; Li, J.; Cheng, S.; Huang, Y.; Tang, Z. Scalable fabrication of carbon-based MEMS/NEMS and their applications: A review. *J. Micromech. Microeng.* **2015**, *25*, 113001. [CrossRef]

11. Katz, E.; Willner, I. Biomolecule-Functionalized Carbon Nanotubes: Applications in Nano-bioelectronics. *ChemPhysChem* **2004**, *5*. [CrossRef]

12. Kis, A.; Zettl, A. Nanomechanics of carbon nanotubes. *Philos. Trans. R. Soc. A* **2008**, *366*, 1591–1611. [CrossRef]

13. Janas, D.; Koziol, K.K. A review of production methods of carbon nanotube and graphene thin films for electrothermal applications. *Nanoscale* **2014**, *6*, 3037. [CrossRef] [PubMed]

14. Zang, X.; Zhou, Q.; Chang, J.; Liu, Y.; Lin, L. Graphene and CNT in MEMS/NEMS applications. *Microelectron. Eng.* **2015**, *132*, 192–206. [CrossRef]

15. Sharma, S.; Sharma, A.; Cho, Y.-K.; Madou, M. Increased Graphitization in Electrospun Single Suspended Carbon Nanowires Integrated with Carbon-MEMS and Carbon-NEMS Platforms. *ACS Appl. Mater. Interfaces* **2012**, *4*, 34–39. [CrossRef] [PubMed]

16. Varney, M.W.; Aslam, D.M.; Janoudi, A.; Chan, H.Y.; Wang, D.H. Polycrystalline-Diamond MEMS Biosensors Including Neural Microelectrode-Arrays. *Biosensors* **2011**, *1*, 118–133. [CrossRef] [PubMed]

17. Choi, J.; Eun, Y.; Pyo, S.; Sim, J.; Kim, J. Vertically aligned carbon nanotube arrays as vertical comb structures for electrostatic torsional actuator. *Microelectron. Eng.* **2012**, *98*, 405–408. [CrossRef]

18. Kim, M.-O.; Lee, K.; Na, H.; Kwon, D.-S.; Choi, J.; Lee, J.-I.; Baek, D.-H.; Kim, J. Highly sensitive cantilever type chemo-mechanical hydrogen sensor based on contact resistance of self-adjusted carbon nanotube arrays. *Sens. Actuators B Chem.* **2014**, *197*, 414–421. [CrossRef]

19. Choi, J.; Kim, J. Defective carbon nanotube-silicon heterojunctions for photodetector and chemical sensor with improved responses. *J. Micromech. Microeng.* **2015**, *25*, 115004. [CrossRef]

20. Schroeder, V.; Savagatrup, S.; He, M.; Lin, S.; Swager, T.M. Carbon Nanotube Chemical Sensors. *Chem. Rev.* **2019**, *119*, 599–663. [CrossRef]

21. Lyshevski, S.E. *MEMS and NEMS—Systems, Devices, and Structures (2001)*; CRC Press LLC Corporate Blvd: Boca Raton, FL, USA, 2000; ISBN 0-8493-1262-0.

22. Loh, O.; Wei, X.; Ke, C.; Sullivan, J.; Espinosa, H.D. Robust carbon-nanotube-based nano-electromechanical devices: Understanding and eliminating prevalent failure modes using alternative electrode materials. *Small* **2011**, *7*, 79–86. [CrossRef]

23. De Haan, S. NEMS—Emerging products and applications of nano-electro-mechanical systems. *Nanotechnol. Percept.* **2006**, *2*, 267–275. [CrossRef]

24. Joshi, R.K.; Kumar, A. Diamond and Related Nanomaterials for MEMS/NEMS Applications. *J. Nanomater.* **2009**, *2009*. [CrossRef]

25. Neuville, S. Quantum electronic mechanisms of atomic rearrangements during growth of hard carbon films. *Surf. Coat. Technol.* **2011**, *206*, 703–726. [CrossRef]

26. Neuville, S. *Refined Raman Spectroscopy Fundamentals for Improved Carbon Material Engineering*; Lambert Academic Publishing: Kaiserslautern, Germany, 2014; 169p, ISBN 978-3-659-48909-9.

27. Neuville, S.; Matthews, A. A perspective on the optimization of hard carbon and related coatings for engineering applications. *Thin Solid Film* **2007**, *515*, 6619–6655. [CrossRef]

28. Neuville, S. Differentiated Carbon Material for Energy Storage and Conversion. *Mater. Day Proc.* **2018**, *5*, 13837–13845. [CrossRef]

29. Neuville, S. Superconductivity described with Electron-Phonon Synchronic Coupling. *Mater. Today Proc.* **2018**, *5*, 13827–13836. [CrossRef]

30. Bosseboeuf, A.; Mathias, H. Experimental techniques for damping Characterization of micro and nanostructures. In *Advances in Multiphysic Simulation and Experimental Testing of MEMS*; Frangi, A., Cercignagni, C., Mukherjee, S., Aluru, N., Eds.; Imperial College Press: London, UK, 2008.

31. Fu, Y.; Du, H.; Huang, W.; Zhang, S.; Hu, M. TiNi-based thin films for MEMS applications. In *Advanced Materials for Micro- and Nano-Systems (AMMNS)*; Singapore-MIT Alliance (SMA): Singapore, 2004.

32. Saito, K.; Iwata, K.; Ishihara, Y.; Sugita, K.; Takato, M.; Uchikoba, F. Miniaturized Rotary Actuators Using Shape Memory Alloy for Insect-Type MEMS Microrobot. *Micromachines* **2016**, *7*, 58. [CrossRef] [PubMed]

33. Karpelson, M.; Whitney, J.P.; Wei, G.Y.; Wood, R.J. Design and fabrication of ultralight high-voltage power circuits for flapping-wing robotic insects. In Proceedings of the 2011 Twenty-Sixth Annual IEEE Applied Power Electronics Conference and Exposition (APEC), Fort Worth, TX, USA, 6–11 March 2011.

34. Zhang, M.; Llaser, N.; Mathias, H.; Dupret, A. High precision measurement of quality factor for MEMS resonators. *Procedia Chem.* **2009**, *1*, 827–830. [CrossRef]

35. Rotake, A.D. Heavy Metal Ion Detection in Water using MEMS Based Sensor. *Mater. Today Proc.* **2018**, *5*, 1530–1536. [CrossRef]

36. Polster, T.; Hofmann, M. Aluminum Nitride based 3D piezoelectric sensors. *Proc. Chem.* **2009**, *1*, 144–147. [CrossRef]

37. Stampfer, C.; Helbling, T.; Obergfell, D.; Schöberle, B.; Tripp, M.K.; Jungen, A.; Roth, S.; Bright, V.M.; Hierold, C. Fabrication of Single-Walled Carbon-Nanotube-Based Pressure Sensors. *Nano Lett.* **2006**, *6*, 233–237. [CrossRef]

38. Stampfer, C.; Jungen, A.; Linderman, R.; Obergfell, D.; Roth, S.; Hierold, C. Nano-Electromechanical Displacement Sensing Based on Single-Walled Carbon Nanotubes. *Nano Lett.* **2006**, *6*, 1449–1453. [CrossRef] [PubMed]

39. Lee, J.-I.; Song, Y.; Jung, H.; Choi, J.; Eun, Y.; Kim, J. Deformable Carbon Nanotube-Contact Pads for Inertial Microswitch to Extend Contact Time. *IEEE Trans. Ind. Electron.* **2012**, *59*, 4914–4920. [CrossRef]

40. Kaul, A.B.; Wong, E.W.; Epp, L.; Hunt, B.D. Electromechanical Carbon Nanotube Switches for High-Frequency Applications. *Nano Lett.* **2006**, *6*, 942–947. [CrossRef] [PubMed]

41. Lee, S.W.; Lee, D.S.; Morjan, R.E.; Jhang, S.H.; Sveningsson, M.; Nerushev, O.A.; Park, Y.W.; Campbell, E.E.B. A Three-Terminal Carbon Nano-relay. *Nano-Lett.* **2004**, *4*, 2027–2030. [CrossRef]

42. Jang, J.E.; Cha, S.N.; Choi, Y.; Amaratunga, G.A.J.; Kang, D.J.; Hasko, D.G.; Jung, J.E.; Kim, J.M. Nanoelectromechanical switches with vertically aligned carbon nanotubes. *Appl. Phys. Lett.* **2005**, *87*, 163114. [CrossRef]

43. Nguyen, N.T.; Huang, X.Y.; Chuan, T.K. MEMS-micropumps: A review. *J. Fluids Eng.* **2002**, *124*, 384–392. [CrossRef]

44. Roy, S.K. A Survey MEMS Micromotors Assemblies and applications. *Int. J. Eng. Res. Gen. Sci.* **2015**, *3*, 2091–2730.

45. Büttgenbach, S. Review Electromagnetic Micromotors-Design, Fabrication and Applications. *Micromachines* **2014**, *5*, 929–942. [CrossRef]

46. Hsieh, J.Y.; Kuo, P.H.; Huang, Y.C.; Huang, Y.-J.; Tsai, R.D.; Wang, T.; Chiu, H.W.; Wang, Y.H.; Lu, S.S. A Remotely-Controlled Locomotive IC Driven by Electrolytic Bubbles and Wireless Powering. *IEEE Trans. Biomed. Circuits Syst.* **2014**, *8*, 787–798. [CrossRef] [PubMed]

47. Mehta, A.M.; Pister, K. Planar two degree-of-freedom legs for walking microrobots. In Proceedings of the IARP Workshop on Micro & Nano Robotics, Paris, France, 23–24 October 2006.

48. Kovac, M. Learning from nature how to land aerial robots. *Science* **2016**, *352*, 895–896. [CrossRef] [PubMed]

49. Penskiy, I.; Bergbreiter, S. Optimized Electrostatic Inchworm Motors using a Flexible Driving Arm. *J. Micromech. Microeng.* **2013**, *23*, 015018. [CrossRef]

50. Chollet, F.; Goedgebuer, J.-P. Improved LiNbO 3 Technology for Reducing Sidelobe Asymmetry in Mode Converter-Based Wavelength Filters. *Jpn. J. Appl. Phys.* **1998**, *37*, 979–981. [CrossRef]

51. Liu, A.Q.; Zhang, X.M.; Murukeshan, V.M.; Zhang, Q.X.; Zou, Q.B.; Uppili, S. Optical Switch Using Draw-Bridge Micromirror for Large Array. In *Transducers '01 EurosensorsXV, Proceedings of the 11th International Conference on Solid-State Sensors and Actuators*; Springer: Berlin/Heidelberg, Germany, 2001; pp. 1296–1299.

52. Liu, A.; Zhao, B.; Chollet, F.; Zou, Q.; Asundi, A.; Fujita, H. Micro-opto-mechanical grating switches. *Sens. Actuators A Phys.* **2000**, *86*, 127–134. [CrossRef]

53. Ostfeld, A.E.; Gaikwad, A.M.; Khan, Y.; Arias, A.C. High-performance flexible energy storage and harvesting system for wearable electronics. *Sci. Rep.* **2016**, *6*. [CrossRef] [PubMed]

54. Arun, A.; Le Poche, H.; Idda, T.; Acquaviva, D.; Badia, M.F.-B.; Pantigny, P.; Salet, P.; Ionescu, A.M. Tunable MEMS capacitors using vertical carbon nanotube arrays grown on metal lines. *Nanotechnology* **2010**, *22*, 025203. [CrossRef] [PubMed]

55. Despont, M.; Brugger, J.; Drechsler, U.; Dürig, U.; Häberle, W.; Lutwyche, M.; Rothuizen, H.; Stutz, R.; Widmer, R.; Binnig, G.; et al. VLSI-NEMS chip for parallel AFM data storage. *Sens. Actuators A Phys.* **2000**, *80*, 100–107. [CrossRef]

56. Waldner, J.P. *Nano-computers and Swarm Intelligence*; ISTE John Wiley &Sons: London, UK, 2008; p. 205. ISBN 978-1-84821-009-7.

57. Shafagh, R.; Vastesson, A.; Guo, W.; van der Wijngaart, W.; Haraldsson, T. E-Beam Nano-structuring and Direct Click Biofunctionalization of Thiol–Ene Resist. *ACS Nano.* **2018**, *12*, 9940–9946. [CrossRef]

58. Draz, M.S.; Kochehbyoki, K.M.; Vasan, A.; Battalapalli, D.; Sreeram, A.; Kanakasabapathy, M.K.; Kallakuri, S.; Tsibris, A.; Kuritzkes, D.R.; Shafiee, H. DNA engineered micromotors powered by metal nanoparticles for motion -based cellphone diagnostics. *Nat. Commun.* **2018**, *9*. [CrossRef]

59. Godssi, R.; Lin, P. *MEMS Materials and Processes Handbook*; Springer: Berlin, Germany, 2011; ISBN 978-0-387-47316-1.

60. Zhang, S. *Handbook of Nanostructured Thin Films and Coatings: Mechanical Properties*; CRC Press Taylor & Francis Group: Boca Raton, FL, USA, 2010; ISBN 13:978-1-4200-9403-9.

61. Madou, M.J. Fundamentals of Microfabrication and Nanotechnology. In *MEMS to Bio-MEMS. Manufacturing Techniques and Applications*; CRC Press: Boca Raton, FL, USA, 2011; Volume III, p. 252, ISBN 143-989-5244.

62. Rao, A.V.; Yadav, S.K. An introduction to Nano Electro Mechanical Systems. *Int. J. Eng. Sci. (IJES)* **2015**, *4*, 6–9.

63. Ventra, M.D.; Evoy, S.; Heflin, J.R., Jr. Introduction to Nanoscale Science and Technology. In *Nanostructure Science and Technology*; Springer: Berlin, Germany, 2004; ISBN 978-1-4020-7720-3.

64. Mahakud, R.; Panigrahi, M.; Rath, S. Various Technological Aspects of Nano Electro Mechanical Systems—A Review. *Int. J. Anal. Exp. Finite Element Anal. (IJAEFEA)* **2014**, *1*, 1–5.

65. Peng, H.; Li, Q.; Chen, T. (Eds.) *Industrial Applications of Carbon Nanotubes*; Elsevier: Amsterdam, The Netherlands, 2017; 508p, ISBN 978-0-323-41481-4.

66. Garcia, J.C.; Justo, J.F. Twisted ultrathin silicon nanowires: A possible torsion electromechanical nanodevice. *EPL Europhys. Lett.* **2014**, *108*, 36006. [CrossRef]

67. Ke, C.; Espinosa, H.D. In Situ Electron Microscopy Electromechanical Characterization of a Bistable NEMS Device. *Small* **2006**, *2*, 1484–1489. [CrossRef] [PubMed]

68. Ernst, T.; Hentz, S.; Arcamone, J.; Agache, V.; Duraffourg, L.; Ouerghi, I.; Ludurczak, W.; Ladner, C.; Ollier, E.; Andreucci, P.; et al. High performance NEMS devices for sensing applications. In Proceedings of the 2015 45th European Solid-State Device Research Conference (ESSDERC), Graz, Austria, 14–18 September 2015; pp. 31–35.

69. Li, M.; Tang, H.X.; Roukes, M.L. Ultra-sensitive NEMS-based cantilevers for sensing, scanned probe and very high-frequency applications. *Nat. Nanotechnol.* **2007**, *2*, 114–120. [CrossRef] [PubMed]

70. Tao, Y.; Boss, J.M.; Moores, B.A.; Degen, C.L. Single-crystal diamond nanomechanical resonators with quality factors exceeding one million. *Nat. Commun.* **2014**, *5*. [CrossRef] [PubMed]

71. Davis, Z.J.; Abadal, G.; Helbo, B.; Hansen, O.; Campabadal, F.; Pérez-Murano, F.; Esteve, J.; Figueras, E.; Ruiz, R.; Barniol, N.; et al. High Mass and Spatial Resolution Mass Sensor based on Resonating Nano-Cantilevers Integrated with CMOS. In *Transducers '01 Eurosensors XV*; Obermeier, E., Ed.; Springer: Berlin/Heidelberg, Germany, 2001.

72. Wang, Y.-H.; Lee, C.-Y.; Chiang, C.-M. A MEMS-based Air Flow Sensor with a Free-standing Micro-cantilever Structure. *Sensors* **2007**, *7*, 2389–2401. [CrossRef] [PubMed]

73. Vajpayee, S.; Kumar, B.; Thakur, R.; Kumar, M. Design and Development of Nano pH Sensor and Interfacing with Arduino. *Int. J. Electron. Electr. Comput. Syst. IJEECS* **2017**, *8*, 6. [CrossRef]

74. Wang, J. *Nanomachines: Fundamental and Application*; Wiley: Hoboken, NJ, USA, 2013.

75. Yadav, V.; Duan, W.; Butler, P.J.; Sen, A. Anatomy of Nanoscale Propulsion. *Annu. Rev. Biophys.* **2015**, *44*, 77–100. [CrossRef]

76. Ren, L.; Zhou, D.; Mao, Z.; Xu, P.; Huang, T.J.; Mallouk, T.E. Rheotaxis of Bimetallic Micromotors Driven by Chemical–Acoustic Hybrid Power. *ACS Nano* **2017**, *11*, 10591–10598. [CrossRef]

77. Das, S.; Garg, A.; Campbell, A.I.; Howse, J.; Sen, A.; Velegol, D.; Golestanian, R.; Ebbens, S.J. Boundaries can steer active Janus spheres. *Nat. Commun.* **2015**, *6*. [CrossRef]

78. Duan, W.; Ibele, M.; Liu, R.; Sen, A. Motion analysis of light-powered autonomous silver chloride nanomotors. *Eur. Phys. J. E* **2012**, *35*. [CrossRef]

79. Wong, F.; Sen, A. Progress Towards Light-Harvesting Self-Electrophoretic Motors: Highly Efficient Bimetallic Nanomotors and Micropumps in Halogen Media. *ACS Nano* **2016**, *10*, 7172–7179. [CrossRef] [PubMed]

80. Duan, W.; Wang, W.; Das, S.; Yadav, V.; Mallouk, T.E.; Sen, A. Synthetic Nano- and Micromachines in Analytical Chemistry: Sensing, Migration, Capture, Delivery, and Separation. *Annu. Rev. Anal. Chem.* **2015**, *8*, 311–333. [CrossRef] [PubMed]

81. Jurado-Sánchez, B.; Wang, J. Micromotors for environmental applications: A review. *Environ. Sci. Nano Issue* **2018**, *5*, 1530–1544. [CrossRef]

82. Ghosh, A.; Dasgupta, D.; Pal, M.; Morozov, K.; Lehshansky, A.; Ghosh, A. Helical Nano-machines as Mobile Viscometers. *Adv. Funct. Mater.* **2018**, *28*, 1705687. [CrossRef]

83. Hindi, S.S.Z. Molecular Machines: I. An Overview of Biological and Synthetic Angstromic Devices. *Nanosci. Nanotechnol. Res.* **2017**, *4*, 98–105.

84. Bamrungsap, S.; Phillips, J.A.; Xiong, X.; Kim, Y.; Wang, H.; Liu, H.; Hebard, A.; Tan, W. Magnetically driven single DNA nanomotor. *Small* **2011**, *7*, 601–605. [CrossRef] [PubMed]

85. Pal, M.; Somalwar, N.; Singh, A.; Bhat, R.; Eswarappa, S.M.; Saini, D.K.; Ghosh, A. Maneuverability of Magnetic Nanomotors Inside Living Cells. *Adv. Mater.* **2018**, *30*, 1800429. [CrossRef] [PubMed]

86. Dietrich-Buchecker, C.O.; Jimenez-Molero, M.C.; Sartor, V.; Sauvage, J.-P. Rotaxanes and catenanes as prototypes of molecular machines and motors. *Pure Appl. Chem.* **2003**, *75*, 1383–1393. [CrossRef]

87. Kudernac, T.; Ruangsupapichat, N.; Parschau, M.; Maciá, B.; Katsonis, N.; Harutyunyan, S.R.; Ernst, K.H.; Feringa, B.L. Electrically driven directional motion of a four-wheeled molecule on a metal surface. *Nature* **2011**, *479*, 208–211. [CrossRef] [PubMed]

88. Coskun, A.; Banaszak, M.; Astumian, R.D.; Stoddart, J.F.; Grzybowski, B.A. Great expectations: Can artificial molecular machines deliver on their promise? *Chem. Soc. Rev.* **2012**, *41*, 19–30. [CrossRef] [PubMed]

89. Bianco, B.; Moggia, E.; Giordano, S.; Rocchia, W.; Chiabrera, A. Friction and noise in quantum mechanics: A model for the interactions between a system and a thermal bath. *Il Nuovo Cim. B* **2001**, *116*, 155–165.

90. Farias, M.B.; Fosco, C.D.; Lombardo, F.C.; Mazzitelli, F.D. Quantum friction between graphene sheets. *Phys. Rev. D* **2017**, *95*, 065012. [CrossRef]

91. Jiang, Q.D.; Wilczek, F. Chiral Casimir forces: Repulsive, enhanced, tunable. *Phys. Rev. B* **2019**, *99*, 125403. [CrossRef]

92. Bennett, S.D.; Clerk, A.A. Laser-like instabilities in quantum nano-electromechanical systems. *Phys. Rev. B* **2006**, *74*, 201301. [CrossRef]

93. Cao, A.; Baskaran, R.; Frederick, M.; Turner, K.; Ajayan, P.; Ramanath, G. Direction-Selective and Length-Tunable In-Plane Growth of Carbon Nanotubes. *Adv. Mater.* **2003**, *15*, 1105–1109. [CrossRef]

94. Choi, J.; Pyo, S.; Baek, D.-H.; Lee, J.-I.; Kim, J. Thickness-, alignment- and defect-tunable growth of carbon nanotube arrays using designed mechanical loads. *Carbon* **2014**, *66*, 126–133. [CrossRef]

95. Hutchison, D.N.; Morrill, N.B.; Aten, Q.; Turner, B.W.; Jensen, B.D.; Howell, L.L.; Vanfleet, R.R.; Davis, R.C. Carbon Nanotubes as a Framework for High-Aspect-Ratio MEMS Fabrication. *J. Microelectromech. Syst.* **2010**, *19*, 75–82. [CrossRef]

96. Choi, J.; Lee, J.-I.; Eun, Y.; Kim, M.-O.; Kim, J. Aligned Carbon Nanotube Arrays for Degradation-Resistant, Intimate Contact in Micromechanical Devices. *Adv. Mater.* **2011**, *23*, 2231–2236. [CrossRef]

97. Choi, J.; Eun, Y.; Kim, J. Investigation of Interfacial Adhesion between the Top Ends of Carbon Nanotubes. *ACS Appl. Mater. Interfaces* **2014**, *6*, 6598–6605. [CrossRef]

98. Eun, Y.; Lee, J.I.; Choi, J.; Song, Y.; Kim, J. Integrated Carbon Nanotube Array as Dry Adhesive for High-Temperature Silicon Processing. *Adv. Mater.* **2011**, *23*, 2231–2236. [CrossRef]

99. Goold, J.; Huber, M.; Riera, A.; Del Rio, L.; Skrzypczyk, P. The role of quantum information in thermodynamics—A topical review. *J. Phys. A Math. Theor.* **2016**, *49*, 143001. [CrossRef]

100. Neuville, S. Transient Transversal Electric Field by Adsorbtion on Semiconductors. *SNB* **2007**, *121*, 436.

101. Caschera, D.; Federici, F.; Zane, D.; Focanti, F.; Curulli, A.; Padeletti, G. Gold nanoparticles modified GC electrodes: Electrochemical behavior dependence of different neurotransmitters and molecules of biological interest on the particles size and shape. *J. Nanoparticle Res.* **2009**, *11*, 1925. [CrossRef]

102. Krishnaraj, C.; Jagan, E.; Rajasekar, S.; Selvakumar, P.; Kalaichelvan, P.; Mohan, N. Synthesis of silver nanoparticles using Acalypha Indica leaf extracts and its antibacterial activity against water borne pathogens. *Colloids Surf. B Biointerfaces* **2010**, *76*, 50–56. [CrossRef] [PubMed]

103. Zhu, X.; Radovic-Moreno, A.F.; Wu, J.; Shi, R.L.J. Nano-medicine in the management of microbial infection overview and perspectives. *Nano Today* **2014**, *9*, 478–498. [CrossRef]

104. Waqar, A.; Abdelbary, E.; Vinod, D.; Karthikeyan, S. Emerging Nanotechnologies in Dentistry (Second Edition), Chapter 18—Carbon nanotubes: Applications in cancer therapy and drug delivery research. *Micro Nano Technol.* **2018**, 371–389. [CrossRef]

105. Sharifi, S.; Behzadi, S.; Laurent, S.; Forrest, M.L.; Stroeve, P.; Mahmoudi, M. Toxicity of nanomaterials. *Chem. Soc. Rev.* **2012**, *41*, 2323–2343. [CrossRef]

106. Ramsden, J.J. Assessing the toxic risks of the nanotechnology industry. *Nanotechnol. Percept.* **2013**, *9*, 119–134. [CrossRef]

107. Madani, S.Y.; Mandel, A.; Seifalian, A.M. A concise review of carbon nanotube's toxicology. *Nano Rev.* **2013**, *4*. [CrossRef]

108. Choi, J.; Kim, J. Batch-processed carbon nanotube wall as pressure and flow sensor. *Nanotechnology* **2010**, *21*, 105502. [CrossRef]

109. Hayamizu, Y.; Yamada, T.; Mizuno, K.; Davis, R.C.; Futaba, D.N.; Yumura, M.; Hata, K. Integrated three-dimensional micro-electromechanical devices from processable carbon nanotube wafers. *Nat. Nanotechnol.* **2008**, *3*, 289–294. [CrossRef]

110. Lee, J.I.; Eun, Y.; Choi, J.; Kwon, D.S.; Kim, J. Using Confined Self-Adjusting Carbon Nanotube Arrays as High-Sensitivity Displacement Sensing Element. *ACS Appl. Mater. Interfaces* **2014**, *6*, 10181–10187. [CrossRef]

111. Silva, S.P.R.; Carey, J.D.; Kahn, R.U.A.; Gerstner, E.G.; Anguita, J.V. Amorphous carbon thin films. In *Handbook of Thin Film Materials*; Nalwa, H.S., Ed.; Academic Press: New York, NY, USA, 2002; Chapter 9; Volume 4, pp. 403–505.

112. Casiraghi, C.; Robertson, J.; Ferrari, A. Review. Diamond Like Carbon for data storage and beer storage. *Mater. Day* **2007**, *10*, 44–53.

113. Huong, P.V. Structural studies of diamond films and ultrahard materials by Raman spectroscopy. *Diam. Relat. Mater.* **1991**, *1*, 31–41. [CrossRef]

114. Bachmann, P.K.; van Enckevort, W. Diamond deposition technologies. *Diam. Relat. Mater.* **1992**, *1*, 1021–1034. [CrossRef]

115. Nimmagadda, R.R.; Joshi, J.; Hsu, L. Role of microstructure on the oxidation behavior of microwave plasma synthetized diamond and diamond like carbon. *J. Mater. Res.* **1990**, *5*, 2445–2450. [CrossRef]

116. McNamara, K.; Gleason, K.; Vestyck, D.; Butler, J.; Butler, J. Evaluation of diamond films by nuclear magnetic resonance and Raman spectroscopy. *Diam. Relat. Mater.* **1992**, *1*, 1145–1155. [CrossRef]

117. Fayette, L.; Marcus, B.; Mermoux, M.; Tourillon, G.; Laffon, K.; Parent, P.; Le Normand, F. Local order in CVD diamond films: Comparative Raman, x-ray-diffraction, and x-ray-absorption near-edge studies. *Phys. Rev. B* **1998**, *57*, 14123–14132. [CrossRef]

118. Hanada, K.; Nishiyama, T.; Yoshitake, T.; Nagayama, K. Time-Resolved Observation of Deposition Process of Ultra-nanocrystalline Diamond/Hydrogenated Amorphous Carbon Composite Films in Pulsed Laser Deposition. *Hindawi J. Nanomater.* **2009**, *2009*, 901241.

119. Hernberg, R.; Lepisto, T.; Mantyla, T.; Stenberg, T.; Vattulainen, J. Diamond film synthesis on Mo in thermal RF plasma. *Diam. Relat. Mater.* **1992**, *1*, 255–261. [CrossRef]

120. Kim, T.-Y.; Lee, C.S.; Lee, Y.J.; Lee, K.-R.; Chae, K.-H.; Oh, K.H. Reduction of the residual compressive stress of tetrahedral amorphous carbon film by Ar background gas during the filtered vacuum arc process. *J. Appl. Phys.* **2007**, *101*, 023504. [CrossRef]

121. Neuville, S. New applications perspectives on ta-C coating. *QScience Connect* **2014**, *2014*, 8. [CrossRef]

122. Reisse, G.; Weissmantel, S.; Rost, D. Preparation of super-hard coatings by pulsed laser deposition. *Appl. Phys. A* **2004**, *79*, 1275–1278. [CrossRef]

123. Shi, X.; Tay, B.K.; Lau, S.P. The Double Bent Filtered Cathodic Arc Technology and its Applications. *Int. Mod. Phys. B* **2000**, *14*, 136–153. [CrossRef]

124. Schultrich, B. *Tetrahedrally Bonded Amorphous Carbon Films I: Basics, Structure and Preparation*; Springer Nature: Berlin, Germany, 2018; 752p, ISBN 978-3-662-55927-7.

125. Charitidis, C.A. Nanomechanical and nanotribological properties of carbon-based thin films: A review. *Int. J. Refract. Met.* **2010**, *28*, 51–70. [CrossRef]

126. Sullivan, J.P.; Friedmann, T.A.; de Boer, M.P.; la Van, D.A.; Hohlfelder, R.J.; Ashby, C.I.H.; Dugger, M.T.; Mitchell, M.; Dunn, R.G.; Magerkurth, A.J. Developing a new material for MEMS: Amorphous diamond. In Proceedings of the Materials Research Society of MRS Fall Meating 2000 Symposium, Boston, MA, USA, USA, 27 November–1 December 2001; Volume 657.

127. Mel'nichenko, V.M.; Sladkov, A.M.; Nikulin, Y.N. Structure of Polymeric Carbon. In *Russian Chemical Reviews*; The British Library: London, UK, 1982; Volume 51.

128. Li, X.; Masters, A.F.; Maschmeyer, T. Polymeric carbon nitride for solar hydrogen production. *Chem. Commun.* **2017**, *53*, 7438–7446. [CrossRef]

129. Bredas, J.L.; Street, G.B. Electronic structure of amorphous hydrogenated carbon films. In Proceedings of the Vol. XVII MRS Strasbourg Meeting, Strasbourg, France, 2 June 1987; Koidl, P., Oelhafen, P., Eds.; Les Editions de la Physique: Paris, France, 1987; pp. 237–250.

130. Wagner, J.; Ramsteiner, M.; Wild, C. Amorphous hydrogenated carbon films. In Proceedings of the Vol. XVII. MRS Strasbourg Meeting, Strasbourg, France, 2 June 1987; Koidl, P., Oelhafen, P., Eds.; Les Editions de la Physique: Paris, France, 1987; pp. 219–228.

131. Santra, T.S.; Bhattacharyya, T.K.; Patel, P.; Tseng, F.G.; Barik, T.K. Diamond, Diamond-Like Carbon (DLC) and Diamond-Like Nanocomposite (DLN) Thin Films for MEMS Applications. *Microelectromech. Syst. D Devices* **2012**, *28*, 459–480.

132. Badzian, A.R.; Bachmann, P.K.; Hartnett, T.; Badzian, T.; Messier, R. Diamond thin films prepared by PACVD processes. In Proceedings of the Vol. XVII MRS Strasbourg Meeting, Strasbourg, France, 2 June 1987; Koidl, P., Oelhafen, P., Eds.; Les Editions de la Physique: Paris, France, 1987; pp. 63–77.

133. Schueller, O.J.A.; Brittain, S.T.; Marzolin, C.; Whitesides, G.M. Fabrication and Characterization of Glassy Carbon MEMS. *Chem. Mater.* **1997**, *9*, 1399–1406. [CrossRef]

134. Chaudhury, S.; Sinha, S.K. Carbon Nanotube and Nanowires for Future Semiconductor Devices Applications. In *Nanoelectronics*; Kaush, B.K., Ed.; Devices, Circuits and Systems; Elsevier Inc.: Amsterdam, The Netherlands, 2018; Chapter 12, pp. 375–398, ISBN 978-0-12-813353.

135. Spencer, J.H.; Nesbitt, J.M.; Trewhitt, H.; Kashtiban, R.J.; Bell, G.; Ivanov, V.G.; Faulques, E.; Sloan, J.; Smith, D.C. Raman Spectroscopy of Optical Transitions and Vibrational Energies of ~1 nm HgTe Extreme Nanowires within Single Walled Carbon Nanotubes. *ACS Nano* **2014**, *8*, 9044–9052. [CrossRef]

136. Lin, Y.Y.; Wei, H.W.; Leou, K.C.; Lin, H.; Tung, C.H.; Wei, M.T.; Lin, C.; Tsai, C.H. Experimental characterization of an inductively coupled acetylene/hydrogen plasma for Carbon Nano Fiber synthesis (CNF). *J. Vac. Sci. Technol. B* **2006**, *24*, 97–112. [CrossRef]

137. Iijima, S. Synthesis of Carbon Nanotubes. *Nature* **1991**, *354*, 56–58. [CrossRef]

138. Endo, M.; Strano, M.S.; Ajayan, P.M. Potential Applications of Carbon Nanotubes. In *Carbon Nanotubes*; Jorio, A., Dresselhaus, G., Dresselhaus, M.S., Eds.; Topics in Applied Physics; Springer: New York, NY, USA, USA, 2007; pp. 13–62, ISBN 978-3-540-72864-1.

139. Endo, M.; Iijima, S.; Dresselhaus, M.S. *Carbon Nanotubes*; Elsevier: Amsterdam, The Netherlands, 1997; 198p, ISBN 9780080426822.

140. Küttel, O.M.; Gröning, O.; Emmenegger, C.; Nilsson, L.; Maillard, E.; Diederich, L.; Schlapbach, L. Field emission from diamond, diamond-like and nanostructured carbon films. *Carbon* **1999**, *37*, 745–752. [CrossRef]

141. Giubileo, F.; di Bartolomeo, A.; Iemmo, L.; Luongo, G.; Urban, F. Field Emission from Carbon Nanostructures. *Appl. Sci.* **2018**, *8*, 526. [CrossRef]

142. Fedoseeva, Y.V.; Bulusheva, L.G.; Okotrub, A.V.; Kanygin, M.A.; Gorodetskiy, D.V.; Asanov, I.P.; Vyalikh, D.V.; Puzyr, A.P.; Bondar, V.S. Field Emission Luminescence of Nano-Diamonds Deposited on the Aligned Carbon Nanotube Array. *Sci. Rep.* **2015**, *5*, 9379. [CrossRef]

143. Bonard, J.M.; Weiss, N.; Kind, H.; Stöckli, T.; Forró, L.; Kern, K.; Châtelain, A. Tuning the Field Emission Properties of Patterned Carbon Nanotube Films. *Adv. Mater.* **2001**, *13*, 184–188. [CrossRef]

144. Clark, I.T.; Yoshimura, M. Electronic properties of carbon nanotube. In *Fabrication of Carbon Nanotubes for High-Performance Scanning Probe Microscopy*; Intechopen: London, UK, 2011; pp. 91–104.

145. Tanaka, K.; Yoshimura, M.; Ueda, K. High-Resolution Magnetic Force Microscopy Using Carbon Nanotube Probes Fabricated Directly by Microwave Plasma-Enhanced Chemical Vapor Deposition. *J. Nanomater.* **2009**, *2009*. [CrossRef]

146. Laocharoensuk, R.; Burdick, J.; Wang, J. Carbon-Nanotube-Induced Acceleration of Catalytic Nanomotors. *ACS Nano* **2008**, *2*, 1069–1075. [CrossRef]

147. Inami, N.; Shikoh, E.; Mohamed, M.A.; Fujiwara, A. Synthesis-condition dependence of carbon nanotube growth by alcohol catalytic chemical vapor deposition method. *Sci. Technol. Adv. Mater.* **2007**, *8*, 292–295. [CrossRef]

148. Hermann, S.; Ecke, R.; Schulz, S.; Gessner, T. Controlling the formation of nanoparticles for definite growth of carbon nanotubes for interconnect applications. *Microelectron. Eng.* **2008**, *85*, 1979–1983. [CrossRef]

149. Costa, S.; Borowiak-Palen, E.; Kruszynska, M.; Bachmatiuk, A.; Kalenczuk, R.J. Characterization of carbon nanotubes by Raman spectroscopy. *Mater. Sci. Pol.* **2008**, *26*, 433–441.

150. Dresselhaus, M.; Dresselhaus, G.; Saito, R.; Jorio, A. Raman spectroscopy of carbon nanotubes. *Phys. Rep.* **2005**, *409*, 47–99. [CrossRef]

151. Malard, L.M.; Pimenta, M.A.; Dresselhaus, G.; Dresselhaus, M.S. Raman spectroscopy in graphene. *Phys. Rep.* **2009**, *473*, 51–87. [CrossRef]

152. Ritter, K.A.; Lyding, J.W. The influence of edge structure on the electronic properties of graphene quantum dots and nanoribbons. *Nat. Mater.* **2009**, *8*, 235–242. [CrossRef] [PubMed]

153. Begliarbekov, M.; Sul, O.; Kalliakos, S.; Yang, E.-H.; Strauf, S. Determination of edge purity in bilayer graphene using μ-Raman spectroscopy. *Appl. Phys. Lett.* **2010**, *97*, 31908. [CrossRef]

154. You, Y.; Ni, Z.; Yu, T.; Shen, Z. Edge chirality determination of graphene by Raman spectroscopy. *Appl. Phys. Lett.* **2008**, *93*, 163112. [CrossRef]

155. Xu, Y.N.; Zhan, D.; Liu, L.; Suo, H.; Ni, Z.H.; Nguyen, T.T.; Zhao, C.; Shen, Z.X. Thermal Dynamics of Graphene Edges Investigated by Polarized Raman Spectroscopy. *ACS Nano* **2011**, *5*, 142–152. [CrossRef] [PubMed]

156. Iakoubovskii, K.; Minami, N.; Ueno, T.; Kazaoui, S.; Kataura, H. Optical Characterization of Double-Wall Carbon Nanotubes: Evidence for Inner Tube Shielding. *J. Phys. Chem. C* **2008**, *112*, 11194–11198. [CrossRef]

157. Mogensen, K.B.; Kutter, J.P. Carbon nanotube based stationary phases for microchip chromatography. *Lab Chip* **2012**, *12*, 1951–1958. [CrossRef] [PubMed]

158. Bachilo, S.M. Structure-Assigned Optical Spectra of Single-Walled Carbon Nanotubes. *Science* **2002**, *298*, 2361–2366. [CrossRef] [PubMed]

159. Cherukuri, P.; Bachilo, S.M.; Litovsky, S.H.; Weisman, R.B. Near-Infrared Fluorescence Microscopy of Single-Walled Carbon Nanotubes in Phagocytic Cells. *J. Am. Chem. Soc.* **2004**, *126*, 15638–15639. [CrossRef]

160. Yum, K.; McNicholas, T.P.; Mu, B.; Strano, M.S. Single-Walled Carbon Nanotube-Based Near-Infrared Optical Glucose Sensors toward In Vivo Continuous Glucose Monitoring. *J. Diabetes Sci. Technol.* **2013**, *7*, 72–87. [CrossRef] [PubMed]

161. He, X.; Wang, X.; Nanot, S.; Cong, K.; Jiang, Q.; Kane, A.A.; Goldsmith, J.E.M.; Hauge, R.H.; Leonard, F.; Kono, J. Photo-thermoelectric p-n junction photodetector with intrinsic broadband polarimetry based on macroscopic carbon nanotube films. *ACS Nano* **2013**, *7*, 7271–7277. [CrossRef] [PubMed]

162. Janas, D.; Herman, A.P.; Boncel, S.; Koziol, K.K. Iodine monochloride as a powerful enhancer of electrical conductivity of carbon nanotube wires. *Carbon* **2014**, *73*, 225–233. [CrossRef]

163. Zhao, Y.; Wei, J.; Vajtai, R.; Ajayan, P.M.; Barrera, E.V. Iodine doped carbon nanotube cables exceeding specific electrical conductivity of metals. *Sci. Rep.* **2011**, *1*. [CrossRef] [PubMed]

164. Subramaniam, C.; Yamada, T.; Kobashi, K.; Sekiguchi, A.; Futaba, D.N.; Yumura, M.; Hata, K. One-hundred-fold increase in current carrying capacity in a carbon nanotube–copper composite. *Nat. Commun.* **2013**, *4*. [CrossRef] [PubMed]

165. Kim, I.T.; Tannenbaum, A.; Tannenbaum, R. Anisotropic conductivity of magnetic carbon nanotubes embedded in epoxy matrices. *Carbon* **2011**, *49*, 54–61. [CrossRef] [PubMed]

166. Wu, Z.-S.; Ren, W.; Xu, L.; Li, F.; Cheng, H.-M. Doped Graphene Sheets as Anode Materials with Superhigh Rate and Large Capacity for Lithium Ion Batteries. *ACS Nano* **2011**, *5*, 5463–5471. [CrossRef]

167. Javey, A.; Guo, J.; Farmer, D.B.; Wang, Q.; Yenilmez, E.; Gordon, R.G.; Lundstrom, M.; Dai, H. Self-Aligned Ballistic Molecular Transistors and Electrically Parallel Nanotube Arrays. *Nano Lett.* **2004**, *4*, 1319–1322. [CrossRef]

168. Hasan, S.; Salahuddin, S.; Vaidyanathan, M.; Alam, M.A. High frequency performance projections for ballistic carbon nano-tube transistors. *IEEE Trans. Nanotechnol.* **2006**, *5*, 14–22. [CrossRef]

169. Ritter, U.; Scharff, P.; Sigmund, C.; Dimmytrenko, O.; Kulish, N.; Prylutskyy, Y.; Belyi, N.; Gubanov, V.; Komarova, L.; Lizunova, S.; et al. Radiation damage to multi-walled carbon nanotubes and their Raman vibrational modes. *Carbon* **2006**, *44*, 2694–2700. [CrossRef]

170. Dresselhaus, M.S.; Jorio, A.; Filho, A.G.S.; Saito, R. Defect characterization in graphene and carbon nanotubes using Raman spectroscopy. *Philos. Trans. R. Soc. A Math. Phys. Eng. Sci.* **2010**, *368*, 5355–5377. [CrossRef] [PubMed]

171. Banhart, F.; Kotakoski, J.; Krasheninnikov, A.V. Structural Defects in Graphene. *ACS Nano* **2011**, *5*, 26–41. [CrossRef] [PubMed]

172. Campos-Delgado, J.; Kim, Y.A.; Hayashi, T.; Morelos-Gomez, A.; Hofmann, M.; Muramatsu, H.; Endo, M.; Terrones, H.; Shull, R.D.; Dresselhaus, M.S.; et al. Thermal stability studies of CVD-grown graphene nanoribbons: Defect annealing and loop formation. *Chem.Phys. Lett.* **2009**, *469*, 177–182. [CrossRef]

173. Huong, P.V.; Marcus, B.; Mermoux, M.; Veirs, D.K.; Rosenblatt, G.M. Diamond-like films prepared by microwave plasma assisted chemical vapor deposition and by magnetron sputtering. *Diam. Relat. Mater.* **1992**, *1*, 869–873. [CrossRef]

174. Flamant, G.; Guillard, R.; Laplaz, D. La synthese des fullerenes par energie solaire. *Le Vide* **2001**, *2–4*, 266.

175. Sato, F. Diamond-like bonds in amorphous hydrogenated carbon films induced by x-ray irradiation. *J. Vac. Sci. Technol. A* **1998**, *16*, 2553–2555. [CrossRef]

176. Balat-Pichelin, M.; Eck, J.; Sans, J. Thermal radiative properties of carbon materials under high temperature and vacuum ultra-violet (VUV) radiation for the heat shield of the Solar Probe Plus mission. *Appl. Surf. Sci.* **2012**, *258*, 2829–2835. [CrossRef]

177. Matsui, Y.; Yabe, H.; Sugimoto, T.; Hirose, Y. The structure of acetylene flames for diamond synthesis. *Diam. Relat. Mater.* **1991**, *1*, 19–24. [CrossRef]

178. Hinneberg, H.-J.; Eck, M.; Schmidt, K. Hot-filament-grown diamond films on Si: Characterization of impurities. *Diam. Relat. Mater.* **1992**, *1*, 810–813. [CrossRef]

179. Smith, D.; Sevillano, E.; Besen, M.; Berkman, V.; Bourget, L. Large area diamond reactor using a high flow velocity microwave plasma jet. *Diam. Relat. Mater.* **1992**, *1*, 814–817. [CrossRef]

180. Schultrich, B.; Scheibe, H.-J.; Drescher, D.; Ziegele, H. Deposition of superhard amorphous carbon films by pulsed vacuum arc deposition. *Surf. Coat. Technol.* **1998**, *98*, 1097–1101. [CrossRef]

181. Liu, D.; Tu, J.; Hong, C.; Gu, C.; Mao, S. Two-phase nanostructured carbon nitride films prepared by direct current magnetron sputtering and thermal annealing. *Surf. Coat. Technol.* **2010**, *205*, 152–157. [CrossRef]

182. Iwaki, M.; Takahashi, K.; Sekiguchi, A. Wear property and structure of implanted glassy carbon. *J. Mater. Res.* **1990**, *5*, 2562–2566. [CrossRef]

183. Wang, Z.; Wang, C.; Wang, Q.; Zhang, J. Annealing effect on the microstructure modification and tribological properties of amorphous carbon nitride films. *J. Appl. Phys.* **2008**, *104*, 073306. [CrossRef]

184. Li, J.; Zheng, W.; Bian, H.J.; Lu, X.Y.; Jiang, Z.G.; Bai, Y.Z.; Jin, Z.S.; Zhao, Y.N. The effect of annealing on the field emission properties of amorphous CNx films. *Acta Phys. Sin.* **2003**, *52*, 1797–1801.

185. Borroni-Bird, C.E.; King, D.A. An ultrahigh vacuum single crystal adsorption micro-calorimeter. *Rev. Sci. Instrum.* **1991**, *62*, 2177–2185. [CrossRef]

186. Sankara-Reddy, K.S.; Satyam, M. Structural ordering of diamond like carbon films by applied electric field. *Solid State Commun.* **1995**, *93*, 797–799. [CrossRef]

187. De Padova, P.; Quaresima, C.; Ottaviani, C.; Sheverdyaeva, P.M.; Moras, P.; Carbone, C.; Topwal, D.; Olivieri, B.; Kara, A.; Oughaddou, H.; et al. Evidence of graphene-like electronic signature in silicene nanoribbons. *Appl. Phys. Lett.* **2010**, *96*, 261905. [CrossRef]

188. Endo, M.; Kim, Y.A.; Matusita, T.; Hayashi, T. From Vapor-Grown Carbon Fibers (VGCFs) to Carbon Nanotubes. In *Carbon Filaments and Nanotubes: Common Origins, Differing Applications?* NATO Science Series (Series E: Applied Sciences); Biró, L.P., Bernardo, C.A., Tibbetts, G.G., Lambin, P., Eds.; Springer: Dordrecht, The Netherlands, 2001; Volume 372, ISBN 978-0-7923-6908.

189. Spear, K.E.; Phelps, A.W.; White, W.B.; Phelps, A. Diamond polytypes and their vibrational spectra. *J. Mater. Res.* **1990**, *5*, 2277–2285. [CrossRef]

190. Phelps, A.W.; Howard, W.; White, W.B.; Spear, K.E.; Huang, D. Diamond, Boron Nitride, Silicon Carbide, and Related Wide Bandgap Semiconductors. In Proceedings of the Materials Research Society Symposium, Boston, MA, USA, 27 Novermber–1 December 1989; Glass, J.T., Messier, R.F., Fujimori, N., Eds.; Materials Research Society: Pittsburgh, PA, USA, 1989.

191. Beeman, D.; Silverman, J.; Lynds, R.; Anderson, M.R. Modeling studies of amorphous carbon. *Phys. Rev. B* **1984**, *30*, 870–875. [CrossRef]

192. Sønderby, S.; Berthelsen, A.; Almtoft, K.; Christensen, B.; Nielsen, L.; Bøttiger, J. Optimization of the mechanical properties of magnetron sputtered diamond-like carbon coatings. *Diam. Relat. Mater.* **2011**, *20*, 682–686. [CrossRef]

193. Ohl, A.; Röpke, J. Mechanism of substrate heating in diamond-forming low-pressure microwave discharges. *Diam. Relat. Mater.* **1992**, *1*, 243–247. [CrossRef]

194. Neuville, S. The enhancement of interconnected sp3 sites by chemical effects during ta-C film growth. *Diam. Relat. Mater.* **2002**, *11*, 1721–1730. [CrossRef]

195. Wang, X.H.; Ma, G.H.M.; Zhu, W.; Glass, J.T.; Bergman, L.; Turner, K.F.; Nemanich, R.J. Effect of Boron doping on the surface morphology of and structural imperfection of diamond films. *Diam. Relat. Mater.* **1992**, *1*, 828–835. [CrossRef]

196. Zhou, C.; Chen, S.; Lou, J.; Wang, J.; Yang, Q.; Liu, C.; Huang, D.; Zhu, T. Graphene's cousin: The present and future of graphane. *Nanoscale Res. Lett.* **2014**, *9*. [CrossRef] [PubMed]

197. Elias, D.C.; Nair, R.R.; Mohiuddin, T.M.; Morozov, S.V.; Blake, P.; Halsall, M.P.; Ferrari, A.C.; Boukhvalov, D.W.; Katsnelson, M.I.; Geim, A.K.; et al. Control of graphene's properties by reversible hydrogenation: Evidence for graphane. *Science* **2009**, *323*, 610–613. [CrossRef] [PubMed]

198. Tolédano, J.C.; Tolédano, P. *The Landau Theory of Phase Transitions*; World Scientific: Singapore, 1987; ISBN 9971-50-025-6.

199. Sofo, J.O.; Chaudhari, A.S.; Barber, G.D. Graphane: A two-dimensional hydrocarbon. *Phys. Rev. B* **2007**, *75*, 153401. [CrossRef]

200. Podlivaev, A.I.; Openov, L.A. Elementary Defects in Graphane. *JETP Lett.* **2017**, *106*, 110–115. [CrossRef]

201. Poh, H.L.; Šaněk, F.; Sofer, Z.; Pumera, M. High-pressure hydrogenation of graphene: Towards graphane. *Nanoscale* **2012**, *4*, 7006–7011. [CrossRef]

202. Liu, A.Y.; Cohen, M.L. Structural properties and electronic structure of low-compressibility materials: β-Si3N4 and hypothetical β-C3N4. *Phys. Rev. B* **1990**, *41*, 10727. [CrossRef]

203. Cao, S.; Yu, J. g-C 3 N 4 -Based Photocatalysts for Hydrogen Generation. *J. Phys. Chem. Lett.* **2014**, *5*, 2101–2107. [CrossRef] [PubMed]

204. Collins, P.G.; Bradley, K.; Ishigami, M.; Zettl, A. Extreme Oxygen Sensitivity of Electronic Properties of Carbon Nanotubes. *Science* **2000**, *287*, 1801–1804. [CrossRef] [PubMed]

205. Al-Mumen, H.; Rao, F.; Li, W.; Dong, L. Singular Sheet Etching of Graphene with Oxygen Plasma. *Nano-Micro Lett.* **2014**, *6*, 116–124. [CrossRef]

206. Czech, B.; Oleszczuk, P.; Wiacek, A.E.; Barczak, M. Cracking of CNT wall by UV. *Environ. Sci. Pollut. Res.* **2015**, *22*, 20198–20206. [CrossRef] [PubMed]

207. Neuville, S. Perspective of low energy Bethe nuclear fusion reactor with quantum electronic atomic rearrangement of carbon. *Condens. Matter. Nucl. Sci.* **2017**, *22*, 1–26.

208. Rouzaud, J.; Oberlin, A.; Beny-Bassez, C. Carbon films: Structure and micro-texture (optical and electron microscopy, Raman spectroscopy). *Thin Solid Film.* **1983**, *105*, 75–96. [CrossRef]

209. Anders, S.; Iii, J.W.A.; Pharr, G.M.; Tsui, T.Y.; Brown, I.G. Heat treatment of cathodic arc deposited amorphous hard carbon films. *Thin Solid Film.* **1997**, *308*, 186–190. [CrossRef]

210. Prawer, S.; Nugent, K.; Lifshitz, Y.; Lempert, G.; Grossman, E.; Kulik, J.; Avigal, I.; Kalish, R. Systematic variation of the Raman spectra of DLC films as a function of sp2: sp3 composition. *Diam. Relat. Mater.* **1996**, *5*, 433–438. [CrossRef]

211. Ferrari, A.C. Determination of bonding in diamond-like carbon by Raman spectroscopy. *Diam. Relat. Mater.* **2002**, *11*, 1053–1061. [CrossRef]

212. Rammamurti, R.; Shanov, V.; Singh, R.N.; Mamedov, S.; Boolchand, P. Raman spectroscopy study of the influence of processing conditions on the structure of polycrystalline diamond films. *J. Vac. Sci. Technol. A* **2006**, *24*, 179. [CrossRef]

213. Tuinstra, F.; Koenig, J.L. Raman spectra of graphite. *J. Phys. Chem.* **1970**, *53*, 1126–1130. [CrossRef]

214. Robertson, J.; O'Reilly, E.P. Electronic and atomic structure of amorphous carbon. *Phys. Rev. B* **1987**, *35*, 2946–2957. [CrossRef] [PubMed]

215. Pauleau, Y. Residual Stresses in DLC Films and Adhesion to Various Substrates. In *Tribology of DLC*; Donnet, C., Erdemir, A., Eds.; Springer: New York, NY, USA, 2008; pp. 102–136.

216. Maultzsch, J.; Reich, S.; Thomsen, C. Double-resonant Raman scattering in graphite: Interference effects, selection rules, and phonon dispersion. *Phys. Rev. B* **2004**, *70*, 155403. [CrossRef]

217. Jorio, A. Raman Spectroscopy in Graphene-Based Systems: Prototypes for Nanoscience and Nanometrology. *Int. Sch. Res. Netw. ISRN Nanotechnol.* **2012**, *2012*, 234216. [CrossRef]

218. Watanabe, E.; Conwill, A.; Tsuya, D.; Koide, Y. Low contact resistance metals for graphene-based devices. *Diam. Relat. Mater.* **2012**, *24*, 171–174. [CrossRef]

219. Reina, A.; Jia, X.; Ho, J.; Nezich, D.; Son, H.; Bulović, V.; Dresselhaus, M.S.; Kong, J. Large Area, Few-Layer Graphene Films on Arbitrary Substrates by Chemical Vapor Deposition. *Nano Lett.* **2009**, *9*, 30–35. [CrossRef] [PubMed]

220. Kumar, P.; Wani, M.F. Synthesis and tribological properties of graphene: A review. *J. Tribol.* **2017**, *13*, 36–71.

221. Penkov, O.; Kim, H.J.; Kim, H.J.; Kim, D.E. Tribology of Graphene: A Review. *Int. J. Precis. Eng. Manuf.* **2014**, *15*, 577–585. [CrossRef]

222. Cançado, C.L.G.; Pimenta, M.A.; Neves, B.R.; Dantas, M.S.; Jorio, A. Influence of the Atomic Structure on the Raman Spectra of Graphite Edges. *Phys. Rev. Lett.* **2004**, *93*, 247401. [CrossRef] [PubMed]

223. Dischler, B. Bonding and hydrogen incorporation in a-C: H studied by infrared spectroscopy. In *EMRS XVII Proceedings*; Koidl, P., Oelhafen, P., Eds.; Les Editions de la Physique: Paris, France, 1987; pp. 189–201.

224. Marcus, B.; Fayette, L.; Mermoux, M.; Abello, L.; Lucazeau, G. Analysis of the structure of multi-component carbon films by resonant Raman scattering. *J. Appl. Phys.* **1994**, *76*, 3463–3470. [CrossRef]

225. Singh, R.C.V. Raman and the Discovery of the Raman Effect. *Phys. Perspect.* **2002**, *4*, 399–420. [CrossRef]

226. Saito, R.; Jorio, A.; Filho, A.G.S.; Dresselhaus, G.; Dresselhaus, M.S.; Pimenta, M.A. Probing Phonon Dispersion Relations of Graphite by Double Resonance Raman Scattering. *Phys. Rev. Lett.* **2001**, *88*, 027401. [CrossRef] [PubMed]

227. Lazzeri, M.; Attaccalite, C.; Wirtz, L.; Mauri, F. Impact of the electron-electron correlation on phonon dispersion: Failure of LDA and GGA DFT functionals in graphene and graphite. *Phys. Rev. B* **2008**, *78*, 081406. [CrossRef]

228. Dallas, T.E.J. Thesis on Structural Phases of Disordered Carbon Materials. Ph.D. Thesis, Texas Tech University, Lubbock, TX, USA, 1996.

229. Chivilikhin, S.A.; Gusarov, V.V.; Yu-Popov, I. Flows in nanostructures: Hybrid classical-quantum models. *Nano-Syst. Phys. Chem. Math.* **2012**, *3*, 7–26.

230. Piscanec, S.; Lazzeri, M.; Mauri, F.; Ferrari, A.C.; Robertson, J. Kohn Anomalies and Electron-Phonon Interactions in Graphite. *Phys. Rev. Lett.* **2004**, *93*, 185503. [CrossRef] [PubMed]

231. Fakrach, B.; Rahmani, A.; Chadli, H.; Sbai, K.; Sauvajol, J.L. Raman spectrum of single-walled boron nitride nanotube. *Physica E* **2009**, *41*, 1800–1805. [CrossRef]

232. Lee, S. Effect of Intrinsic Ripples on Elasticity of the Graphene Monolayer. *Nanoscale Res. Lett.* **2015**, *10*, 1704. [CrossRef]

233. Nicholl, R.J.T.; Conley, H.J.; Lavrik, N.V.; Vlassiouk, I.; Puzyrev, Y.S.; Sreenivas, V.P.; Pantelides, S.T.; Bolotin, K.I. The effect of intrinsic crumpling on the mechanics of free-standing graphene. *Nat. Commun.* **2015**, *6*. [CrossRef]

234. Thompson-Flagg, R.C.; Moura, M.J.B.; Marder, M. Rippling of graphene. *EPL (Europhys. Lett.)* **2009**, *85*. [CrossRef]

235. Tanigaki, K.; Ogi1, H.; Sumiya, H.; Kusakabe, K.; Nakamura, N.; Hirao, M.; Ledbetter, H. Observation of higher stiffness in nano-polycrystal diamond than monocrystal diamond. *Nat. Commun.* **2013**. [CrossRef]

236. Gómez-Navarro, C.; Weitz, R.T.; Bittner, A.M.; Scolari, M.; Mews, A.; Burghard, M.; Kern, K. Electronic Transport Properties of Individual Chemically Reduced Graphene Oxide Sheets. *Nano Lett.* **2007**, *7*, 3499–3503. [CrossRef] [PubMed]

237. Childres, I.; A Jauregui, L.; Tian, J.; Chen, Y.P. Effect of oxygen plasma etching on graphene studied using Raman spectroscopy and electronic transport measurements. *New J. Phys.* **2011**, *13*, 25008. [CrossRef]

238. Janas, D. Towards monochiral carbon nanotubes: A review of progress in the sorting of single-walled carbon nanotubes. *Mater. Chem. Front.* **2018**, *2*, 36–63. [CrossRef]

239. Green, A.A.; Duch, M.C.; Hersam, M.C. Isolation of single-walled carbon nanotube enantiomers by density differentiation. *Nano Res.* **2009**, *2*, 69–77. [CrossRef]

240. Neuville, S. Prebiotic mechanisms of life appearance near deep sea vents with early earth synthetized L-handed amino acids being adsorbed on SWCNT. *Q. Phys. Rev.* **2018**, *4*, 1–36.

241. De Volder, M.F.L.; Tawfick, S.H.; Baughman, R.H.; Hart, A.J. Carbon Nanotubes: Present and Future Commercial Applications. *Science* **2013**, *339*, 535–539. [CrossRef] [PubMed]

242. Li, Z.; Guo, Q.; Li, Z.; Fan, G.; Xiong, D.B.; Su, Y.; Zhang, J.; Zhang, D. Enhanced Mechanical Properties of Graphene (Reduced Graphene Oxide)/Aluminum Composites with a Bioinspired Nano-laminated Structure. *Nano Lett.* **2015**, *15*, 8077–8083. [CrossRef]

243. Zhang, M.; Knoch, D.; Pascual-Leone, A.; Meyer, K.; Treyer, V.; Fehr, E. Strong, Transparent, Multifunctional, Carbon Nanotube Sheets. *Science* **2005**, *309*, 1215–1219. [CrossRef]

244. Coleman, J.N.; Khan, U.; Blau, W.J.; Gun'Ko, Y.K. Small but strong: A review of the mechanical properties of carbon nanotube–polymer composites. *Carbon* **2006**, *44*, 1624–1652. [CrossRef]
245. Kim, K.H.; Oh, Y.; Islam, M.F.; Islam, M. Graphene coating makes carbon nanotube aerogels superelastic and resistant to fatigue. *Nat. Nanotechnol.* **2012**, *7*, 562–566. [CrossRef]
246. Poelma, R.H.; Morana, B.; Vollebregt, S.; Schlangen, E.; Van Zeijl, H.W.; Fan, X.; Zhang, G.Q. Tailoring the Mechanical Properties of High-Aspect-Ratio Carbon Nanotube Arrays using Amorphous Silicon Carbide Coatings. *Adv. Funct. Mater.* **2014**, *24*, 5737–5744. [CrossRef]
247. Watcharotone, S.; Dikin, D.A.; Stankovich, S.; Piner, R.; Jung, I.; Dommett, G.H.B.; Evmenenko, G.; Wu, S.E.; Chen, S.F.; Liu, C.P.; et al. Graphene–Silica Composite Thin Films as Transparent Conductors. *Nano Lett.* **2007**, *7*, 1888–1892. [CrossRef] [PubMed]
248. Susantyoko, R.A.; Alkindi, T.S.; Kanagaraj, A.B.; An, B.; Alshibli, H.; Choi, D.; Aldahmani, S.; Fadaq, H.; Almheiri, S. Performance optimization of freestanding MWCNT-LiFePO4 sheets as cathodes for improved specific capacity of lithium-ion batteries. *RSC Adv.* **2018**, *8*, 16566–16573. [CrossRef]
249. Lalwani, G.; Kwaczala, A.T.; Kanakia, S.; Patel, S.C.; Judex, S.; Sitharaman, B. Fabrication and characterization of three-dimensional macroscopic all-carbon scaffolds. *Carbon* **2013**, *53*, 90–100. [CrossRef] [PubMed]
250. Van Der Zande, A.M.; Verbridge, S.S.; Frank, I.W.; Tanenbaum, D.M.; Parpia, J.M.; Craighead, H.G.; McEuen, P.L.; Bunch, J.S. Electromechanical Resonators from Graphene Sheets. *Science* **2007**, *315*, 490–493.
251. Xiao, L.; Chen, Z.; Feng, C.; Liu, L.; Bai, Z.-Q.; Wang, Y.; Qian, L.; Zhang, Y.; Li, Q.; Jiang, K.; et al. Flexible, Stretchable, Transparent Carbon Nanotube Thin Film Loudspeakers. *Nano Lett.* **2008**, *8*, 4539–4545. [CrossRef]
252. Zhang, J.-C.; Tang, Y.-J.; Yi, Y.; Ma, K.-F.; Zhou, M.-J.; Wu, W.-D.; Wang, C.-Y. Large-scale synthesis of novel vertically-aligned helical carbon nanotube arrays. *New Carbon Mater.* **2016**, *31*, 568–573. [CrossRef]
253. Aliev, A.E.; Oh, J.; Kozlov, M.E.; Kuznetsov, A.A.; Fang, S.; Fonseca, A.F.; Ovalle, R.; Lima, M.D.; Haque, M.H.; Gartstein, Y.N.; et al. Giant-Stroke, Superelastic Carbon Nanotube Aerogel Muscles. *Science* **2009**, *323*, 1575–1578. [CrossRef]
254. Du, S. A microelectromechanical system based inertial system with rotating accelerometers and gyroscope for Land navigation systems. *Int. J. Distrib. Sens. Netw.* **2017**, *13*. [CrossRef]
255. Miller, D.C.; Foster, R.R.; Zhang, Y.; Jen, S.H.; Bertrand, J.A.; Lu, Z.; Seghete, D.; O'Patchen, J.L.; Lee, R.Y.Y.C.; George, S.M.; et al. The mechanical robustness of atomic-layer- and molecular-layer-deposited coatings on polymer substrates. *J. Appl. Phys.* **2009**, *105*. [CrossRef]
256. Berman, D.; Erdemir, A.; Sumant, A.V. Reduced wear and friction enabled by graphene layers on sliding steel surfaces in dry nitrogen. *Carbon* **2013**, *59*, 167–175. [CrossRef]
257. Yan, C.; Kim, K.S.; Lee, S.K.; Bae, S.H.; Hong, B.H.; Kim, J.H.; Lee, H.J.; Ahn, J.H. Mechanical and Environmental Stability of Polymer Thin-Film-Coated Graphene. *ACS Nano* **2012**, *6*, 2096–2103. [CrossRef] [PubMed]
258. Marchetto, D.; Held, C.; Hausen, F.; Wählisch, F.; Dienwiebel, M.; Bennewitz, R. Friction and Wear on Single-Layer Epitaxial Graphene in Multi-Asperity Contacts. *Tribol. Lett.* **2012**, *48*, 77–82. [CrossRef]
259. Naraghi, M.; Filleter, T.; Moravsky, A.; Locascio, M.; Loutfy, R.O.; Espinosa, H.D. A Multiscale Study of High-Performance Double-Walled Nanotube–Polymer Fibers. *ACS Nano* **2010**, *4*, 6463–6476. [CrossRef] [PubMed]
260. Yildirim, T.; Gülseren, O.; Kılıç, C.; Ciraci, S. Pressure-induced interlinking of carbon nanotubes. *Phys. Rev. B* **2000**, *62*. [CrossRef]
261. Zhang, R.; Zhang, Y.; Zhang, Q.; Xie, H.; Qian, W.; Wei, F. Growth of Half-Meter Long Carbon Nanotubes Based on Schulz–Flory Distribution. *ACS Nano* **2013**, *7*, 6156–6161. [CrossRef] [PubMed]
262. Motta, M.; Moisala, A.; Kinloch, I.A.; Windle, A.H. High Performance Fibers from 'Dog Bone' Carbon Nanotubes. *Adv. Mater.* **2007**, *19*, 3721–3726. [CrossRef]
263. Dalton, A.B.; Collins, S.; Muñoz, E.; Razal, J.M.; Ebron, V.H.; Ferraris, J.P.; Coleman, J.N.; Kim, B.G.; Baughman, R.H. Super-tough carbon-nanotube fibers. *Nature* **2003**, *423*. [CrossRef]
264. Yang, Y.; Chen, X.; Shao, Z.; Zhou, P.; Porter, D.; Knight, D.P.; Vollrath, F. Toughness of Spider Silk at High and Low Temperatures. *Adv. Mater.* **2005**, *17*, 84–88. [CrossRef]
265. Behabtu, N.; Young, C.C.; Tsentalovich, D.E.; Kleinerman, O.; Wang, X.; Ma, A.W.K.; Bengio, E.A.; Ter Waarbeek, R.F.; De Jong, J.J.; Hoogerwerf, R.E.; et al. Strong, Light, Multifunctional Fibers of Carbon Nanotubes with Ultrahigh Conductivity. *Science* **2013**, *339*, 182–186. [CrossRef]

266. Liu, G.; Zheng, H.; Jiang, Z.; Zhao, J.; Wang, Z.; Pan, B.; Xing, B. Formation and Physicochemical Characteristics of Nano Biochar: Insight into Chemical and Colloidal Stability. *Environ. Sci. Technol.* **2018**, *52*, 10369–10379. [CrossRef] [PubMed]

267. Giorcelli, M.; Savi, P.; Khan, A.; Tagliaferro, A. Analysis of biochar with different pyrolysis temperatures used as filler in epoxy resin composites. *Biomass Bioenergy* **2019**, *122*, 466–471. [CrossRef]

268. Srivastava, N.; Joshi, R.; Banerjee, K. Carbon nanotube interconnects: Implications for performance, power dissipation and thermal management. In Proceedings of the IEEE International Electron Devices Meeting, Washington, DC, USA, 5 December 2005; IEDM Technical Digest. IEEE: Piscataway, NJ, USA, 2005; p. 249, ISBN 0-7803-9268-X.

269. Coiffic, J.C.; Fayolle, M.; Maitrejean, S.; Torres, L.E.F.F.; Le Poche, H. Conduction regime in innovative carbon nanotube via interconnect architectures. *Appl. Phys. Lett.* **2007**, *91*, 252107. [CrossRef]

270. Bauerdick, S.; Linden, A.; Stampfer, C.; Helbling, T.; Hierold, C. Direct wiring of carbon nanotubes for integration in nanoelectromechanical systems. *J. Vac. Sci. Technol. B Microelectron. Nanometer. Struct.* **2006**, *24*, 3144–3147. [CrossRef]

271. Banerjee, K.; Srivastava, N. Are carbon nanotubes the future of VLSI interconnections? In Proceedings of the 39th conference on Design automation—DAC '02 2006, San Francisco, CA, USA, 24–28 July 2006; p. 809.

272. Naeemi, A.; Meindl, J.D. Carbon nanotube interconnects. In Proceedings of the 2007 International Symposium on Physical Design, ISPD '07, Austin, TX, USA, 18–21 March 2007; p. 77.

273. Chai, Y.; Chan, P.C.H. High electro-migration-resistant copper/carbon nanotube composite for Interconnect application. In Proceedings of the 2008 IEEE Electron Devices Meeting, San Francisco, CA, USA, 15–17 December 2008.

274. Kordás, K.; TóTh, G.; Moilanen, P.; Kumpumäki, M.; Vähäkangas, J.; Uusimäki, A.; Vajtai, R.; Ajayan, P.M. Chip cooling with integrated carbon nanotube microfilm architectures. *Appl. Phys. Lett.* **2007**, *90*, 123105. [CrossRef]

275. Li, J.; Ye, Q.; Cassell, A.; Ng, H.T.; Stevens, R.; Han, J.; Meyyappan, M. Bottom-up approach for carbon nanotube interconnects. *Appl. Phys. Lett.* **2003**, *82*, 2491–2493. [CrossRef]

276. Collins, P.G. Nanotubes for Electronics. *Sci. Am.* **2000**, *283*, 67–69. [CrossRef]

277. Westervelt, R.M. APPLIED PHYSICS: Graphene Nanoelectronics. *Science* **2008**, *320*, 324–325. [CrossRef]

278. Potsma, H.W.C.; Teepen, T.; Yao, Z.; Grifoni, M.; Dekker, C. Carbon Nanotube Single-Electron Transistors at Room temperature. *Science* **2001**, *293*, 76–79.

279. Chen, Z.; Sippel-Oakley, J.; Tang, J.; Wind, S.J.; Solomon, P.M.; Appenzeller, J.; Lin, Y.-M.; Rinzler, A.G.; Avouris, P. An Integrated Logic Circuit Assembled on a Single Carbon Nanotube. *Science* **2006**, *311*, 1735. [CrossRef]

280. Wang, C.; Chien, J.-C.; Takei, K.; Takahashi, T.; Nah, J.; Niknejad, A.M.; Javey, A. Extremely Bendable, High-Performance Integrated Circuits Using Semiconducting Carbon Nanotube Networks for Digital, Analog, and Radio-Frequency Applications. *Nano Lett.* **2012**, *12*, 1527–1533. [CrossRef] [PubMed]

281. Appenzeller, J.; Lin, Y.-M.; Knoch, J.; Chen, Z.; Avouris, P. Comparing Carbon Nanotube Transistors—The Ideal Choice: A Novel Tunneling Device Design. *IEEE Trans. Electron Devices* **2005**, *52*, 2568–2576. [CrossRef]

282. Parnell, A.J. Nano-technology and the potential for a renewable solar future. *Nanotechnol. Percept.* **2011**, *7*, 180–187. [CrossRef]

283. Rasmussen, L. *Electroactivity in Polymeric Materials*; Springer Science & Business Media: Berlin, Germany, 2012; 162p.

284. Jørgensen, M.; Carlé, J.E.; Søndergaard, R.R.; Lauritzen, M.; Dagnæs-Hansen, N.A.; Byskov, S.L.; Andersen, T.R.; Larsen-Olsen, T.T.; Böttiger, A.P.; Andreasen, B.; et al. The state of organic solar cells—A meta-analysis. *Sol. Energy Mater. Sol. Cells* **2013**, *119*, 84–93. [CrossRef]

285. Chauhan, A.K.; Gusain, A.; Jha, P.; Koiry, S.P.; Saxena, V.; Veerender, P.; Aswal, D.K.; Gupta, S.K. Graphene composite for improvement in the conversion efficiency of flexible poly 3-hexyl-thiophene: [6,6]-phenyl C71 butyric acid methyl ester polymer solar cells. *Appl. Phys. Lett.* **2014**, *104*. [CrossRef]

286. Guldi, D.M.; Rahman, G.M.A.; Prato, M.; Jux, N.; Qin, S.; Ford, W. Single-Wall Carbon Nanotubes as Integrative Building Blocks for Solar-Energy Conversion. *Angew. Chem.* **2005**, *117*, 2051–2054. [CrossRef]

287. Neuville, S. Advanced ta-C coatings with up-dated fundamentals for energy production efficiency. *Mater. Day Proc.* **2018**, *5*, 13816–13826.

288. Casas, C.D.L.; Li, W. A review of application of carbon nanotubes for lithium ion battery anode material. *J. Power Sources* **2012**, *208*, 74–85. [CrossRef]

289. Banerjee, A.; Ziv, B.; Levi, E.; Shilina, Y.; Luski, S.; Aurbac, D. Single-Wall Carbon Nanotubes Embedded in Active Masses for High-Performance Lead-Acid Batteries. *J. Electrochem. Soc.* **2016**, *163*, A1518–A1526. [CrossRef]

290. Fu, K.; Yildiz, Ö.; Bhanushali, H.; Wang, Y.; Stano, K.; Xue, L.; Zhang, X.; Bradford, P.D. Aligned Carbon Nanotube-Silicon Sheets: A Novel Nano-architecture for Flexible Lithium Ion Battery Electrodes. *Adv. Mater.* **2013**, *25*, 5109–5114. [CrossRef]

291. Barghi, S.H.; Tsotsis, T.T.; Sahimi, M. Chemisorption, physisorption and hysteresis during hydrogen storage in carbon nanotubes. *Int. J. Hydrog. Energy* **2014**, *39*, 1390–1397. [CrossRef]

292. Dillon, A.C.; Jones, K.M.; Bekkedahl, T.A.; Kiang, C.H.; Bethune, D.S.; Heben, M.J. Storage of hydrogen in single-walled carbon nanotubes. *Nature* **1997**, *386*, 377–379. [CrossRef]

293. Safa, S.; Mojtahedzadeh, L.M.; Fathollahi, V.; Kakuee, O.R. Investigating hydrogen storage behavior of carbon nanotubes at ambient temperature by ion beam analysis. *Nano* **2010**, *5*, 341–347. [CrossRef]

294. Yuca, N.; Karatepe, N. Hydrogen storage in single-walled carbon nanotubes purified by microwave digestion method. *World Acad. Sci. Eng. Technol.* **2011**, *79*, 605–610.

295. Valenti, G.; Boni, A.; Melchionna, M.; Cargnello, M.; Nasi, L.; Bertoni, G.; Gorte, R.J.; Marcaccio, M.; Rapino, S.; Bonchio, M.; et al. Co-axial heterostructures integrating palladium/titanium dioxide with carbon nanotubes for efficient electrocatalytic hydrogen evolution. *Nat. Commun.* **2016**, *7*. [CrossRef] [PubMed]

296. Gong, K.; Du, F.; Xia, Z.; Durstock, M.; Dai, L. Nitrogen-Doped Carbon Nanotube Arrays with High Electrocatalytic Activity for Oxygen Reduction. *Science* **2009**, *323*, 760–764. [CrossRef] [PubMed]

297. Zhang, S.J.; Shao, T.; Kose, H.S.; Karanfil, T. Adsorption of aromatic compounds by carbonaceous absorbents: A comparative study on granular activated carbon, activated carbon fibers, and carbon nanotubes. *Environ. Sci. Technol.* **2010**, *44*, 6377–6383. [CrossRef] [PubMed]

298. Apul, O.G.; Karanfil, T. Adsorption of synthetic organic contaminants by carbon nanotubes: A critical review. *Water Res.* **2015**, *68*, 34–55. [CrossRef] [PubMed]

299. Polikarpova, N.P.; Zaporotskova, I.V.; Vilkeeva, D.E.; Polikarpov, D.I. Sensor properties of carboxyl-modified carbon nanotubes. *Nanosyst. Phys. Chem. Math.* **2014**, *5*, 101–106.

300. Pötschke, P.; Andres, T.; Villmow, T.; Pegel, S.; Brünig, H.; Kobashi, K.; Fischer, D.; Häussler, L. Liquid sensing properties of fibers prepared by melt spinning from polylactic acid containing multi-walled carbon nanotubes. *Compos. Sci. Technol.* **2010**, *70*, 343–349. [CrossRef]

301. Hashim, D.P.; Narayanan, N.T.; Romo-Herrera, J.M.; Cullen, D.A.; Hahm, M.G.; Lezzi, P.; Suttle, J.R.; Kelkhoff, D.; Muñoz-Sandoval, E.; Ganguli, S.; et al. Covalently bonded three-dimensional carbon nanotube solids via boron induced nanojunctions. *Sci. Rep.* **2012**, *2*. [CrossRef] [PubMed]

302. Mikael, P.E.; Nukavarapu, S.P. Functionalized Carbon Nanotube Composite Scaffolds for Bone Tissue Engineering: Prospects and Progress. *J. Biomater. Tissue Eng.* **2011**, *1*, 76–85. [CrossRef]

303. Balavoine, F.; Schultz, P.; Richard, C.; Mallouh, V.; Ebbesen, T.W.; Mioskowski, C. Helical Crystallization of Proteins on Carbon Nanotubes: A First Step towards the Development of New Biosensors. *Angew. Chem. Int. Ed.* **1999**, *38*, 1912–1915. [CrossRef]

304. Fasano, M.; Chiavazzo, E.; Asinari, P. Water transport control in carbon nanotube arrays. *Nanoscale Res. Lett.* **2014**, *9*. [CrossRef] [PubMed]

305. Hesami, M.; Bagheri, R.; Masoomi, M. Combination effects of carbon nanotubes, MMT and phosphorus flame retardant on fire and thermal resistance of fiber-reinforced epoxy composites. *Iran. Polym. J.* **2014**, *23*, 469–476. [CrossRef]

306. Lv, J.; Leong, E.S.; Jiang, X.; Kou, S.; Dai, H.; Lin, J.; Liu, Y.J.; Si, G. Plasmon-Enhanced Sensing: Current Status and Prospects. *Hindawi J. Nanomater.* **2015**, *16*. [CrossRef]

307. Lamb, H. *Sandvik Reveals 3D Printed Diamond Composite*; Engineering and Technology: Hertfordshire, UK, 2019.

 *micromachines* 

Review

Functionalized Carbon Materials for Electronic Devices: A Review

Urooj Kamran, Young-Jung Heo, Ji Won Lee and Soo-Jin Park *

Department of Chemistry, Inha University, 100 Inharo, Incheon 22212, Korea; malikurooj9@gmail.com (U.K.); heoyj1211@naver.com (Y.-J.H.); lj529@naver.com (J.W.L.)
* Correspondence: sjpark@inha.ac.kr; Tel.: +82-32-876-7234

Received: 27 February 2019; Accepted: 1 April 2019; Published: 3 April 2019

Abstract: Carbon-based materials, including graphene, single walled carbon nanotubes (SWCNTs), and multi walled carbon nanotubes (MWCNTs), are very promising materials for developing future-generation electronic devices. Their efficient physical, chemical, and electrical properties, such as high conductivity, efficient thermal and electrochemical stability, and high specific surface area, enable them to fulfill the requirements of modern electronic industries. In this review article, we discuss the synthetic methods of different functionalized carbon materials based on graphene oxide (GO), SWCNTs, MWCNTs, carbon fibers (CFs), and activated carbon (AC). Furthermore, we highlight the recent developments and applications of functionalized carbon materials in energy storage devices (supercapacitors), inkjet printing appliances, self-powered automatic sensing devices (biosensors, gas sensors, pressure sensors), and stretchable/flexible wearable electronic devices.

Keywords: carbon nanotubes; graphene; carbon fibres; functionalization; supercapacitors; sensors; inkjet printer inks; flexible wearable devices; electronics

---

1. Introduction

Carbon is a crucial element, as it is considered the sixth most abundant element in the universe, fourth most ordinary species in the planetary system, as well as the seventeenth most dominant component in the earth's crust [1]. The overall assessed relative abundance of carbon is 180–270 ppm [2]. The existence of some carbon allotropes, such as graphene, fullerenes (C60), carbon nanotubes (CNTs), amorphous carbon, diamonds (NDs), and londsaleite [3], has attracted the interest of scientists in recent years due to their brilliant properties. These include wide specific surface area, good electrical conductance, high electro-chemical stabilization ability, good thermal conductivity, and high mechanical resistance, which can fulfill the energy-storing needs in electronic-based industries [4–8]. A brief outline of carbon allotropes, their essential physio-chemical properties, and applications in various fields are mentioned in Figure 1.

The progress in electronic industries has reorganized expertise in electro-communications, automation, and computational industries, all of which have a tremendous influence on various aspects of human lives [9]. Previously, the silicon industry was established to develop more effective, compressed, dense, rapid, and stable electronic devices [10]. These advances in silicon technology were synchronized with the simultaneous minimization of electronic devices in order to make them more effective. Despite this progress in silicon-based devices, further advancement is inadequate to satisfy the various scientific and technical aspects for electronic devices because of resource limitations [11]. Recently, researchers have found that carbon-based materials are an alternative to silicon technology and have proven to be very effective to fabricate electronic devices with superior performance.

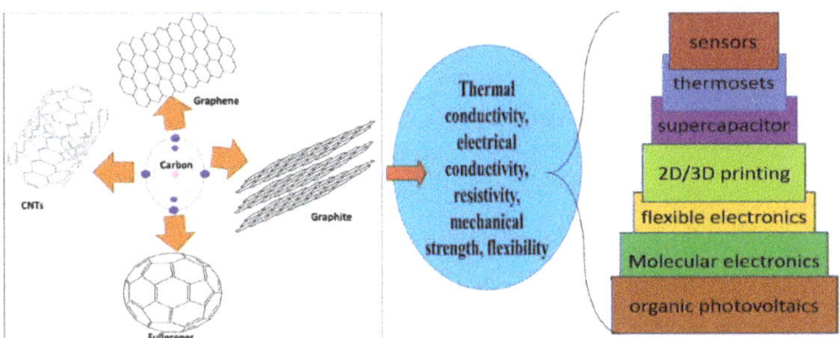

Figure 1. Brief description of various carbon allotropes, their electronic properties and applications in electronic industries.

The importance of carbon-based materials in electrical devices is due to their tremendous properties. Graphene exhibits low weight, efficient electric as well as thermal conductivity, high specific tunable surface area (up to 2675 $m^2 \cdot g^{-1}$), superior mechanical properties, with a Young's modulus (YMs) of almost 1 TPa, and high chemical stability [12–14]. CNTs exhibit efficient mechanical characteristics, with a YMs in the range of 0.27–1 TPa and high mechanical strength (11–130 GPa) [15,16]. A literature survey indicated that CNTs possessed thermal and electrical conductance up to 3,000 $Wm \cdot K^{-1}$ [17] and 1800 $S \cdot cm^{-1}$ [18], respectively. NDs exhibit sp^2 and sp^3 carbon layers, which contribute to their high durability [19,20]. These tremendous properties of carbon-based materials result in various applications in an electronic devices, energy production (Li-ion batteries, solar-fuel cells) [21], energy storage devices, and environment-protection-related appliances [22–34]. Therefore, the utilization of carbon-based materials instead of silicon provides an innovative and new technological platform in the electronic industry. Recently, graphene, CNTs, fullerene derivatives (FDs), carbon nanofibers (CFs), NDs, 2D graphene layers, and carbon foams, have been widely used in electronics; their applications include supercapacitors, thermosets, molecular electronics, organic photovoltaics, gas-sensor, biosensors, 2D or 3D printers, flexible electronic wearable electronics, wireless telecommunication systems, electromagnetic-wave (EM) absorber materials in reverberation chambers, fillers in EM shielding cementitious materials, EM wave propagating tuner materials [35–54]. However, the direct use of carbon-based materials without any modification and functionalization lead to frequent aggregation, and insolublilization sometimes limits their industrial applications. Therefore, from a practical perspective for industrial applications, it is extremely necessary to fabricate functionalized materials with tuneable physio-chemical surface properties in order to improve their performance. For this purpose, many methodologies have been proposed to produce doped and functionalized carbon-based materials for energy and electronic applications [55–57].

Herein, we focused on the strategies of functionalization or modification of carbon-based materials and highlighted their applications in electronic devices, including supercapacitors, inks, inkjet/3D printing appliances, biosensors, gas sensors, and flexible electronics wearable.

2. Methodology

2.1. Functionalized Carbon-Based Electrode Materials

2.1.1. Activation Method

A carbon-based electrode hybrid material is obtained by polymer functionalization, metal oxide (MO) doping, or by activation with activating agents. The polymer-based functionalization of activated carbon (AC) for supercapacitor electrodes has been reported in literature. Lee et al. [58] synthesized tubular polypyrrole-(T-Ppy) functionalized pitch-based AC to obtain a hybrid composite material (AC/Ppy) for a supercapacitor electrode; the synthetic layout is mentioned in Figure 2a. Different contents ratios (1:0.5, 1:1, 1:2, 1:4) of T-Ppy were loaded onto AC [59,60]. The electrode was prepared by dipping a small nickel piece in a slurry of 80% Ppy/AC hybrid composite and 10% conductive material, with 10% (PVDF:NMP) as a binder solvent. Vighnesha et al. [61] fabricated activated AC (the AC was derived from coconut shell) and functionalized by polyaniline. KOH-activated AC was obtained by polymerization of aniline [62]. A hybrid electrode was prepared by mechanical mixing of AC and PANi with various ratios and N-methyl pyrrolidone and PVDF as a binder, followed by loading on a current-collector steel.

2.1.2. Electrospinning Technology

Polypyrrole (Ppy) and polyacrylonitrile (PAN) have also been used to functionalize CFs. In this regard, Tao et al. [63] fabricated a composite hybrid material for flexible supercapacitors by in-situ growth followed by conductive wrapping. The fabrication steps included the electrodeposition of MnO_2 nanoparticles on cleaned CFs, which was followed by the wrapping of a conductive Ppy layer. The supercapacitor electrode was prepared by fixing the as-synthesized hybrid material ($Ppy-MnO_2$-CFs) to a supercapacitor in a sandwich structure, with PVA/H_3PO_4 as the membrane as well as an electrolyte solution between the electrodes. Furthermore, electrospinning technology was used to prepare PAN-functionalized CFs (APCFs) [64], as mentioned in Figure 2b. The pristine PAN-CFs material was chemically activated by KOH at different temperatures (600–1000 °C) under N_2 atmosphere. The capacitance test was performed in a 1.0 M Na_2SO_4 electrolyte and the coated working electrode was prepared in the same way as mentioned in Ref. [58].

2.1.3. Greener and Rotational Hydrothermal Method

Recently, Cakici et al. [65] utilized carbon fiber fabric (CFF) to prepare a hybrid composite with MnO_2, named $CFF-MnO_2$. A greener hydrothermal approach was used for the doping of MnO_2 nanoparticles on CFF. Briefly, the polymer sizing on CFF was first eliminated by annealing at a specified temperature under Ar atmosphere. Fine CFF was then autoclaved with a 6 mM $KMnO_4$ solution at 175 °C. Different samples were prepared under same conditions with varying times. The electrochemical test was obtained using a 3-electrode electrochemical cell system, with a 1 M Na_2SO_4 electrolyte, Ag/AgCl as a reference electrode, and the prepared fabric material (CFF/MnO_2) itself as a working electrode.

Graphene-based modified hybrid electrode materials have also been reported. Ke et al. [66] demonstrated different approaches for fabricating a reduced graphene oxide (rGO)-doped Fe_3O_4 nanocomposite electrode material based on strong electrostatic interactions between oppositely charged graphene oxide (GO) and Fe_3O_4 via a rotational hydrothermal approach with some modifications. A modified Hummers' process was used to convert graphite into GO. However, Li et al. [67] fabricated a hydrothermally reduced graphene-(HRG) decorated MnO_2 composite on the basis of previous reports [68–72] via a hydrothermal approach. In this case, GO was prepared by a modified Hummers' method [73,74]. The influence of polyaniline-functionalized graphene sheets (GS) doped with various MO (RuO_2, TiO_2, and Fe_3O_4) on the supercapacitance was also studied [75]. The graphite was converted into graphitic oxide by Hummers' method. Hydrogen thermal exfoliation of graphitic oxide lead to the formation of hydrogen-exfoliated graphene sheet (HEG), which was further functionalized by HNO_3. Sawangphruk et al. [76] invented a hybrid pseudo-supercapacitor based on double layer graphene functionalized with polyaniline and doped with silver nanoparticles deposited on flexile type CFs paper. Wang et al. [77] successfully fabricated q hybrid material containing $Ni(OH)_2$, GS, and GO; GS was obtained by an exfoliation-expansion method and GO was obtained by a modified Hummers' method. GO/DMF and GS/DMF were identically mixed with a $Ni(Ac)_2$ solution for hydrothermal autoclave treatment at 80 °C for 10 h and coated on a pseudo-capacitive electrode. The composite material was dispersed in an ethanol-PTFE solution and hydrated into nickel foam through compression at 80 °C. Electrochemical measurements were conducted on the foam sample. Similarly, Zhang et al. [78] produced GO nanofibers (GONFs) and GO nanoribbons (GONRs) from carbon nanofibers (CNFs) using Hummers' method. The growth of crystalline $Ni(OH)_2$ nanoplates and reduction of prepared GONFs and GONRs were obtained via a hydrothermal approach. Specifically, the electrode composite material was synthesized by mixing $N_2H_4 \cdot H_2O$ (35%), $NH_3 \cdot H_2O$ (28 wt%), and $Ni(NO_3)_2$ (0.4 M) with 2 mg·mL^{-1} aqueous solution containing either GONF, CNF, or GONR. After sonication, the mixture was hydrothermally annealed under certain conditions. The composites were named rGONF/Ni(OH)$_2$, rGONR/Ni(OH)$_2$, and CNF/Ni(OH)$_2$. The general schematic layout and morphology of the composites are mentioned in Figure 2c.

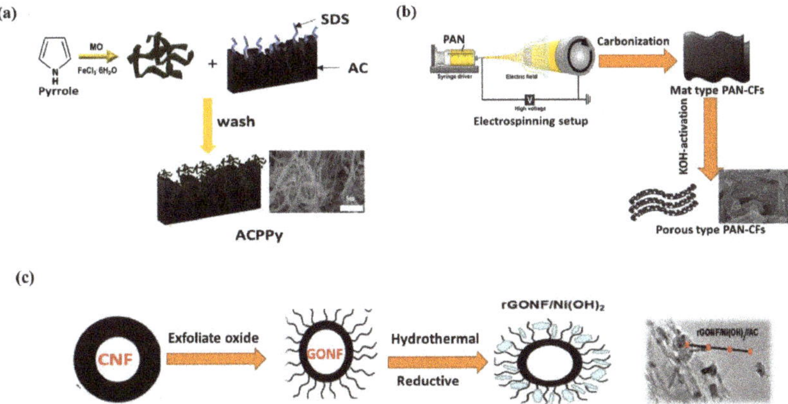

Figure 2. (**a**) Schematic layout of synthesis of activated carbon/polypyrrole (AC-PPy) composites, reproduced with permission from [58], published by Elsevier, 2018; (**b**) synthetic scheme of porous polymer-based carbon fibers (PPCFs) preparation process by electrospinning technique, reproduced with permission from [64], published by Elsevier, 2019; and (**c**) general synthetic layout for rGONF/Ni(OH)$_2$ fabrication, reproduced with permission from [78], published by ACS Publications, 2016.

2.2. Functionalized Carbon-Based Inks and Inkjet Printer Materials

Solvent Exfoliated Method

According to a literature survey, carbon allotropes are used in formation of various inks and hybrid materials within inkjet printing devices [79,80]. For this purpose, Secor et al. [81] formulated ink from graphene by using the solvent exfoliated method [82]. Briefly, graphene was dispersed efficiently in ethanol with a stabilizing polymer, ethyl cellulose (EC). The obtained pure graphene-EC powder was mixed with terpineol to improve the dispersion and obtain smooth printing. The formulated graphene-based ink was utilized for gravure printing by using a flooding doctoring printing system. Similarly, Torrisi et al. [83] formulated a graphene-based printable ink by ultrasonic exfoliation of graphite flakes in an NMP solvent; the obtained mixture was centrifuged to eliminate graphite flakes. A pristine GS-based ink was formulated by Gao et al. [84] through the exfoliation of graphite flakes in a mixture of EC and cyclohexane through ultrasonication under supercritical CO_2. On the other hand, CNTs are also used to synthesize inks. Homenick et al. [85] fabricated an ink from SWCNTs, which was used in the formation of a SWCNT transistor on a test chip. The scintillation vial was charged by ultrasonication with IsoSol-S100 and toluene. The polymer (PFDD) and SWCNTs were mixed in appropriate ratios and the obtained ink was deposited on a silicon chip through optimized print waveform (20 μm) fitted in an inkjet printer. Multiple layers of the SWCNT-functionalized ink were printed by rinsing the chip several time with toluene within the inkjet printer. In the same way, a SWCNT transistor was prepared on SiO_2 wafers. The easiest approach for this, using a water-based ink, was reported by Han et al. [86], with a mixture of SWCNTs and sodium dodecyl-benzene sulfonate (SDBS). The SWCNTs and SDBS surfactant with ratio of 1:2 were mixed in 10% distilled water by ultrasonication and probe sonication for 2 h. The fabricated inks were put in a cartridge for a month to obtain a well-dispersed ink. The writing efficiency of the water-based SWCNTs ink was checked on a paper and on a curved cup, as shown in Figure 3a. The primary resistivity of the SWCNTs-ink-coated filter paper was under 20 kV·cm^{-1}.

Further improvements on three-dimensional (3D) printed designs were obtained by Foster et al. [87]. They invented a 3D printed design with the help of a Rep-Rap printer by utilizing a graphene/PLA filament named "black magic", as shown in Figure 3b; this filament possessed a conductance of 2.13 S·cm^{-1}. The electrochemical test was conducted using a 3-electrode-based system with a printed 3D graphene/PLA anode as the working electrode and calomel and platinum electrodes as the reference electrodes. You et al. [88] fabricated a well-dispersed graphene-based ink for 3D printers. This ink was prepared by ultrasonication of graphene with EGB to achieve a homogenized graphene dispersion, which was further mixed with a solution of DBP and PVB in ethanol medium (2.5:1 ratio), followed by volatilization of ethanol to obtained 3D slurry inks for printing. They also prepared graphene-based filaments, which were used as primary blocks for 3D structures. The different 3D morphological structures with the graphene filaments were prepared via a 3-axis robocasting system and are mentioned in Figure 3e,f. Sarapuk et al. [89] investigated the influence of dispersing agents on the properties of inks. Graphene nanoplates were added into a mixture of glycol and ethanol (1:1) followed by sonication. A functional polymer (AKM-0531) was then added as a dispersing agent to improve the effectiveness of the prepared inks.

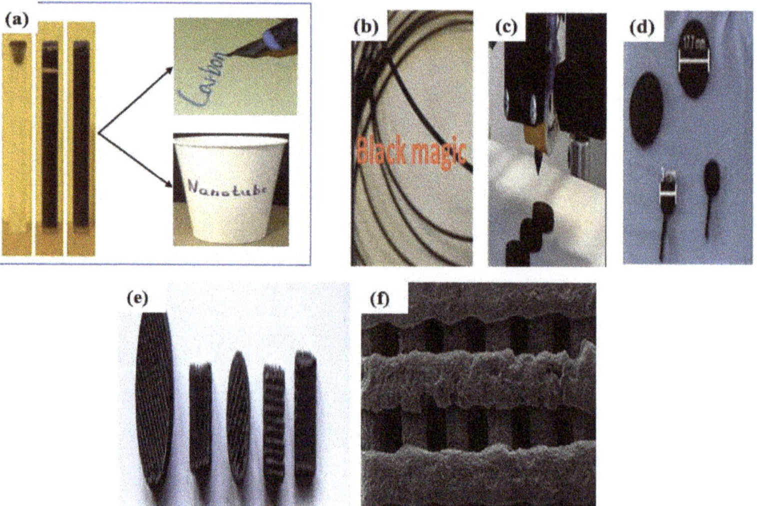

Figure 3. (a) Water-based carbon nanotubes ink dispersion in cartridges after 7 days, and direct writing above paper and cup using fountain pen, reproduced with permission from [86], published by Elsevier, 2014; (b) 3D printable graphene/PLA, (c) 3D printing process, (d) various printed 3D electrodes, reproduced with permission from [87], published by Nature, 2017; (e) morphology, (f) and FESEM images of different 3D graphene-based structure, reproduced with permission from [88], published by Elsevier, 2018.

2.3. Functionalized Carbon Hybrids for Sensors

2.3.1. Immobilization, Direct and In-Situ Methods

Li et al. [90] fabricated a composite material for biosensor electrodes through a facile method. The GO was produced from graphite and reduced to rGO using NaOH (Hummers' method) [91].

The rGO/Pd hybrid was obtained by mutual sonication of the rGO suspension and palladium acetate [92]. The immobilization of lacasse was realized through self-polymerization of dopamine, leading to poly-dopamine, and subsequently, the formation of a PDA-Lac-rGO-Pd hybrid composite. The modified biosensor electrode was developed by polishing a glass carbon electrode (GCE) with alumina and coating the GCE with the PDA-Lac-rGO-Pd hybrid material. Hassan et al. [93] also prepared a functionalized graphene-based glucose sensor through the immobilization method. The GO was prepared by a modified Hummers' method using graphite flakes, while the functionalization and reduction of GO were performed by adding PIL and hydrazine [94]. The obtained GO sheet was coated on a hot Pt wire by dipping [95,96]. The glucose oxidase (GOD) enzyme was electrostatically introduced above the GO-metal wire. The electrochemical test of the prepared GOD-graphene biosensor was conducted with Ag/AgCl as a reference electrode.

Hemanth et al. [97] fabricated a bio-functionalized graphene 3D carbon electrode biosensor. The reduction and functionalization of Hummers-modified GO was obtained by treatment with branched polyethylenimine (PEI) in a single reaction [98]. Further modification and induction of biosensor ability in PEI-GO was carried out by adding ferrocene carboxylic acid; after combing with a biosensing enzyme, the hybrid material was coated on an electrode. A brief schematic layout is shown in Figure 4a, where the biosensor is prepared by coating WE area of 2D and 3D micro-electrodes chip with a solution of the hybrid material (ferrocene-modified RGO-PEI). A similar method was used to prepare a glucose-based biosensing electrode with a GOD (glucose oxidase) enzyme solution. In the case of carbon-based gas sensors, Liu et al. [99] investigated a ZnO-rGO hybrid material for NO_2 gas sensors. The rGO was prepared from graphite flakes powder by Hummers' method, and was in-situ decorated with ZnO nanoparticles. The hybrid ZnO-rGO composite material was reduced by hydrazine hydrate. The composite hybrid was mixed with DMF suspension to obtain a DMF-ZnO-rGO hybrid suspension, which was then coated on ceramic substrate electrodes; this was used to test the NO_2 gas-sensing properties with a static test system.

2.3.2. Thermal Annealing and Hydrothermal Methods

Novikov et al. [100] used graphene films to fabricate a NO_2-based gas sensor. The graphene film was grown by annealing a 4H-SiC crystal under Ar atmosphere at 1700 °C. Laser photolithography was used to form a bar-shaped design on the graphene surface. The minimum and suitable resistance contacts were obtained via double step metallization, which was carried out by e-beam evaporation and lift-off photolithography [101]. The prepared graphene-based sensor chip was placed in a prototype portable device i.e., above a holder having a platinum resistor as a heater (platinum was chosen due to its minimum thermal inertia). The sample was subjected to rapid thermal cycling, as mentioned in Figure 4b, and the NO_2 gas adsorption was monitored. Huang and Hu [102] also investigated a rGO-based polyaniline hybrid (rGO-PANI) for the detection of ammonia (NH_3). A single layer of GO was obtained by ultra-sonication of GO paper in distilled water. The monomer for aniline polymerization ($GO-MnO_2$) was prepared by direct reaction with $KMnO_4$ and thermal annealing [103,104]. The electrodes for the sensor device were synthesized via a standard microfabrication approach, in which a micro syringe was utilized for the deposition of a rGO-PANI hybrid ethanol solution within the electrode gap, generating a rGO-PANI bridge. Ye and Tai [105] successfully invented a novel rGO-based TiO$_2$ thin-film sensor; the brief schematic layout is shown in Figure 4c. The hybrid reduced graphene oxide based TiO_2 composite ($rGO-TiO_2$) was prepared by ultrasonic dispersion of rGO and TiO_2 with a ratio of 1:96 in a hydrothermally treated mixture of TIP, HCl, and DI water. The prepared $rGO-TiO_2$ composite material was deposited on a micro-electrode. This $rGO-TiO_2$ thin-film sensor was applied for the detection of NH_3.

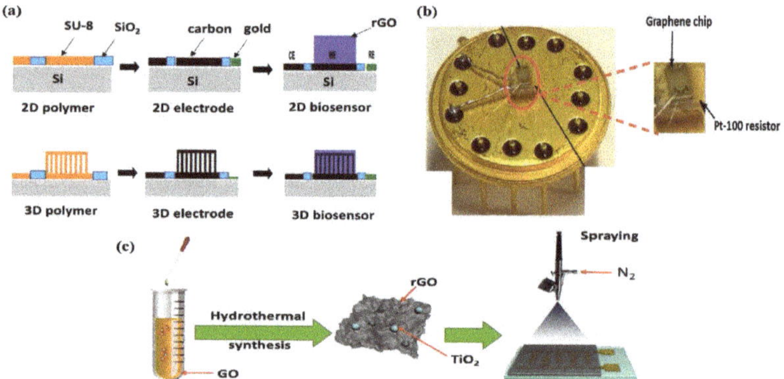

Figure 4. (**a**) Schematic illustration of 2D and 3D carbon-based biosensors, reproduced with permission from [97], published by MPDI, 2018; (**b**) graphene-based sensor chip above holder reproduced with permission from [100], published by Elsevier, 2016; and (**c**) synthetic layout of rGO-TiO$_2$ thin-film sensor, reproduced with permission from [105], published by Elsevier, 2017.

2.4. Carbon-Based Materials for Wearables Electronics

2.4.1. Impregination, Thermal Annealing, and Spray Deposition Methods

The applications of carbon allotropes in the field of wearable, flexible, and stretchable electronics are noteworthy. Previously, SWCNT-tissue-paper-based flexible wearable pressure sensor was synthesized by impregnation and thermal annealing [106]. The fabrication procedure involved the mixing of a SWCNT suspension with tissue paper, evaporating the water, and coating the dried tissue paper/SWCNTs solid on a substrate to obtain a resistance of up to 12.6 kΩ. Then, a prepared PDMS layer was poured onto a glass slide and cured by heating. The PDMS layer (300 μm) was deposited on a PI-coated Ti/Au electrode via metal shadow masking. The pressure sensor device was prepared by coating the active SWCNTs/tissue paper on the PDMS/PI surface deposited on an alumina wire; this was connected to an electrode pad with Ag epoxy for electric networking. On the other hand, Wang and Loh [107] synthesized a type of nanocomposite sensing material based on MWCNTs. A thin film was prepared via airbrushing; this technique has been reported in literature [108,109]. Briefly, the dispersion (MWCNTs:distilled water) and Poly(sodium 4-styrenesulfonate):N-Methyl-2-pyrrolidone (PSS:NMP) were mixed to obtain a latex solution. This ink solution was sprayed onto a glass microscopic slide and annealed at a specified temperature to improve its electrical and mechanical characteristics. The annealed thin film was sandwiched among layers of two-sided Fe on an adhesive fabric and pressed by an iron heat presser. The prepared MWCNTs-sensor was cut into small pieces for electrochemical experimental tests. Li et al. [110] invented a multifunctional wristband fabricated from flexible carbon-sponge polydimethylsiloxane composites (CS/PDMS). The brief synthetic outline is clearly mentioned in Figure 5a, the CS conductive material was synthesized from tissue paper waste through ultrasonication in distilled water followed by freeze-drying and pyrolytic heat treatment. A polydimethylsiloxane resin and curing agent were mixed with the prepared CS. After mixing and vacuum drying, the CS was cut into sheets and coated with aluminum foil and silver paste to obtained conductive electrodes, which were then encapsulated by PDMS elastomers. The sensors were finally fixed on wristbands and sports shoes to detect their performance.

2.4.2. In-Situ Chemical Reduction and Full-Solution Methods

Karim et al. [111] fabricated a rGO-based wearable e-textile device. The brief synthetic protocol is presented in Figure 5c. The in-situ chemical reduction method was used to reduce GO; it involved the addition of PSS in a GO suspension followed by stirring. The prepared mixture was transferred to a round-bottom flask that was fixed in an oil bath, then ammonia and $Na_2S_2O_4$ were added for the reduction of GO. A dispersion of rGO ink was formed with distilled water. The e-textile wearable device was prepared by the pad-drying method, where the textile fabric is dipped/padded several times (1 to 10) in the rGO suspension (pick-up was 80% of textile fabric) in order to improve the electrical conductance of the graphene-based textile fiber. Wu et al. [112] also synthesized a smart e-textile wearable electronic generator based on polyester/Ag nanowires/graphene core-shell nanocomposites via a full-solution process. Plasma-treated textiles were strained with a tension of approximately 20 N. Then, a Ag nanowire solution was pipetted and a blade was used to coat the textile surface via a bottom-up approach. After that, the textile-Ag nanowire was coated with a dispersed GO (prepared by modified Hummers' method) solution. The prepared e-textile was then treated with hydrazine-hydrated vapor for the reduction of the coated GO. For the tribo-electric generator formation, a PMMA/chloroform suspension was blade-coated onto the textile. Similarly, a PI/e-textile was prepared and annealed under Ar atmosphere for the polymerization of polyamic acid within the layers. The PDMS film was obtained by solidification of the e-textile/PDMS.

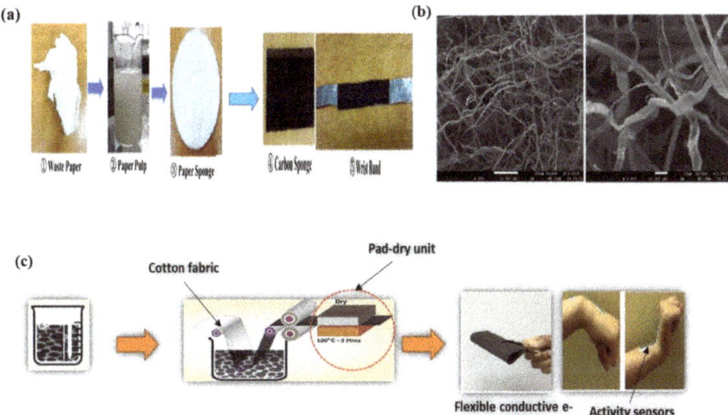

Figure 5. (**a**) Schematic illustration of fabrication of multifunctional wrist band made of CS/PDMS composite, (**b**) SEM images of carbon sponge prepared from paper waste at different magnifications, reproduced with permission from [110], published by ACS Publications, 2016; and (**c**) schematic illustration of scalable formation of reduced graphene oxide-based wearable e-textile, reproduced with permission from [111], published by ACS Publications, 2017.

3. Results and Discussion

3.1. Supercapacitors

Supercapacitors play a very important role in applications of energy storage devices, for example, in energy storage chips, implantable electronic devices, wire free electronic-sensors, and many others [113–117]. The different electrode configurations of mini-supercapacitors include sandwich, roll-shaped, as well as inter-digital [118]. The tremendous merits of supercapacitors are its higher power density, lengthy life cycle, and inexpensiveness. For commercial applications of supercapacitor, the electrodes are required to exhibit high ionic sorption ability at solid liquid interfaces and rapid charge transfer. For this purpose, scientists have explored various carbon-based materials (AC, CFs, graphene etc).

Lee et al. [58] prepared an AC/Ppy hybrid composite for a supercapacitor electrode. The prepared AC/Ppy composite had a maximum specific capacitance of 82.3 $F{\cdot}g^{-1}$, determined by performing a charge/discharge galvanostatic test. The primary stability loss was observed after 1000 cycles due to the instable nature of Ppy. In comparison, Vighnesha et al. [61] fabricated an AC-PANi composite electrode material; the maximum specific capacitance recorded for this composite at 0.5 m $A \cdot g^{-1}$ was 99.6 $F \cdot g^{-1}$ by a charge-/discharge test, which was higher than the previously reported value by Lee et al. [58]. A flexible hybrid electrode composite, Ppy-MnO_2-CFs, displayed a specific capacitance of 69.3 $F \cdot cm^3$ at 0.1 $A \cdot cm^3$ with an energy density (ED) of 0.00616 $Wh \cdot cm^3$, obtained via cyclic voltammetry and a charge/discharge galvanostatic testing system using a PVA/H_3PO_4 electrolyte. It shows electro-chemical stability up-to 1000 cycles [63]. Furthermore, the influence of the specific surface area and pore size on a composite material was studied by Heo et al. [64]. In this work, a CF-based composite activated at 1000 °C had a high specific surface area (1886 $m^2 \cdot g^{-1}$) and reliable pore size (0.021–1.196 $cm^3 \cdot g^{-1}$). The APCFs-1000 sample showed a specific capacitance of 103 $F \cdot g^{-1}$ (1 $A \cdot g^{-1}$) in a 1 M Na_2SO_4 electrolyte with stability up to 3000 cycles. This enhanced supercapacitance occurred due to the high specific surface area, suitability of pore size, and effect of heteroatoms among the enhanced double layers. Further improvements in the specific capacitance properties of an electrode material was reported by Cakici and their co-workers for a CFF/MnO_2 composite that acted as a flexible electrode in a supercapacitor. Three tests were used to verify the electrochemical performance: cyclic voltammetry, charge/discharge galvanostatic test, and electro-chemical impedance spectroscopy. The cyclic voltammetry analysis showed that the prepared materials had a significantly efficient specific capacitance, with a value of 467 $F \cdot g^{-1}$ (1 $A \cdot g^{-1}$). The materials also exhibited high electro-chemical stability after 5000 cycles. Additionally, the as-synthesized device showed a very high energy density (approximately 20 $Wh \cdot kg^{-1}$) at 0.176 $kWh \cdot kg^{-1}$. This means that this carbon-based composite hybrid material can be used as a potential electrode for commercial energy-saving devices with supercapacitors [65].

Many researchers have worked on synthesizing graphene-based materials for supercapacitor electrodes. Ke et al. [66] fabricated a Fe_3O_4-rGO nanocomposite electrode material that exhibits a specific supercapacitance of 169 $F \cdot g^{-1}$ at 1 $A \cdot g^{-1}$ (obtained by cyclic voltammeter and galvanostatic charge/discharge tests). The electrode material showed a capacitance retention of above 88% after 100 cycles. In contrast, Li et al. [67] fabricated GO-decorated MnO_2 (GO/MnO_2). The supercapacitor fabricated with the GO/MnO_2 composite showed a specific capacitance of 211.5 $F \cdot g^{-1}$ at 2 $mV \cdot s^{-1}$. The charge/discharge studies showed that 75% capacitance stability was maintained after 1000 cycles by utilizing 1 M Na_2SO_4 as an electrolytic solution. Further improvements in hybrid materials with high-specific capacitance has been achieved by developing various MO-doped (RuO_2, TiO_2, and Fe_3O_4) and polyaniline-functionalized GS. The highest specific capacitance recorded was 375 $F \cdot g^{-1}$ with 1 M sulphuric acid (as an electrolytic solution) and a sweep-voltage speed range of 10–100 $Mv \cdot S^{-1}$. Despite the high voltage rate, 85% of the capacitance was maintained, proving that this composite material can be used as an excellent supercapacitor [75]. Sawangphruk et al. [76] invented a hybrid pseudo-supercapacitor (AgNP/PANI-Graphene-CFP) that showed a high specific capacitance in a 1 M Na_2SO_4 electrolyte, approximately 828 $F \cdot g^{-1}$ at 1.5 $A \cdot g^{-1}$ with a capacitive stability up to 97% after 3000 cycles, obtained via a charge-discharge test.

It is known that exfoliated GO sheets exhibit a wider specific surface area and that they need to be restacked. In order to increase the efficiency of GO sheets, a famous approach is to decorate GO layers using oxide- or hydroxide-based nanomaterials. This technique enables the prepared materials to have enhanced functional applications, for example, increased pseudo-capacitance and also act as a spacer among the GO layers. In this regard, Wang et al. [77] fabricated a hybrid electrode based on Ni(OH)$_2$ nanoplates deposited on graphene. The maximum specific capacitance of the Ni(OH)$_2$/graphene electrode, as shown in Figure 6a, is approximately 1335 $F \cdot g^{-1}$ with a 2.8 $A \cdot g^{-1}$ current density. The capacitance was stable after 2000 charge/discharge cycles at the highest current density tested (28.6 $A \cdot g^{-1}$), as shown in Figure 6b. A literature survey showed that so far, the most efficient pseudo-capacitive hybrid electrode composite was invented by Zhang et al. [78], named rGONF/Ni(OH)$_2$. This composite has the highest recorded specific supercapacitance (1433 $F \cdot g^{-1}$) at 5 $mV \cdot s^{-1}$ and maintained approximately 90% of the capacitance after 2000 cycles; this is clearly shown in Figure 6c,d.

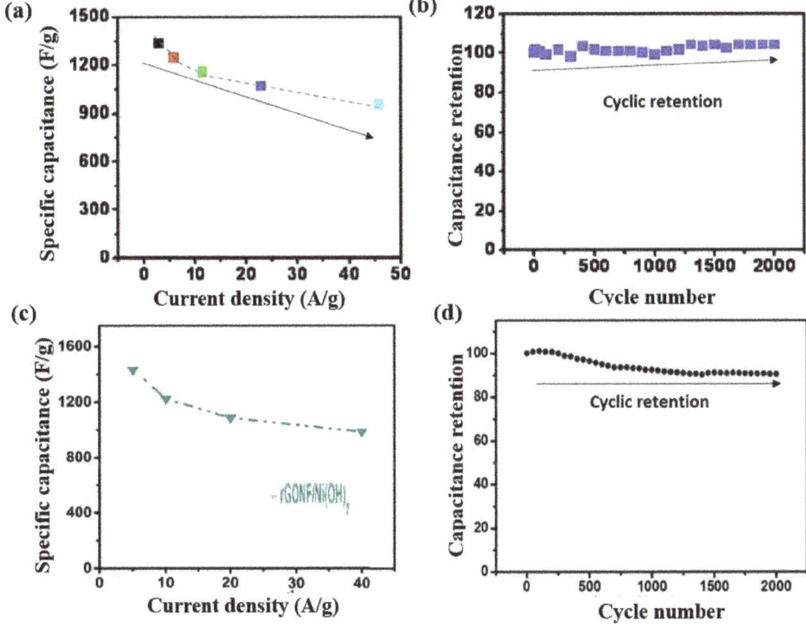

Figure 6. (a) Average specific capacitance (SC) at different charge-discharge current densities of 28.6 A·g^{-1}, (b) capacitance stability versus cycle number of Ni(OH)$_2$/GS, reproduced with permission from [77], published by ACS Publications, 2010; (c) SC of rGONF/Ni(OH)$_2$ hybrid electrode composites and (d) cycling retention of rGONF/Ni(OH)$_2$ material, reproduced with permission from [78], published by ACS Publications, 2016.

The rGONF/Ni(OH)$_2$ hybrid exhibited efficient energy density, and can be utilized at an industrial level for supercapacitor applications. Besides these reported works, as mentioned in Table 1, different modified carbon-based composites have also been studied for supercap, the acitor applications.

Table 1. Different carbon forms, their modification and application in supercapcitors. GNS: graphene nanosheet, KTP: Korean traditional paper, pErGO: porous electrochemically reduced graphene oxide, Cuf; copper foil, AC; activated carbon.

Carbon Forms	Functionalization	Specific Capacitance (F/g)	Electrolyte (M)	Retention Cycle	References
MWCNTs	PANI (20%)	670	H_2SO_4 (1)	-	[119]
MWCNTs	Ppy	427	Na_2SO_4 (1)	-	[120]
CNTs	M-PANI	1030	H_2SO_4 (1)	5000	[121]
MCNTs	PEDOT	120	H_2SO_4 (1)	20000	[122]
Graphene	PANI(80%)	320	H_2SO_4 (2)	-	[123]
Graphene	PANI	1126	H_2SO_4 (1)	1000	[124]
GNS	PANI	1130	H_2SO_4 (1)	1000	[125]
pErGO	Cuf/Cu wire	81	PVA/H_3PO_4	5000	[126]
AC	Fe_3O_4	37.9	KOH (6)	500	[127]
CNTs	RuO_2-TiO_2	50	KOH (1)	1000	[128]
Carbon black	Fe_3O_4	5.3	Na_2SO_4 (1)	10000	[129]
3D GO	PANI	1341	H_2SO_4 (1)	5000	[130]
N-doped-rGO	PANI	610	H_2SO_4 (1)	1000	[131]
rGO	PANI-Co_3O_4	1063	KOH (6)	2500	[132]
rGO	PANI, ZrO_2	1360	H_2SO_4 (1)	1000	[133]
B-doped rGO	PANI	406	H_2SO_4 (1)	10000	[134]
MWCNT	Ni_3S_2	55.8	KOH (2)	5000	[135]
graphene	MoS_2	268	Na_2SO_4 (1)	1000	[136]
CNTs	CuS	112	KOH (2)	1000	[137]

3.2. Inks and Inkjet Printing Devices

In the previous decades, electronics related to the printing technology have attracted increasing attraction due to the potential of manufacturing inexpensive and large-scale electronic circuits. It is necessary to develop reliable functionalized inks, highly mobile printable semiconductors, conducting inks that consume less energy, and higher-magnification and uniform printing tools, in contrast to conventional techniques [138]. The first work related to the production of conductive inks for inkjet printing based on graphene was conducted by Torrisi and Coleman [139]. They obtained a large amount of GS rapidly by liquid-phase exfoliation in easily printable solvents, including water and other organic liquids. The resulting ink showed stability and easy processability at room temperature and exhibited high batch reproducibility with efficient rheological characteristics for printing. Secor et al. [81] reported on gravure printing of graphene to rapidly form conductive patterns on a stretchable substrate; they fabricated reliable inks and presented printing parameters permitting the synthesis of patterns under magnifications up to 30 μm. The inclusion of a mild heating process resulted in conductive lines with high regularity. These results provide an effective approach for the integration of graphene into large printed areas as well as in flexible electronics. Torrisi et al. [83] investigated a feasible approach for broad-scale synthesis of graphene-based devices with inkjet printing. The researchers successfully prepared graphene-contained inks, which were formed by liquid-based exfoliation of graphite powder with N-methyl-pyrrolidine. These functional inks were used to print thin-film transistors which had a mobility of approximately 95 $cm^2 \cdot V \cdot s^{-1}$. The transmittance of these inks was approximately 80% of the transmittance of graphene and the resistance was approximately 30 $k\Omega \cdot cm^{-2}$. These inventions paved the way for the fabrication of printed, flexible, and transparent graphene-based electronics on a uniform substrate. Pristine GO ink was found to exhibit stability above 9 months at a 1 $mg \cdot mL^{-1}$ concentration, with well-suited fluidic properties for effective and viable ink-jet printing devices [84]. A conductance of 9.24×10^3 $S \cdot m^{-1}$ was obtained after 30 printing stages at 300 °C. The resistance of the electrode printed on a flexible substrate was raised by < 5% after 1000 bending cycles and by 5.3% under a 180° bending angle. This reported technique for developing inks and conductive electrodes has proven to be promising for applications related to graphene-based flexible electronics. Homenick et al. [85] formulated a SWCNT-based ink via hybrid extraction-adsorption. The SWCNTs concentration, amount of ink incorporated, and comparative ratio of the polymer to SWCNTs were controlled using the SWCNTs' network density. The optimized inkjet printing parameters were identified on Si/SiO$_2$, where an ink with a polymer: SWCNTs ratio of 6:1 and 50 $mg \cdot L^{-1}$ SWCNTs concentration printed at drops spaced 20 μm apart leads to a mobility of thin film transistor of approximately 25 $cm^2 \cdot V \cdot s^{-1}$ with on-off voltage ratios greater than 10^5. The mentioned conditions produced an efficient network regularity and was utilized in an additive process to synthesize TFT on a PET substrate, with motilities greater than 5 $cm^2 \cdot V \cdot s^{-1}$. The ink-jet printing encapsulation layer effectively resulted in a TFT sample with a mobility of more than 1 $cm^2 \cdot V \cdot s^{-1}$; the use of inverter circuits resulted in stable and efficient operation conditions. Han et al. [86] reported the fabrication of a water-based conductive SWCNTs ink with sodium dodecyl-benzene sulfonate (surfactant). Direct writing on a paper with the prepared ink was performed with an off-the-shelf nib as well as a cartridge in the jet pen handwriting device. The lighter weight and moveable nature of the device meant that writing on a curved substrate was possible. The paper was obtained by a wetting approach. Double-sided and many-layered paper circuit boards were established via direct writing. The printed writing showed regularity and renewability. The extraordinary adhesion of SWCNTs on the cellulose paper showed efficient robustness against different mechanical loads or stress.

Additionally, Foster et al. [87] reported on using graphene-based polylactic acid filaments to print 3D disc electrodes with a Rep-Rep FDM 3-D printer. The prepared 3D electrodes were characterized electrochemically and physiochemically. The prepared 3D electrode was applied as freestanding anodes in a Li$^+$ battery and used in hydrogen generation via H-evolution reaction and exhibited tremendous catalytic activity. The results suggested that 3D printing of graphene-based conductive filaments enables the facile synthesis of energy storage devices. Recently, You et al. [88] demonstrated a direct ink-writing technique by preparing a 3D structure with stacked layers based on light-mass cellular interlinked networks. A homogenized graphene dispersion was formed via ultrasonication in ethanol. The speed of the printer and nozzle size were organized in such a way as to form 3D graphene; the material exhibited morphological stability, with 50% of the graphene contents in the filament, as mentioned in Figure 7a–c, and maintained the tremendous properties of graphene. This material was utilized as a 3D material in 3D printers. For the purpose of improving the printing properties, Sarapuk et al. [89] worked on the properties of graphene ink (including viscosity, strain, and shear ink rate) and presented the results via model-based calculations for an inkjet printer nozzle. The experimental study was conducted by preparing inks with or without a dispersing agent to validate the model. The results showed that the dispersing agent played a valuable role in improving the viscosity, printing power, and path with minimum resistivity, as shown in Figure 7d–f. Furthermore, stable graphene-based inks with a dispersing agent can produce effective prints.

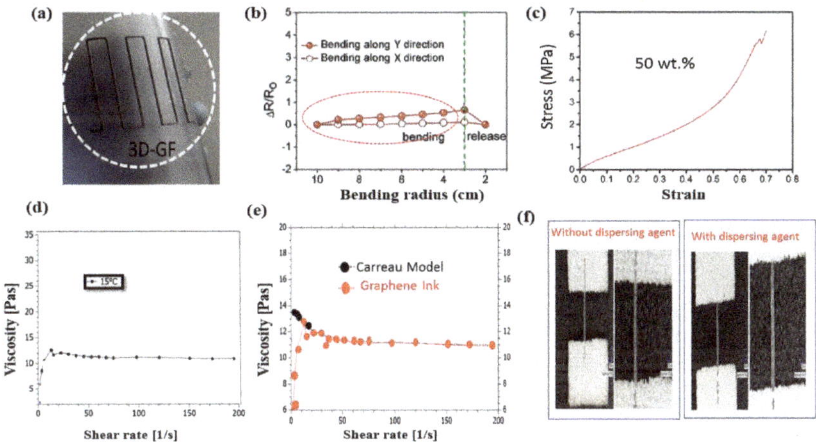

Figure 7. (**a**) Macrograph of 3D printed individual graphene filament, (**b**) electrical resistance of flexible circuit with 50 wt.% graphene content at various bending radius, and (**c**) typical stress strain curves of printed structure of 50 wt.% graphene loading, reproduced with permission from [88], published by Elsevier, 2018; (**d**) viscosity curves of graphene based ink with dispersing agent at 15 °C temperatures, (**e**) graphene ink with dispersing agent viscosity curves with fitted Carreau Model curve, and (**f**) graphene printed paths produced with or without dispersing agent graphene ink were printed via piezoelectric inkjet printer at following parameters, nozzle diameter (50 m), 40–50 V, pulse length (150–200 s), reproduced with permission from [89], published by MDPI, 2018.

3.3. Biosensors

A biosensor is a machine that can detect molecules with a specific transducer to produce a detectable signal from the sample [140,141]. The schematic illustration is shown in Figure 8, indicating a general platform and the interface between a bio-receptor and a transducer.

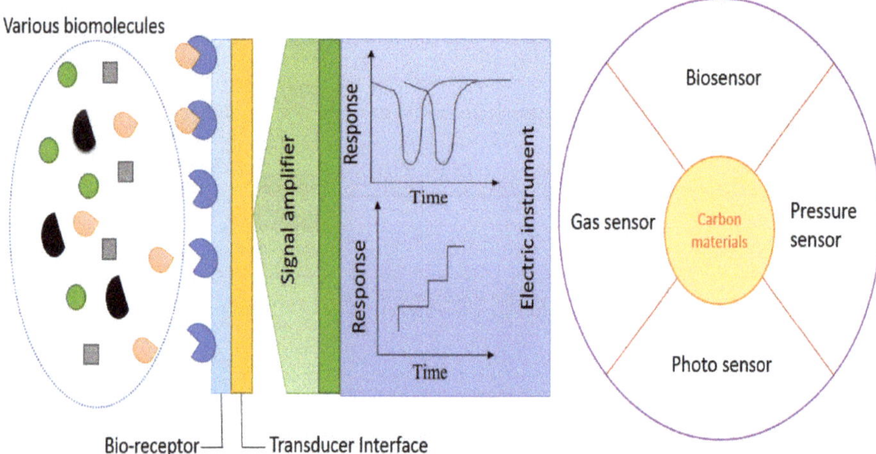

Figure 8. General layout of a typical biosensors system, and use of carbon materials in various sensors applications.

The electrochemical-based biosensor contains 2–3 electrodes in a cell that are responsible for the transformation of a biological condition to electrochemical signals. It mostly contains biomolecules on an electrode that interact with analytes and lead to the generation of electrochemical signals [142]. Li et al. [90] fabricated a PDA-Lac-rGO-Pd composite for use as a sensing material. An electrochemical biosensor was prepared using composite to detect catechol. Under optimum conditions, the prepared electrochemical biosensor exhibited linearity in the range of 0.1–263 µM, with a sensitivity of 18.4 µAm·M^{-1} at the minimum detection range of 0.03 µM. Furthermore, the electrochemical sensor possessed efficient repeatability, renewability, as well as stability. An identical electrochemical biosensor containing Lac exhibited the ability to detect minute amounts of catechol in unique water conditions. On the other hand, Hassan et al. [93] investigated a unique, effective glucose biosensor above functionalized rGO. This biosensor had a glucose-dependent electro-chemical behavior with Ag-AgCl as the standard electrode. This graphene-based biosensor with an enzyme possessed a very wide range of detection of glucose concentrations (up to 100 Mm), with sensitiveness of 5.59 µA·D^{-1}. These types of biosensors can be used as an efficient indicator of sugar level for biological diagnoses. Hemanth et al. [97] also reported an enzyme-focused electro-chemical biosensor. This biosensor was fabricated using a 3D pyrolytic carbon micro-electrode that was deposited with biologically functionalized rGO. The glucose-sensing properties of the 3D rGO-based biosensor were compared with those of 2D electrodes using cyclic voltammetry. The results showed that the 3D biosensor exhibited twice the sensing ability of the 2D electrodes i.e., with a value of 23.56 µAm·(M·cm^2)$^{-1}$ instead of 10.19 µAm·(Mcm^2)$^{-1}$. The stability analysis of the enzyme above 3D rGO presented renewable results above 1 week. The prepared biosensor showed higher glucose selectivity over uric/ascorbic acid and cholesterol.

Furthermore, CNT-based electrochemical biosensors also proved to be very efficient in sensing applications because of their tremendous merits, including great sensitiveness, rapid responsivity, ease of handling, and reliable transportability. A literature survey [143,144] indicated that CNT-based electrochemical biosensors have been investigated for the detection of biologically essential analytes by electro-chemical reaction catalyzed via different bio-enzymes [145], including glucose oxidases (GODs) [146], horse radish peroxidases [147], lactases [148], and malate dehydrogenases [149].

3.4. Gas Sensors

Gas sensors are very important components to sense the nature and amount of gas. A gas sensor can alter gas components and concentrations to obtain information based on the electric measurements of gas [150]. In 2014, a hybrid composite with rGO coated with ZnO nanomaterials was fabricated through a redox reaction [99]. The gas-sensing activities indicated that the response of the hybrid composites towards 5 ppm NO_2 gas was 25.6% with a speed of 165 s and a reproducible time of 499 s. Furthermore, Liu et al. [151] reported on gas sensors with bimetallic nanoparticles (SnO_2, ZnO) above graphene having a 3D porous morphology (pore size of 3–10 nm). This $G/SnO_2/ZnO$ composite gas sensor showed rapid and efficient NO_2 gas adsorption response at various concentrations, with a response time of <1 min and efficient reproducibility (within 1 min). Similarly, Novikov et al. [100] successfully fabricated the cheapest ultrasensitive gas sensor based on an epitaxial graphene/silicon carbide composite. This sensor was functional under minimum operating concentrations (<1 ppb) with high sensitivity towards NO_2 in an air mixture. A prototype of a replaceable electronic device to check surrounding levels of NO_2 that uses a mixture of gases at ambient temperature was built. The prepared sensor can be recovered at room temperature and resulted in very rapid and reproducible analysis of NO_2 (5–50 ppb).

On the other hand, Huang and Hu [102] reported on the use of a graphene-doped polyaniline composite as a NH_3 gas sensor. They found that the prepared hybrid sensor was capable of NH_3 sensing with a response of 59.2% at 50 ppm concentration. The sensor also possessed an effective response towards H_2, which was investigated by Zou et al. [152]. A t_{res} of 20 s and t_{rec} of 50 s were found when the prepared PANi-GO-based sensor was exposed to 1% (by volume) of H_2 at ambient temperature. Recently, a hybrid composite was prepared with TiO_2 (due to its high specific surface area and inexpensiveness) and graphene. This type of composite sensor proved to be very efficient to sense NH_3 due to the existence of a large number of active adsorption sites [105]. A detailed literature survey on carbon-based composites and their functionalization for biosensors, gas sensors, and many other sensors are summarized in Table 2.

Table 2. Carbon-based hybrids, their functionalization/modification as sensors.

Carbon Material	Modification	Analyte	Detection Limit	References
MWCNTs	COOH	O_2	0.3%	[153]
MWCNTs	maleic acid, acetylene	NH_3	10 ppm	[154]
SWCNTs	Pd doping/sputtering	H_2	0.5%	[155]
SWCNTs	$LaFeO_3$	methanol	1 ppm	[156]
SWCNTs	Pd nanoparticles	glucose	0.2 mM	[157]
MWCNTs	Pt nanoparticles	glucose	1×10^{-5} mol/L	[158]
Multi-layered graphene	Poly(vinylpyrrolidone), glucose oxidase	glucose	2 mM	[159]
rGO	Sulfophenyl, ethylenediamine	NO_2	3.6 ppm	[160]
rGO	Au-Pt alloy, chitosan-glucose oxidase	glucose	5 mM	[161]
Graphene foam	α-Fe_2O_3	NO_2	0.12 mM	[162]
GO	poly(3,4-ethylenedioxythiophene)	dopamine	0.33 mM	[163]
rGO	PNF-AgNPs	H_2O_2	10.4 µM.	[164]
GO	peptide-AgNPs	H_2O_2	0.13 mM	[165]
Graphene foam	CuO nanoflower	ascorbic acid	0.43 mM	[166]
rGO	CeO_2/GCE	NO_2	9.6 nM	[167]
rGO	$AuFe_3O_4$/Pt	H_2O_2	0.1 nM	[168]
GO	Au@Pt@Au NPs	H_2O_2	0.02 nM	[169]

3.5. Wearable Electronic Devices

There are broad varieties of flexible and stretchable materials available for utilization in different wearable appliances, including replaceable sensors and flexible electrodes and circuits. In the healthcare sector, wearable electronics have proven to be very efficient to check human health, movements, and thermo-therapy. Dinh et al. [170] designed CNT-based thermal sensors as wearable electronics for humans by utilizing lightweight high-strength stretchable CNT yarns as a heating wire, a graphitic pencil as an electrode, and minimum weight, durable, and bio-recyclable paper as a stretchable substrate. This CNT-based device was used as a sensor by fixing on human skin to monitor real-time respiration rates and to detect respiratory-related diseases. In addition, they also fixed a temperature detector within a similar sensor to analyze human body temperature in a contactless mode. This invention paved the way for the utilization of CNT yarns for the development of a broad range of eco-friendly, inexpensive, and lighter weight stretchable and wearable electronics related to temperature and breathing detectors. Zhan et al. [106] reported an identical, inexpensive SWCNT-tissue paper based stretchable wearable pressure sensor. This wearable pressure sensor showed tremendous performance and various advantages, such as greater sensitiveness for a wide range of pressures with minimum energy utilization (6–10 W), with real-time monitoring of various physical muscle activities. Additionally, the application of this pressure sensor to detect the response of force and pressure under synthetic robotic skin was also investigated. Wearable devices related to the detection of human movements and physical actions of athletes have also been developed. In this regard, Wang and Loh [107] fabricated multi-functional wearable sensors with a CNT-based fabric for monitoring the bending movements of human fingers in order to detect the commercial strain-sensing capability of the sensors. In addition, the CNT-fabric based sensor was fixed on a chest band to monitor the human breathing rate. This invented CNT-fabric-based sensor exhibited several merits, including flexibility, facile synthetic approach, low weight, inexpensiveness, and reliability for human use. Li et al. [110] fabricated a multi-functional wearable wristband prepared from CS/PDMS composites that exhibited high flexibility, confirmed by stress-strain curve as shown in Figure 9a. This wristband could act as a heater for thermotherapy, a biosensor for human blood pulse, and a breathing (shown in Figure 9c) and movements detector. The wristband could act as a heater below 15 V with a constant temperature variation of 20 °C. As a strain sensor, it exhibited rapid, reproducible response and efficient stability within a strain range of 0–20% and an employed frequency range of 0.01 to 10 Hz. The efficient flexibility, intermediate conductivity, efficient strain-sensing ability, and inexpensive nature make the multifunctional wearable CS/PDMS band a potential candidate for healthcare devices.

rGO-based e-textile wearables exhibit a wide range of benefits over conventional metal-based approaches. These conventional methods are complex and unsuitable for a wide range of practical applications. Therefore, Karim et al. [111] reported a facile, inexpensive technique to fabricate rGO-doped wearable devices through a facile pad-dry approach. This technique enabled the efficient manufacturing of conductive rGO e-textiles at an industrial fabrication rate of approximately 150 m·min^{-1}. The rGO-based e-textile electronic wearable devices possessed reliable softness, stability, and washability. The use of rGO increased the flexible nature and tensile strength of cotton fabrics by increasing the percentage of strain at the highest weight. The activity of the prepared rGO e-textile sensor was monitored by human wrist upward-downward motions and the results are shown in Figure 9d.

This invented rGO e-textile wearable device was tested as a commercial sensor and heating appliance. The wearable tribo-electric sensor was compatible with an intelligent setup. However, the low conductivity, stability, and compatibility of e-textile electronics prohibited the fabrication of reliable incorporated generators for human clothing. To overcome these drawbacks, Wu et al. [112] fabricated wearable electronic generators based on polyester/Ag nanowires/graphene core-shell hybrids impregnated on an efficient, opaque, and smart e-textile via an eco-friendly full solution technique. The prepared smart e-textile device showed tremendous conductivity under a 20 Ω square and was efficient, flexible, stretchable, bendable, and washable. It can act as an electrode and wearable device, however due to its wearability, the smart e-textile generator was easily fitted in gloves to demonstrated the mechanical power produced by movements of fingers. The maximum recorded power by a single generator-based glove due to finger movement was 7 nW·cm^{-2}. The properties of this composite-based smart e-textile prove that it as an efficient candidate for practical wearable clothing.

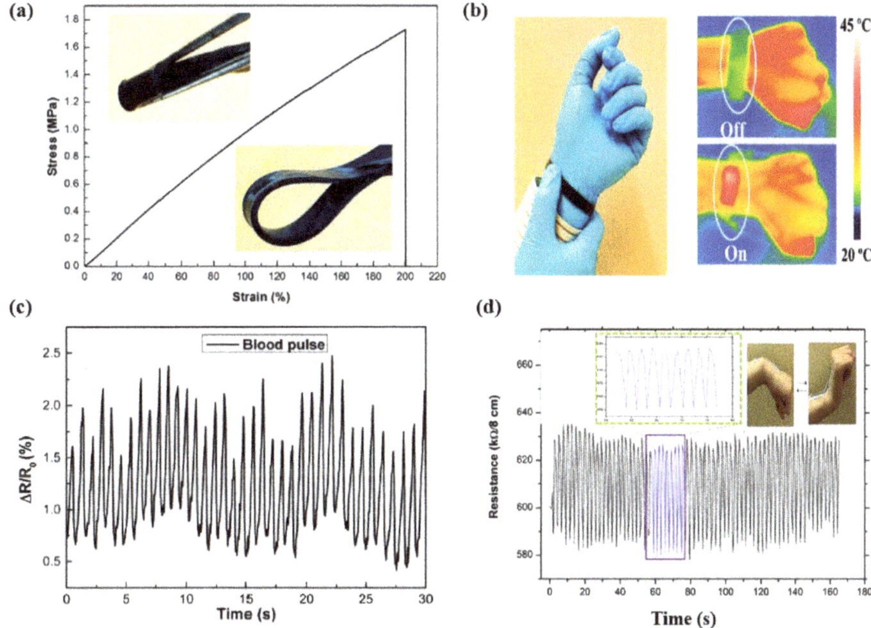

Figure 9. (**a**) Typical tensile strain-stress curve of CS/PDMS composite, (**b**) Infrared images of multifunctional wrist band worn by a volunteer as a wearable heater at on and off modes, (**c**) relative change of resistance (RCR) response of the wrist band to blood pulse of an adult volunteer; monitoring of the breathing, reproduced with permission from [110], published by ACS Publications, 2016; and (**d**) the upward/downward motion of wrist recorded by rGO-based cotton fabric sensors, and inset shows the magnified version of blue box in image (**d**) in the time range of 57–77 s, reproduced with permission from [111], published by ACS Publications, 2017.

4. Conclusions

In this review, we have summarized a large number of research papers related to the fabrication of nanocomposites materials based on chemically functionalized GO/rGO/GS, SWCNTs, MWCNTs, AC, and CFs. The outstanding properties of these functionalized carbon allotrope-based composites, including high specific surface area, high durability, high thermal and electrical conductivity, high resistivity, and elastic flexibility, have resulted in remarkable applications in the electronic industry. MO-doped, hybrids, and composites of carbon materials have improved the electronic properties of devices. This review also elaborated the recent development of carbon allotrope composites and their applications in energy storage devices, supercapacitors, ink formulation, inkjet-printing devices, bio-gas and pressure sensors, as well as stretchable/flexible wearable electronics. Overall, the results indicated that the properties of carbon allotropes can be improved and polished by fabricating composites and hybrids, which broaden their applications in the commercial electronic industry.

5. Future Outlooks

As mentioned in the literature survey, different hybrid materials, methodologies, and techniques have been applied to produce energy storage electronics that exhibit high specific capacitance, energy density, and power density. It is well known that carbon-based hybrids are an efficient contender for electrodes in supercapacitors; however, they require supportive dynamic materials to increase their efficiency. The surface modification of electrodes by carbon allotropes (CNTs,

AC, CFs, and rGO) functionalized by conducting materials, such as polymers, MO, and magnetic nanoparticles, significantly enhances the power and energy density, specific capacitance, and reusability of supercapacitor electrodes. If further research efforts are devoted to this field, then functionalized carbon allotrope-based supercapacitors can be expected to be a novel innovation in the field of electronics.

On the other hand, important developments have been made in inkjet 3D printing by using graphene- and SWCNT-based inks. Many newly designed inkjet 3D printers exhibit performance superior to that of earlier-generation printers. Conventional approaches have not been frequently adopted by a large number of industries because of practical issues related to efficiency, price, and reliability of bulk material formation. For solvent-based direct-writing techniques in inkjet printers, the rheology of inks is essential. In order to manufacture reliable inks, adequate filler contents with a homogeneous dispersion are mandatory. Furthermore, a small concentration of graphene-based inks has been utilized to prepare 3D graphene-based tools, but has recently been deemed inappropriate for broad-range production because of its harsh print settings. Therefore, the proper selection of printing technique and materials are still existing issues. The use of carbon-based inks still requires optimized conditions in order to improve the homogeneity, stabilization, and fine-printing ability at an industrial level.

Furthermore, electrodes functionalized with SWCNTs, MWCNTs, or different graphene forms with changeable band-gaps seem to be more reliable for the fabrication of biosensors. These sensors are able to sense different biological components, including protein, glucose, and enzymes, with great sensitivity because of their high flexibility and specific surface area. These carbon-based sensors also have the ability of detecting gases (O_2, N_2, NH_3, and H_2). However, a facile approach for controlled fabrication and easy handling of graphene should be an important research topic in the upcoming years. Recent chemical methodologies to functionalize graphene with biological compounds have proven to be efficient in enhancing the detecting efficiency of the targeted components, although the refining of the detecting surface material is required in order to restrict the adsorption of non-relevant species during the sensing performance. In addition, miniaturized compact-type biosensor devices having high durability, accuracy, sensitivity, and cost efficiency are highly demanded for the detection of viruses and bacteria to monitor the human health situation.

A large number of applications of functionalized carbon-based stretchable devices for the monitoring of human health and body movements have been developed. This includes the observation of nerve pulse, respiratory rate, and body heat. The development of multifunctional wearable electronics remains limited and should be given greater consideration. In the near future, the simultaneous use of functionalized carbon-based composites and biocompatible components for efficient wearable devices may be possible. Smart electronic devices have also attracted attention recently. In the near future, more progress in advanced carbon-based stretchable devices will highly contribute to the development of more efficient devices for the health-care and medical improvement sectors.

Author Contributions: U.K. as first author conducted the literature search. All four authors were involved in writing and editing the review manuscript.

Acknowledgments: This work was supported by the Technological Innovation R&D Program (S2460853) funded by the Small and Medium Business Administration (SMBA, Korea) and the Commercialization Promotion Agency for R&D Outcomes (COMPA) funded by the Ministry of Science and ICT (MSIT) [2018_RND_002_0064, Development of 800 mA·h·g^{-1} pitch carbon coating material].

References

1. Zhang, Y.; Yin, Q.-Z. Carbon and other light element contents in the earth's core based on first-principles molecular dynamics. *Proc. Natl. Acad. Sci. USA* **2012**, *109*, 19579–19583. [CrossRef] [PubMed]
2. Allègre, C.J.; Poirier, J.-P.; Humler, E.; Hofmann, A.W. The chemical composition of the earth, earth planet. *Sci. Lett.* **1995**, *134*, 515–526.

3. Karthik, P.S.; Himaja, A.L.; Singh, S.P. Carbon-allotropes: Synthesis methods, applications and future perspectives. *Carbon Lett.* **2014**, *15*, 219–237. [CrossRef]

4. Shin, H.K.; Rhee, K.Y.; Park, S.J. Effects of exfoliated graphite on the thermal properties of erythritol-based composites used as phase-change materials. *Compos. Part B Eng.* **2016**, *96*, 350–353. [CrossRef]

5. Zhang, Y.; Park, S.J. In situ shear-induced mercapto group-activated graphite nanoplatelets for fabricating mechanically strong and thermally conductive elastomer composites for thermal management applications. *Compos. Part A Appl. Sci. Manuf.* **2018**, *112*, 40–48. [CrossRef]

6. Zhang, Y.; Park, S.J. Imidazolium-optimized conductive interfaces in multilayer graphene nanoplatelet/epoxy composites for thermal management applications and electroactive devices. *Polym. J.* **2019**, *168*, 23–60. [CrossRef]

7. Zhang, Y.; Heo, Y.J.; Son, Y.R.; In, I.; An, K.H.; Kim, B.J.; Park, S.J. Recent advanced thermal interfacial materials: A review of conducting mechanisms and parameters of carbon materials. *Carbon* **2018**, *142*, 445–460. [CrossRef]

8. Zhang, Y.; Choi, J.R.; Park, S.J. Interlayer polymerization in amine-terminated macromolecular chain-grafted expanded graphite for fabricating highly thermal conductive and physically strong thermoset composites for thermal management applications. *Compos. Part A Appl. Sci. Manuf.* **2018**, *109*, 498–506. [CrossRef]

9. Mowery, D.C.; Nelson, R.R. *Sources of Industrial Leadership*; Cambridge University Press: Cambridge, UK, 1999.

10. Walsh, S.T.; Boylan, R.L.; McDermott, C.; Paulson, A. The semiconductor silicon industry roadmap: Epochs driven by the dynamics between disruptive technologies and core competencies. *Technol. Forecast. Soc. Chang.* **2005**, *72*, 213–236. [CrossRef]

11. Schulz, M. The end of the road for silicon? *Nature* **1999**, *399*, 729–730. [CrossRef]

12. Lee, J.H.; Marroquin, J.; Rhee, K.Y.; Park, S.J.; Hui, D. Cryomilling application of graphene to improve material properties of graphene/chitosan nanocomposites. *Comp. Part B-Eng.* **2013**, *45*, 682–687. [CrossRef]

13. Mittal, G.; Rhee, K.Y.; Park, S.J.; Hui, D. Generation of the pores on graphene surface and their reinforcement effects on the thermal and mechanical properties of chitosan-based composites. *Comp. Part B Eng.* **2017**, *114*, 348–355. [CrossRef]

14. Kalosakas, G.; Lathiotakis, N.N.; Galiotis, C.; Papagelis, K. In-plane force fields and elastic properties of graphene. *J. Appl. Phys.* **2013**, *113*, 134307. [CrossRef]

15. Yu, M.-F.; Lourie, O.; Dyer, M.J.; Moloni, K.; Kelly, T.F.; Ruoff, R.S. Strength and breaking mechanism of multiwalled carbon nanotubes under tensile load. *Science* **2000**, *287*, 637–640. [CrossRef] [PubMed]

16. Lee, C.; Wei, X.; Kysar, J.W.; Hone, J. Measurement of the elastic properties and intrinsic strength of monolayer graphene. *Science* **2008**, *321*, 385–388. [CrossRef] [PubMed]

17. Kim, P.; Shi, L.; Majumdar, A.; McEuen, P. Thermal transport measurements of individual multiwalled nanotubes. *Phys. Rev. Lett.* **2001**, *87*, 215502. [CrossRef] [PubMed]

18. Ando, Y.; Zhao, X.; Shimoyama, H.; Sakai, G.; Kaneto, K. Physical properties of multiwalled carbon nanotubes. *Int. J. Inorg. Mater.* **1999**, *1*, 77–82. [CrossRef]

19. Mochalin, V.N.; Shenderova, O.; Ho, D.; Gogotsi, Y. The properties and applications of nanodiamonds. *Nat. Nanotechnol.* **2012**, *7*, 11–23. [CrossRef]

20. Osswald, S.; Yushin, G.; Mochalin, V.; Kucheyev, S.O.; Gogotsi, Y. Control of sp^2/sp^3 carbon ratio and surface chemistry of nanodiamond powders by selective oxidation in air. *J. Am. Chem. Soc.* **2006**, *128*, 11635–11642. [CrossRef]

21. Simon, P.; Gogotsi, Y. Materials for electrochemical capacitors. *Nat. Mater.* **2008**, *7*, 845–854. [CrossRef]

22. Liang, J.J.; Huang, Y.; Zhang, L.; Wang, Y.; Ma, Y.F.; Guo, T.Y.; Chen, Y. Molecular level dispersion of graphene into poly (vinyl alcohol) and effective reinforcement of their nanocomposites. *Adv. Funct. Mater.* **2009**, *19*, 2297–2302. [CrossRef]

23. Zhang, Y.; Rhee, K.Y.; Hui, D.; Park, S.J. A critical review of nanodiamond based nanocomposites: Synthesis, properties and applications. *Comp. Part B-Eng.* **2018**, *143*, 19–27. [CrossRef]

24. Zhang, Y.; Park, S.J. Influence of the nanoscaled hybrid based on nanodiamond@ graphene oxide architecture on the rheological and thermo-physical performances of carboxylated-polymeric composites. *Compos. Part A Appl. Sci. Manuf.* **2018**, *112*, 356–364. [CrossRef]

25. Zhang, Y.; Rhee, K.Y.; Park, S.J. Nanodiamond nanocluster-decorated graphene oxide/epoxy nanocomposites with enhanced mechanical behavior and thermal stability. *Compos. Part B-Eng.* **2017**, *114*, 111–120. [CrossRef]

26. Schedin, F.; Geim, A.K.; Morozov, S.V.; Hill, E.W.; Blake, P.; Katsnelson, M.I.; Novoselov, K.S. Detection of individual gas molecules adsorbed on graphene. *Nat. Mater.* **2007**, *6*, 652–655. [CrossRef]

27. He, Q.Y.; Sudibya, H.G.; Yin, Z.Y.; Wu, S.X.; Li, H.; Boey, F.; Huang, W.; Chen, P.; Zhang, H. Centimeter long and large-scale micro patterns of reduced graphene oxide films: Fabrication and sensing applications. *ACS Nano* **2010**, *4*, 3201–3208. [CrossRef]

28. Liang, J.J.; Xu, Y.F.; Huang, Y.; Zhang, L.; Wang, Y.; Ma, Y.F.; Ma, Y.; Li, F.; Guo, T.; Chen, Y. Infrared-triggered actuators from graphene-based nanocomposites. *J. Phys. Chem. C* **2009**, *113*, 9921–9927. [CrossRef]

29. Park, S.; An, J.; Suk, J.W.; Ruoff, R.S. Graphene-based actuators. *Small* **2010**, *6*, 210–212. [CrossRef]

30. Xie, X.J.; Qu, L.T.; Zhou, C.; Li, Y.; Zhu, J.; Bai, H.; Shi, G.; Dai, L. An asymmetrically surface-modified graphene film electrochemical actuator. *ACS Nano* **2010**, *4*, 6050–6054. [CrossRef]

31. Liang, J.J.; Huang, Y.; Oh, J.; Kozlov, M.; Sui, D.; Fang, S.; Baughman, R.H.; Ma, Y.; Chen, Y. Electromechanical actuators based on graphene and graphene/Fe_3O_4 hybrid paper. *Adv. Funct. Mater.* **2011**, *21*, 3778–3784. [CrossRef]

32. Becerril, H.A.; Mao, J.; Liu, Z.; Stoltenberg, R.M.; Bao, Z.; Chen, Y. Evaluation of solution-processed reduced graphene oxide films as transparent conductors. *ACS Nano* **2008**, *2*, 463–470. [CrossRef] [PubMed]

33. Brownson, D.A.; Kampouris, D.K.; Banks, C.E. An overview of graphene in energy production and storage applications. *J. Phys. Chem. C* **2011**, *196*, 4873–4885. [CrossRef]

34. Allen, M.J.; Tung, V.C.; Kaner, R.B. Honeycomb carbon: A review of graphene. *Chem. Rev.* **2010**, *110*, 132–145. [CrossRef] [PubMed]

35. Kim, K.S.; Park, S.J. Influence of multi-walled carbon nanotubes on the electrochemical performance of graphene nanocomposites for supercapacitor electrodes. *Electrochim. Acta* **2011**, *56*, 1629–1635. [CrossRef]

36. Seo, M.K.; Park, S.J. Electrical resistivity and rheological behaviors of carbon nanotubes-filled polypropylene composites. *Chem. Phys. Lett.* **2004**, *395*, 44–48. [CrossRef]

37. Lee, S.Y.; Park, S.J. Isothermal exfoliation of graphene oxide by a new carbon dioxide pressure swing method. *Carbon* **2014**, *68*, 112–117. [CrossRef]

38. Akaike, K.; Kumai, T.; Nakano, K.; Abdullah, S.; Ouchi, S.; Uemura, Y.; Ito, Y.; Onishi, A.; Yoshida, H.; Tajima, K.; et al. Effects of molecular orientation of a fullerene derivative at the donor/acceptor interface on the device performance of organic photovoltaics. *Chem. Mater.* **2018**, *30*, 8233–8243. [CrossRef]

39. Zhang, Y.; Park, S.J. Incorporation of RuO_2 into charcoal-derived carbon with controllable microporosity by CO_2 activation for high-performance supercapacitor. *Carbon* **2017**, *122*, 287–297. [CrossRef]

40. Zhang, Y.; Park, S.J. Enhanced interfacial interaction by grafting carboxylated-macromolecular chains on nanodiamond surfaces for epoxy-based thermosets. *J. Polym. Sci. B* **2017**, *55*, 1890–1898. [CrossRef]

41. Elhaes, H.; Fakhry, A.; Ibrahim, M. Carbon nano materials as gas sensors. *Mater Today Proc.* **2016**, *3*, 2483–2492. [CrossRef]

42. Rabti, A.; Raouafi, N.; Merkoçi, A. Bio (sensing) devices based on ferrocene–functionalized graphene and carbon nanotubes. *Carbon* **2016**, *108*, 481–514. [CrossRef]

43. Kim, K.; Park, J.; Suh, J.H.; Kim, M.; Jeong, Y.; Park, I. 3D printing of multiaxial force sensors using carbon nanotube (CNT)/thermoplastic polyurethane (TPU) filaments. *Sens. Actuator B-Chem.* **2017**, *263*, 493–500. [CrossRef]

44. Shin, S.R.; Farzad, R.; Tamayol, A.; Manoharan, V.; Mostafalu, P.; Zhang, Y.S.; Akbari, M.; Jung, S.M.; Kim, D.; Comotto, M.; et al. A bioactive carbon nanotube-based ink for printing 2D and 3D flexible electronics. *Adv. Mater.* **2016**, *28*, 3280–3289. [CrossRef]

45. Liu, L.; Ye, D.; Yu, Y.; Liu, L.; Wu, Y. Carbon-based flexible micro-supercapacitor fabrication via mask-free ambient micro-plasma-jet etching. *Carbon* **2017**, *111*, 121–127. [CrossRef]

46. Huang, G.; Zhang, Y.; Wang, L.; Sheng, P.; Peng, H. Fiber-based MnO_2/carbon nanotube/polyimide asymmetric supercapacitor. *Carbon* **2017**, *125*, 595–604. [CrossRef]

47. Paulo, S.; Palomares, E.; Martinez-Ferrero, E. Graphene and carbon quantum dot-based materials in photovoltaic devices: From synthesis to applications. *Nanomaterials* **2016**, *6*, 157. [CrossRef]

48. Chen, L.; Liu, X.; Wang, C.; Lv, S.; Chen, C. Amperometric nitrite sensor based on a glassy carbon electrode modified with electrodeposited poly (3,4-ethylenedioxythiophene) doped with a polyacenic semiconductor. *Microchim. Acta* **2017**, *184*, 2073–2079. [CrossRef]

49. Sarma, S.D.; Adam, S.; Hwang, E.H.; Rossi, E. Electronic transport in two-dimensional graphene. *Rev. Mod. Phys.* **2011**, *83*, 407. [CrossRef]

50. Kim, H.J.; Lee, S.Y.; Yeo, C.S.; Son, Y.R.; Cho, K.R.; Song, Y.; Ju, S.; Shin, M.K.; Park, S.J.; Park, S. Maximizing volumetric energy density of all-graphene-oxide-supercapacitors and their potential applications for energy harvest. *J. Power Sources* **2017**, *346*, 113–119. [CrossRef]

51. Pastore, R.; Delfini, A.; Micheli, D.; Vricella, A.; Marchetti, M.; Santoni, F.; Piergentili, F. Carbon foam electromagnetic mm-wave absorption in reverberation chamber. *Carbon* **2019**, *144*, 63–71. [CrossRef]

52. Micheli, D.; Vricella, A.; Pastore, R.; Delfini, A.; Morles, R.B.; Marchetti, M.; Santoni, F.; Bastianelli, L.; Moglie, F.; Primiani, V.M.; Corinaldesi, V. Electromagnetic properties of carbon nanotube reinforced concrete composites for frequency selective shielding structures. *Constr. Build. Mater.* **2017**, *131*, 267–277. [CrossRef]

53. Mazzoli, A.; Corinaldesi, V.; Donnini, J.; Di Perna, C.; Micheli, D.; Vricella, A.; Pastore, R.; Bastianelli, L.; Moglie, F.; Primiani, V.M. Effect of graphene oxide and metallic fibers on the electromagnetic shielding effect of engineered cementitious composites. *J. Build. Eng.* **2018**, *18*, 33–39. [CrossRef]

54. Micheli, D.; Pastore, R.; Vricella, A.; Marchetti, M. Matter's electromagnetic signature reproduction by graded-dielectric multilayer assembly. *IEEE Trans. Microw. Theory Tech.* **2017**, *65*, 2801–2809. [CrossRef]

55. Cheng, J.; Wang, B.; Xin, H.L.; Kim, C.; Nie, F.; Li, X.; Yang, G.; Huang, H. Conformal coating of TiO_2 nanorods on a 3-D CNT scaffold by using a CNT film as a nanoreactor: A free-standing and binder-free Li-ion anode. *J. Mater. Chem. A* **2014**, *2*, 2701–2707. [CrossRef]

56. Cheng, J.; Wang, B.; Park, C.M.; Wu, Y.; Huang, H.; Nie, F. CNT@ Fe_3O_4@ C coaxial nanocables: One-pot, additive-free synthesis and remarkable lithium storage behavior. *Chem. Eur. J.* **2013**, *19*, 9866–9874. [CrossRef] [PubMed]

57. Takenaka, S.; Miyamoto, H.; Utsunomiya, Y.; Matsune, H.; Kishida, M. Catalytic activity of highly durable Pt/CNT catalysts covered with hydrophobic silica layers for the oxygen reduction reaction in PEFCs. *J. Phys. Chem. C* **2014**, *118*, 774–783. [CrossRef]

58. Lee, J.W.; Lee, H.I.; Park, S.J. Facile synthesis of petroleum-based activated carbons/tubular polypyrrole composites with enhanced electrochemical performance as supercapacitor electrode materials. *Electrochim. Acta* **2018**, *263*, 447–453. [CrossRef]

59. Zhang, X.; Jin, B.; Li, L.; Cheng, T.; Wang, H.; Xin, P.; Jiang, Q. (De) Lithiation of tubular polypyrrole-derived carbon/sulfur composite in lithium-sulfurbatteries. *J. Electroanal. Chem.* **2016**, *780*, 26–31. [CrossRef]

60. Choi, J.Y.; Park, S.J. Effect of manganese dioxide on supercapacitive behaviors of petroleum pitch-based carbons. *J. Ind. Eng. Chem.* **2015**, *29*, 408–413. [CrossRef]

61. Vighnesha, K.M.; Sangeetha, D.N.; Selvakumar, M. Synthesis and characterization of activated carbon/conducting polymer composite electrode for supercapacitor applications. *J. Mater. Sci. Mater. Electron.* **2018**, *29*, 914–921. [CrossRef]

62. Sudhakar, Y.N.; Hemant, H.; Nitinkumar, S.S.; Poornesh, P.; Selvakumar, M. Green synthesis and electrochemical characterization of rGO–CuO nanocomposites for supercapacitor applications. *Ionics* **2017**, *23*, 1267–1276. [CrossRef]

63. Tao, J.; Liu, N.; Ma, W.; Ding, L.; Li, L.; Su, J.; Gao, Y. Solid-state high performance flexible supercapacitors based on polypyrrole-MnO_2-carbon fiber hybrid structure. *Sci. Rep.* **2017**, *3*, 2286. [CrossRef] [PubMed]

64. Heo, Y.J.; Lee, H.I.; Lee, J.W.; Park, M.; Rhee, K.Y.; Park, S.J. Optimization of the pore structure of PAN-based carbon fibers for enhanced supercapacitor performances via electrospinning. *Compos. B Eng.* **2019**, *161*, 10–17. [CrossRef]

65. Cakici, M.; Kakarla, R.R.; Alonso-Marroquin, F. Advanced electrochemical energy storage supercapacitors based on the flexible carbon fiber fabric-coated with uniform coral-like MnO_2 structured electrodes. *J. Chem. Eng.* **2017**, *309*, 151–158. [CrossRef]

66. Ke, Q.; Tang, C.; Liu, Y.; Liu, H.; Wang, J. Intercalating graphene with clusters of Fe_3O_4 nanocrystals for electrochemical supercapacitors. *Mater. Res. Express* **2014**, *1*, 025015. [CrossRef]

67. Li, Z.; Wang, J.; Liu, S.; Liu, X.; Yang, S. Synthesis of hydrothermally reduced graphene/MnO_2 composites and their electrochemical properties as supercapacitors. *J. Power Sources* **2011**, *196*, 8160–8165. [CrossRef]

68. Cheng, C.; Wen, Y.; Xu, X.; Gu, H. Tunable synthesis of carboxyl-functionalized magnetite nanocrystal clusters with uniform size. *J. Mater. Chem.* **2009**, *19*, 8782–8788. [CrossRef]

69. Chen, S.; Zhu, J.; Wu, X.; Han, Q.; Wang, X. Graphene oxide-MnO_2 nanocomposites for supercapacitors. *ACS Nano* **2010**, *4*, 2822–2830. [CrossRef]

70. Wu, Z.S.; Ren, W.; Wang, D.W.; Li, F.; Liu, B.; Cheng, H.M. High-energy MnO_2 nanowire/graphene and graphene asymmetric electrochemical capacitors. *ACS Nano* **2010**, *4*, 5835–5842. [CrossRef] [PubMed]

71. Chen, S.; Zhu, J.; Wang, X. From graphene to metal oxide nanolamellas: Aphenomenon of morphology transmission. *ACS Nano* **2010**, *4*, 6212–6218. [CrossRef] [PubMed]

72. Yan, J.; Fan, Z.; Wei, T.; Qian, W.; Zhang, M.; Wei, F. Fast and reversiblesurface redox reaction of graphene-MnO$_2$ composites as supercapacitor electrodes. *Carbon* **2010**, *48*, 3825–3833. [CrossRef]

73. Hummers, W.S.; Offeman, R.E. Preparation of graphitic oxide. *J. Am. Chem. Soc.* **1958**, *80*, 1339. [CrossRef]

74. Liu, S.; Wang, J.; Zeng, J.; Ou, J.; Li, Z.; Liu, X.; Yang, S. "Green" electrochemical synthesis of Pt/graphene sheet nanocomposite film and its electrocatalytic property. *J. Power Sources* **2010**, *195*, 4628–4633. [CrossRef]

75. Mishra, A.K.; Ramaprabhu, S. Functionalized graphene-based nanocomposites for supercapacitor application. *J. Phys. Chem. C* **2011**, *115*, 14006–14013. [CrossRef]

76. Sawangphruk, M.; Suksomboon, M.; Kongsupornsak, K.; Khuntilo, J.; Srimuk, P.; Sanguansak, Y.; Klunbud, P.; Suktha, P.; Chiochan, P. High-performance supercapacitors based on silver nanoparticle–polyaniline–graphene nanocomposites coated on flexible carbon fiber paper. *J. Mater. Chem. A* **2013**, *1*, 9630–9636. [CrossRef]

77. Wang, H.; Casalongue, H.S.; Liang, Y.; Dai, H. Ni(OH)$_2$ nanoplates grownon graphene as advanced electrochemical pseudocapacitor materials. *J. Am. Chem. Soc.* **2010**, *132*, 7472–7477. [CrossRef] [PubMed]

78. Zhang, C.; Chen, Q.; Zhan, H. Supercapacitors based on reduced graphene oxide nanofibers supported Ni (OH)$_2$ nanoplates with enhanced electrochemical performance. *ACS Appl. Mater. Interfaces* **2016**, *8*, 22977–22987. [CrossRef]

79. O'Mahony, C.; Haq, E.U.; Sillien, C.; Tofail, S.A. Rheological issues in carbon-based inks for additive manufacturing. *Micromachines* **2019**, *10*, 99. [CrossRef]

80. Hu, G.; Kang, J.; Ng, L.W.; Zhu, X.; Howe, R.C.; Jones, C.G.; Hersam, M.C.; Hasan, T. Functional inks and printing of two-dimensional materials. *Chem. Soc. Rev.* **2018**, *47*, 3265–3300. [CrossRef]

81. Secor, E.B.; Lim, S.; Zhang, H.; Frisbie, C.D.; Francis, L.F.; Hersam, M.C. Gravure printing of graphene for large-area flexible electronics. *Adv. Mater.* **2014**, *26*, 4533–4538. [CrossRef]

82. Secor, E.B.; Prabhumirashi, P.L.; Puntambekar, K.; Geier, M.L.; Hersam, M.C. Inkjet printing of high conductivity, flexible graphene patterns. *J. Phys. Chem. Lett.* **2013**, *4*, 1347–1351. [CrossRef]

83. Torrisi, F.; Hasan, T.; Wu, W.; Sun, Z.; Lombardo, A.; Kulmala, T.S.; Hsieh, G.W.; Jung, S.; Bonaccorso, F.; Paul, P.J.; et al. Inkjet-printed graphene electronics. *ACS Nano* **2012**, *6*, 2992–3006. [CrossRef]

84. Gao, Y.; Shi, W.; Wang, W.; Leng, Y.; Zhao, Y. Inkjet printing patterns of highly conductive pristine graphene on flexible substrates. *Ind. Eng. Chem. Res.* **2014**, *53*, 16777–16784. [CrossRef]

85. Homenick, C.M.; James, R.; Lopinski, G.P.; Dunford, J.; Sun, J.; Park, H.; Jung, Y.; Cho, G.; Malenfant, P.R. Fully printed and encapsulated SWCNT-based thin film transistors via a combination of R2R gravure and inkjet printing. *ACS Appl. Mater. Interfaces* **2016**, *8*, 27900–27910. [CrossRef]

86. Han, J.W.; Kim, B.; Li, J.; Meyyappan, M. Carbon nanotube ink for writing on cellulose paper. *Mater. Res. Bull.* **2014**, *50*, 249–253. [CrossRef]

87. Foster, C.W.; Down, M.P.; Zhang, Y.; Ji, X.; Rowley-Neale, S.J.; Smith, G.C.; Kelly, P.J.; Banks, C.E. 3D printed graphene based energy storage devices. *Sci. Rep.* **2017**, *7*, 42233. [CrossRef]

88. You, X.; Yang, J.; Feng, Q.; Huang, K.; Zhou, H.; Hu, J.; Dong, S. Three-dimensional graphene-based materials by direct ink writing method for lightweight application. *Int. J. Lightweight Mater. Manuf.* **2018**, *1*, 96–101. [CrossRef]

89. Dybowska-Sarapuk, L.; Kielbasinski, K.; Arazna, A.; Futera, K.; Skalski, A.; Janczak, D.; Sloma, M.; Jakubowska, M. Efficient inkjet printing of graphene-based elements: Influence of dispersing agent on ink viscosity. *Nanomaterials* **2018**, *8*, 602. [CrossRef]

90. Li, D.W.; Luo, L.; Lv, P.F.; Wang, Q.Q.; Lu, K.Y.; Wei, A.F.; Wei, Q.F. Synthesis of polydopamine functionalized reduced graphene oxide-palladium nanocomposite for laccase based biosensor. *Bioinorg. Chem. Appl.* **2016**, *2016*, 5360361. [CrossRef]

91. Fan, X.; Peng, W.; Li, Y.; Li, X.; Wang, S.; Zhang, G.; Zhang, F. Deoxygenation of exfoliated graphite oxide under alkaline conditions: A green route to graphene preparation. *Adv. Mater.* **2008**, *20*, 4490–4493. [CrossRef]

92. Karousis, N.; Tsotsou, G.-E.; Evangelista, F.; Rudolf, P.; Ragoussis, N.; Tagmatarchis, N. Carbon nanotubes decorated with palladium nanoparticles: Synthesis, characterization, and catalytic activity. *J. Phys. Chem. C* **2008**, *112*, 13463–13469. [CrossRef]

93. Ul Hasan, K.; Asif, M.H.; Nur, O.; Willander, M. Needle-type glucose sensor based on functionalized graphene. *J. Biosens. Bioelectron.* **2012**, *3*, 114. [CrossRef]

94. Ul Hasan, K.; Sandberg, M.; Nur, O.; Willander, M. Polycation stabilization of graphene suspensions. *Nanoscale Res. Lett.* **2011**, *6*, 493. [CrossRef] [PubMed]

95. Penicaud, A.; Valles, C. *Graphene Solutions*; Centre National De La Recherche Scientifique—CNRS: Paris, France, 2011.

96. Zhou, H.; Wang, X.; Yu, P.; Chen, X.; Mao, L. Sensitive and selective voltammetric measurement of Hg^{2+} by rational covalent functionalization of graphene oxide with cysteamine. *Analyst* **2011**, *137*, 305–308. [CrossRef] [PubMed]

97. Hemanth, S.; Halder, A.; Caviglia, C.; Chi, Q.; Keller, S. 3D Carbon microelectrodes with bio-functionalized graphene for electrochemical biosensing. *Biosensors* **2018**, *8*, 70. [CrossRef] [PubMed]

98. Chi, Q.; Han, S.; Halder, A.; Zhu, N.; Ulstrup, J. Graphene-Polymer-Enzyme Hybrid Nanomaterials for Biosensors. WIPO Patent 2016083204, 6 February 2016.

99. Liu, S.; Yu, B. Enhancing NO_2, gas sensing performances at room temperature based on reduced graphene oxide-ZnO nanoparticles hybrids. *Sens. Actuators B Chem.* **2014**, *202*, 272–278. [CrossRef]

100. Novikov, S.; Lebedeva, N.; Satrapinski, A.; Walden, J.; Davydov, V.; Lebedev, A. Graphene based sensor for environmental monitoring of NO_2. *Sens. Actuator B-Chem.* **2016**, *236*, 1054–1060. [CrossRef]

101. Novikov, S.; Hämäläinen, J.; Walden, J.; Iisakka, I.; Lebedeva, N.; Atrapinski, A. Characterization of epitaxial and CVD graphene with double metal-graphene contacts for gas sensing. In Proceedings of the 16th International Congress of Metrology, Paris, France, 7–10 October 2013; p. 13003.

102. Huang, X.; Hu, N. Reduced graphene oxide–polyaniline hybrid: Preparation, characterization and its applications for ammonia gas sensing. *J. Mater. Chem.* **2012**, *22*, 22488–22495. [CrossRef]

103. Sathish, M.; Mitani, S.; Tomai, T.; Honma, I. MnO_2 assisted oxidative polymerization of aniline on graphene sheets: Superior nanocomposite electrodes for electrochemical supercapacitors. *J. Mater. Chem. A* **2011**, *21*, 16216–16222. [CrossRef]

104. Pan, L.J.; Pu, L.; Shi, Y.; Song, S.Y.; Xu, Z.; Zhang, R.; Zheng, Y.D. Synthesis of polyaniline nanotubes with a reactive template of manganese oxide. *Adv. Mater.* **2007**, *19*, 461–464. [CrossRef]

105. Ye, Z.; Tai, H. Excellent ammonia sensing performance of gas sensor based on graphene/titanium dioxidehybrid with improved morphology. *Appl. Surf. Sci.* **2017**, *419*, 84–90. [CrossRef]

106. Zhan, Z.; Lin, R.; Tran, V.T.; An, J.; Wei, Y.; Du, H.; Tran, T.; Lu, W. Paper/carbon nanotube-based wearable pressure sensor for physiological signal acquisition and soft robotic skin. *ACS Appl. Mater. Interfaces* **2017**, *9*, 37921–37928. [CrossRef]

107. Wang, L.; Loh, K.J. Wearable carbon nanotube-based fabric sensors for monitoring human physiological performance. *Smart Mater. Struct.* **2017**, *26*, 055018. [CrossRef]

108. Wang, L.; Loh, K.J. Spray-coated carbon nanotube-latex strain sensors. *Sci. Lett. J.* **2016**, *5*, 234.

109. Wang, L.; Loh, K.J.; Brely, L.; Bosia, F.; Pugno, N.M. An experimental and numerical study on the mechanical properties of carbon nanotube-latex thin films. *J. Eur. Ceram. Soc.* **2016**, *36*, 2255–2262. [CrossRef]

110. Li, Y.Q.; Zhu, W.B.; Yu, X.G.; Huang, P.; Fu, S.Y.; Hu, N.; Liao, K. Multifunctional wearable device based on flexible and conductive carbon sponge/polydimethylsiloxane composite. *ACS Appl. Mater. Interfaces* **2016**, *8*, 33189–33196. [CrossRef] [PubMed]

111. Karim, N.; Afroj, S.; Tan, S.; He, P.; Fernando, A.; Carr, C.; Novoselov, K.S. Scalable production of graphene-based wearable e-textiles. *ACS Nano* **2017**, *11*, 12266–12275. [CrossRef] [PubMed]

112. Wu, C.; Kim, T.W.; Li, F.; Guo, T. Wearable electricity generators fabricated utilizing transparent electronic textiles based on polyester/Ag nanowires/graphene core–shell nanocomposites. *ACS Nano* **2016**, *10*, 6449–6457. [CrossRef]

113. Beidaghi, M.; Wang, C. Micro-supercapacitors based on interdigital electrodes of reduced graphene oxide and carbon nanotube composites with ultrahigh power handling performance. *Adv. Funct. Mater.* **2012**, *22*, 4501–4510. [CrossRef]

114. Liu, C.; Yu, Z.; Neff, D.; Zhamu, A.; Jang, B.Z. Graphene-based supercapacitor with an ultrahigh energy density. *Nano Lett.* **2010**, *12*, 4863–4868. [CrossRef]

115. El-Kady, M.F.; Kaner, R.B. Scalable fabrication of high-power graphene micro-supercapacitors for flexible and on-chip energy storage. *Nat. Commun.* **2013**, *4*, 1475. [CrossRef] [PubMed]

116. Salanne, M.; Rotenberg, B.; Naoi, K.; Kaneko, K.; Taberna, P.L.; Grey, C.P.; Dunn, B.; Simon, P. Efficient storage mechanisms for building better supercapacitors. *Nat. Energy* **2016**, *1*, 16070. [CrossRef]

117. Zhong, C.; Deng, Y.; Hu, W.; Qiao, J.; Zhang, L.; Zhang, J. A review of electrolyte materials and compositions for electrochemical supercapacitors. *Chem. Soc. Rev.* **2015**, *44*, 7484–7539. [CrossRef] [PubMed]

118. Shen, C.; Wang, X.; Zhang, W.; Kang, F. A high-performancethree-dimensional micro supercapacitor based on self-supporting composite materials. *J. Power Sources* **2011**, *196*, 10465–10471. [CrossRef]

119. Khomenko, V.; Frackowiak, E.; Beguin, F. Determination of the specific capacitance of conducting polymer/nanotubes composite electrodes using different cell configurations. *Electrochim. Acta* **2005**, *50*, 2499–2506. [CrossRef]

120. Fang, Y.; Liu, J.; Yu, D.J.; Wicksted, J.P.; Kalkan, K.; Topal, C.O.; Flanders, B.N.; Wu, J.; Li, J. Self-supported supercapacitor membranes: Polypyrrole-coated carbon nanotube networks enabled by pulsed electrodeposition. *J. Power Sources* **2010**, *195*, 674–679. [CrossRef]

121. Zhang, H.; Cao, G.; Wang, Z.; Yang, Y.; Shi, Z.; Gu, Z. Tube-covering-tube nanostructured polyaniline/carbon nanotube array composite electrode with high capacitance and superior rate performance as well as good cycling stability. *Electrochem. Commun.* **2008**, *10*, 1056–1059. [CrossRef]

122. Lota, K.; Khomenko, V.; Frackowiak, E. Capacitance properties of poly (3,4-ethylenedioxythiophene)/carbon nanotubes composites. *J. Phys. Chem. Solids* **2004**, *65*, 295–301. [CrossRef]

123. Heo, Y.J.; Lee, J.W.; Son, Y.R.; Lee, J.H.; Yeo, C.S.; Lam, T.D.; Park, S.Y.; Park, S.J.; Sinh, L.H.; Shin, M.K. Large-scale conductive yarns based on twistable korean traditional paper (Hanji) for supercapacitor applications: Toward high-performance paper supercapacitors. *Adv. Energy Mater.* **2018**, *8*, 1801–1854. [CrossRef]

124. Wang, H.; Hao, Q.; Yang, X.; Lu, L.; Wang, X. A nanostructured graphene/polyaniline hybrid material for supercapacitors. *Nanoscale* **2010**, *2*, 2164–2170. [CrossRef]

125. Li, J.; Xie, H.; Li, Y.; Liu, J.; Li, Z. Electrochemical properties of graphene nanosheets/polyaniline nanofibers composites as electrode for supercapacitors. *J. Power Sources* **2011**, *196*, 10775–10781. [CrossRef]

126. Purkait, T.; Singh, G.; Kumar, D.; Singh, M.; Dey, R.S. High-performance flexible supercapacitors based on electrochemically tailored three-dimensional reduced graphene oxide networks. *Sci. Rep.* **2018**, *8*, 640. [CrossRef]

127. Wang, S.Y.; Wu, N.L. Operating characteristics of aqueous magnetite electrochemical capacitors. *J. Appl. Electrochem.* **2003**, *33*, 345–348. [CrossRef]

128. Wang, Y.G.; Wang, Z.D.; Xia, Y.Y. An asymmetric supercapacitor using RuO_2/TiO_2 nanotube composite and activated carbon electrodes. *Electrochim. Acta* **2005**, *50*, 5641–5646. [CrossRef]

129. Du, X.; Wang, C.; Chen, M.; Jiao, Y.; Wang, J. Electrochemical performances of nanoparticle Fe_3O_4/activated carbon supercapacitor using KOH electrolyte solution. *J. Phys. Chem. C* **2009**, *113*, 2643–2646. [CrossRef]

130. Yu, M.; Huang, Y.; Li, C.; Zeng, Y.; Wang, W.; Li, Y.; Fang, P.; Lu, X.; Tong, Y. Building three-dimensional graphene frameworks for energy storage and catalysis. *Adv. Funct. Mater.* **2015**, *25*, 324–330. [CrossRef]

131. Luo, J.; Zhong, W.; Zou, Y.; Xiong, C.; Yang, W. Preparation of morphology-controllable polyaniline and polyaniline/graphene hydrogels for high performance binder-free supercapacitor electrodes. *J. Power Sources* **2016**, *319*, 73–81. [CrossRef]

132. Li, S.; Wu, D.; Cheng, C.; Wang, J.; Zhang, F.; Su, Y.; Feng, X. Polyaniline-coupled multifunctional 2D metal oxide/hydroxide graphene nanohybrids. Angew. *Chem. Int. Ed.* **2013**, *52*, 12105–12109. [CrossRef]

133. Giri, S.; Ghosh, D.; Das, C.K. Growth of vertically aligned tunable polyaniline on graphene/ZrO_2 nanocomposites for supercapacitor energy-storage application. *Adv. Funct. Mater.* **2014**, *24*, 1312–1324. [CrossRef]

134. Hao, Q.; Xia, X.; Lei, W.; Wang, W.; Qiu, J. Facile synthesis of sandwich-like polyaniline/boron-doped graphene nano hybrid for supercapacitors. *Carbon* **2015**, *81*, 552–563. [CrossRef]

135. Dai, C.S.; Chien, P.Y.; Lin, J.Y.; Chou, S.W.; Kai Wu, W.; Li, P.H.; Wu, K.Y.; Lin, T.W. Hierarchically structured Ni_3S_2/carbon nanotube composites as high performance cathode materials for asymmetric supercapacitors. *ACS Appl. Mater. Interfaces* **2013**, *5*, 12168–12174. [CrossRef]

136. Zhu, T.; Xia, B.; Zhou, L.; Lou, X.W.D. Arrays of ultrafine CuS nanoneedles supported on a CNT backbone for application in supercapacitors. *J. Mater. Chem. A* **2012**, *22*, 7851–7855. [CrossRef]

137. Yang, M.; Jeong, J.M.; Huh, Y.S.; Choi, B.G. High-performance supercapacitor based on three-dimensional MoS2/graphene aerogel composites. *Comp. Sci. Technol.* **2015**, *121*, 123–128. [CrossRef]

138. Baeg, K.J.; Caironi, M.; Noh, Y.Y. Toward printed integrated circuits based on unipolar or ambi polar polymer semiconductors. *Adv. Mater.* **2013**, *25*, 4210–4244. [CrossRef]

139. Torrisi, F.; Coleman, J.N. Electrifying inks with 2D materials. *Nat. Nanotechnol.* **2014**, *9*, 738–739. [CrossRef]

140. Touhami, A. Biosensors and nanobiosensors design and applications. In *Nanomedicine*; One Central Press (OCP): Cheshire, UK, 2014; pp. 374–403.

141. Turner, A.P. Biosensors: Sense and sensibility. *Chem. Soc. Rev.* **2013**, *42*, 3184–3196. [CrossRef]

142. Gan, N.; Jin, H.; Li, T.; Zheng, L. Fe_3O_4/Au magnetic nanoparticle amplification strategies for ultrasensitive electrochemical immunoassay of alfa-fetoprotein. *Int. J. Nanomed.* **2011**, *6*, 3259–3269. [CrossRef]

143. Sireesha, M.; Jagadeesh Babu, V.; Kranthi Kiran, A.S.; Ramakrishna, S. A review on carbon nanotubes in biosensor devices and their applications in medicine. *Nanocomposites* **2018**, *4*, 36–57. [CrossRef]

144. Gupta, S.; Murthy, C.N.; Prabha, C.R. Recent advances in carbon nanotube based electrochemical biosensors. *Int. J. Biol. Macromol.* **2018**, *108*, 687–703. [CrossRef]

145. Pandey, P.; Datta, M.; Malhotra, B.D. Prospects of nanomaterials in biosensors. *Anal. Lett.* **2008**, *41*, 159–209. [CrossRef]

146. Cai, C.; Chen, J. Direct electron transfer of glucose oxidase promoted by carbon nanotubes. *Anal. Biochem.* **2004**, *332*, 75–83. [CrossRef]

147. Lee, Y.M.; Kwon, O.Y.; Yoon, Y.J.; Ryu, K. Immobilization of horse radish peroxidase on multi-wall carbon nanotubes and its electrochemical properties. *Biotechnol. Lett.* **2006**, *28*, 39–43. [CrossRef] [PubMed]

148. Rubianes, M.D.; Rivas, G.A. Enzymatic biosensors based on carbon nanotubes paste electrodes. *Electroynalysis* **2005**, *17*, 73–78. [CrossRef]

149. Ruhal, A.; Rana, J.S.; Kumar, S.; Kumar, A. Immobilization of malate dehydrogenase on carbon nanotubes for development of malate biosensor. *Cell. Mol. Biol.* **2012**, *58*, 15–20. [PubMed]

150. Zhang, X.; Wang, Y. Research and development of gas sensors based on nanomaterials. *Sens. Microsyst.* **2013**, *32*, 1–5.

151. Liu, S.; Wang, Z. A Resistive Gas Sensor Based on Graphene/Tin Dioxide/Zinc Oxide Composite, Its preparation and Application. Patent CN105891271A, 24 August 2016.

152. Zou, Y.; Wang, Q. Doping composite of polyaniline and reduced graphene oxide with palladium nanoparticles for room-temperature hydrogen-gas sensing. *Int. J. Hydrogen Energy* **2016**, *41*, 5396–5404. [CrossRef]

153. Rajavel, K.; Lalitha, M.; Radhakrishnan, J.K.; Senthilkumar, L.; Thangavelu, R.; Kumar, R. Multiwalled carbon nanotube oxygen sensor: Enhanced oxygen sensitivity at room temperature and mechanism of sensing. *ACS Appl. Mater. Interfaces* **2015**, *7*, 23857–23865. [CrossRef]

154. Bannov, A.G.; Jašek, O.; Manakhov, A.; Márik, M.; Nečas, D.; Zajíčková, L. High-performance ammonia gas sensors based on plasma treated carbon nanostructures. *IEEE Trans. Plasma Sci.* **2017**, *17*, 1964–1970. [CrossRef]

155. Sayago, I.; Terrado, E.; Lafuente, E.; Horrillo, M.C.; Maser, W.K.; Benito, A.M.; Navarro, R.; Urriolabeitia, E.P.; Martinez, M.T.; Gutierrez, J. Hydrogen sensors based on carbon nanotubes thin films. *Synth. Met.* **2005**, *148*, 15–19. [CrossRef]

156. Zhang, J.; Zhu, Q.; Zhang, Y.; Zhu, Z.; Liu, Q. Methanol gas-sensing properties of SWCNT-MIP composites. *Nanoscale Res. Lett.* **2016**, *11*, 522. [CrossRef]

157. Meng, L.; Jin, J.; Yang, G.; Lu, T.; Zhang, H.; Cai, C. Non enzymatic electrochemical detection of glucose based on palladium-single-walled carbon nanotube hybrid nanostructures. *Anal. Chem.* **2009**, *81*, 7271–7280. [CrossRef] [PubMed]

158. Zhao, K.; Zhuang, S.; Chang, Z.; Songm, H.; Dai, L.; He, P.; Fang, Y. Amperometric glucose biosensor based on platinum nanoparticles combined aligned carbon nanotubes electrode. *Electroanalysis* **2007**, *19*, 1069–1074. [CrossRef]

159. Shan, C.; Yang, H.; Song, J.; Han, D.; Ivaska, A.; Niu, L. Direct electrochemistry of glucose oxidase and biosensing for glucose based on graphene. *Anal. Chem.* **2009**, *81*, 2378–2382. [CrossRef]

160. Yuan, W.; Liu, A.; Huang, L.; Li, C.; Shi, G. High-performance NO_2 Sensors Based on Chemically Modified graphene. *Adv. Mater.* **2013**, *25*, 766–771. [CrossRef]

161. Xuan, X.; Yoon, H.S.; Park, J.Y. A wearable electrochemical glucose sensor based on simple and low-cost fabrication supported micro-patterned reduced graphene oxide nanocomposite electrode on flexible substrate. *Biosens. Bioelectron.* **2018**, *109*, 75–82. [CrossRef]

162. Ma, Y.; Song, X.; Ge, X.; Zhang, H.; Wang, G.; Zhang, Y.; Zhao, H. In situ growth of Fe_2O_3 nanorod arrays on 3D carbon foam as an efficient binder-free electrode for highly sensitive and specific determination of nitrite. *J. Mater. Chem. A* **2017**, *5*, 4726–4736. [CrossRef]

163. Xu, G.; Jarjes, Z.A.; Desprez, V.; Kilmartin, P.A.; Travas-Sejdic, J. Sensitive, selective, disposable electrochemical dopamine sensor based on PEDOT-modified laser scribed graphene. *Biosens. Bioelectron.* **2018**, *107*, 184–191. [CrossRef]

164. Wang, H.; Zhao, X.J.; Li, J.F.; Kuang, X.; Fan, Y.Q.; Wei, G.; Su, Z. Electrostatic assembly of peptide nanofiber-biomimetic silver nanowires onto graphene for electrochemical sensors. *ACS Macro Lett.* **2014**, *3*, 529–533. [CrossRef]

165. Wang, L.; Lin, J. Phenylalanine-rich peptide mediated binding with graphene oxide and bioinspired synthesis of silver nanoparticles for electrochemical sensing. *Appl. Sci.* **2017**, *7*, 160. [CrossRef]

166. Ma, Y.; Zhao, M.; Cai, B.; Wang, W.; Ye, Z.; Huang, J. 3D graphene foams decorated by CuO nanoflowers for ultrasensitive ascorbic acid detection. *Biosens. Bioelectron.* **2014**, *59*, 384–388. [CrossRef]

167. Hu, F.X.; Xie, J.L.; Bao, S.J.; Yu, L.; Li, C.M. Shape-controlled ceria-reduced graphene oxide nanocomposites toward high sensitivein situ detection of nitric oxide. *Biosens. Bioelectron.* **2015**, *70*, 310–317. [CrossRef] [PubMed]

168. Wang, L.; Zhang, Y.; Cheng, C.; Liu, X.; Jiang, H.; Wang, X. Highly sensitive electrochemical biosensor for evaluation of oxidative stress based on the nano interface of graphene nanocomposites blended with gold, Fe_3O_4, and platinum nanoparticles. *ACS Appl. Mater. Interfaces* **2015**, *7*, 18441–18449. [CrossRef] [PubMed]

169. Li, X.; Xu, M.; Chen, H.; Xu, J. Bimetallic Au@Pt@Au core–shell nanoparticles on graphene oxide nanosheets for high performance H_2O_2 bi-directional sensing. *J. Mater. Chem. B* **2015**, *3*, 4355–4362. [CrossRef]

170. Dinh, T.; Phan, H.P.; Nguyen, T.K.; Qamar, A.; Foisal, A.R.M.; Viet, T.N.; Tran, C.D.; Zhu, Y.; Nguyen, N.T.; Dao, D.V. Environment-friendly carbon nanotube based flexible electronics for noninvasive and wearable healthcare. *J. Mater. Chem. C* **2016**, *4*, 10061–10068. [CrossRef]

Review

Carbon-Based Materials for Humidity Sensing: A Short Review

Jean-Marc Tulliani [1,2,*], Barbara Inserra [1,2] and Daniele Ziegler [1,2]

[1] Department of Applied Science and Technology (DISAT), Politecnico di Torino, C.so Duca degli Abruzzi, 24-10129 Torino, Italy; barbara.inserra@polito.it (B.I.); daniele.ziegler@polito.it (D.Z.)

[2] INSTM Research Unit PoliTO, LINCE Laboratory, C.so Duca degli Abruzzi, 24-10129 Torino, Italy

* Correspondence: jeanmarc.tulliani@polito.it; Tel.: +39-(0)11-090-4700

Received: 20 February 2019; Accepted: 27 March 2019; Published: 31 March 2019

Abstract: Humidity sensors are widespread in many industrial applications, ranging from environmental and meteorological monitoring, soil water content determination in agriculture, air conditioning systems, food quality monitoring, and medical equipment to many other fields. Thus, an accurate and reliable measurement of water content in different environments and materials is of paramount importance. Due to their rich surface chemistry and structure designability, carbon materials have become interesting in humidity sensing. In addition, they can be easily miniaturized and applied in flexible electronics. Therefore, this short review aims at providing a survey of recent research dealing with carbonaceous materials used as capacitive and resistive humidity sensors. This work collects some successful examples of devices based on carbon nanotubes, graphene, carbon black, carbon fibers, carbon soot, and more recently, biochar produced from agricultural wastes. The pros and cons of the different sensors are also discussed in the present review.

Keywords: humidity sensor; carbon-based materials; carbon nanotubes; graphene; carbon black; carbon fibers; carbon soot; biochar; flexible electronics

1. Introduction

Gas sensors are miniaturized analytical devices that can deliver real-time and on-line information on the presence of a target gas. Humidity sensors are largely used in many fields where accurate and reliable measurements of water content in different environments and materials are of paramount importance, as depicted in Figure 1. These sensors can also constitute a cheap alternative to a laboratory's analytical technique, so moisture (the water content of any material) sensors and humidity (the water vapor content in gases) sensors are used in many different areas of human activity, such as food quality monitoring, conditioning systems, meteorology, agriculture, manufacturing and process control, medical equipment, and so forth [1]. When air is fully saturated with water, the pressure exerted by the contained water vapor is defined as the saturation water vapor pressure (Ps) that is dependent on the temperature. Thus, the ratio of the current water vapor pressure to the saturation water vapor pressure at a specific temperature is a common way to quantify the amount of water vapor contained in the air [1], and it represents the relative humidity (RH).

The sensor's response is often determined as the relative changes in a measured physical parameter like the impedance (Z), resistance (R), current (I), conductance ($G = \frac{I}{V}$), capacitance (C), power gain, or resonant frequency (f_0) value of the device with respect to time. Different conventions have been adopted for plotting these measurements: $\frac{\Delta X}{X_0}$, $\frac{X}{X_0}$, or simply ΔX (where $X = Z, R, I, G, C, f_0$ or power gain) [2]. Materials that show a resistivity decrease are classified as n-type semiconductors, while p-type ones are those presenting an increase of the resistivity when the target reducing gas concentration rises. In fact, humidity usually exhibits reducing characteristics, even if the oxidizing effect of water at 300 °C was studied in a recent paper by Staerz et al. on the WO_3 surface [3]. The limit of detection (LOD) is the lowest amount of target gas that can be detected at a known confidence level [2]. The calibration curve in Figure 2 shows the relationship between the sensor's response and the concentration of the target gas. The sensitivity of the sensor is given by the slope of the calibration curve (Figure 2). The drift is defined as the change in the sensor's response over time, independently of the gas concentration (Figure 3). In addition, the selectivity is the ability of the sensor to discriminate the target gas among other species [2].

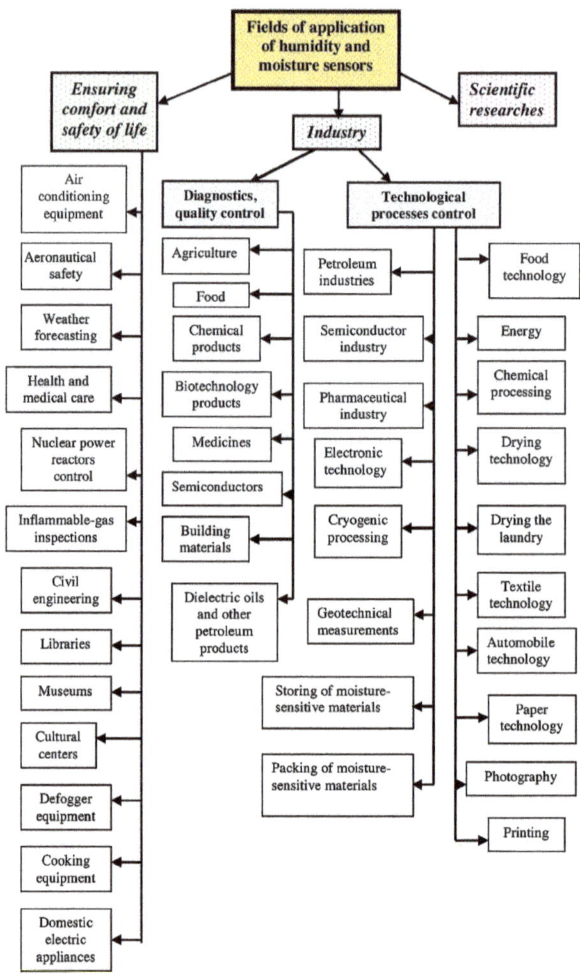

Figure 1. Fields of application of humidity and moisture sensors. Reproduced with permission from [1], published by Elsevier, 2016.

The requirements for effective humidity sensors that change impedance after exposure to humidity in view of practical applications are the following: good sensitivity (i.e., the capability to discriminate small differences in concentration of the analyte over a wide range of RH values, Figure 2), a short response time (the time that a sensor needs to reach usually 90% of the total impedance or capacitance change during adsorption of gas) and recovery time (the time taken for a sensor to achieve usually 90% of the total impedance or capacitance change in the case of gas desorption) (Figure 3), good reproducibility (small standard deviation in sensor response of devices realized with the same material), great repeatability (small standard deviation in sensor response of the same device under a definite humidity concentration), very small hysteresis (the difference between the impedance value during adsorption and desorption cycles for the same RH value), negligible temperature dependence, low cost of fabrication and maintenance, resistance to contaminants, a linear response, an easy fabrication process, and durability [1]. Finally, a low weight and compatibility with a microprocessor are also required features for some specific applications (e.g., portable devices) [1].

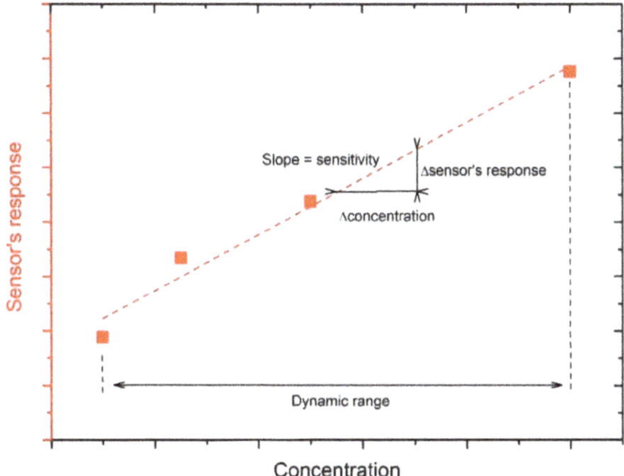

Figure 2. Calibration curve of a sensor exposed to increasing concentrations of an analyte. Elaboration from [2], published by ACS Publishing, 2019.

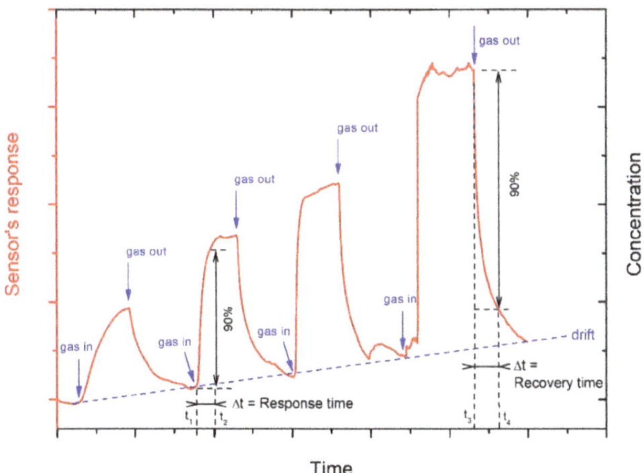

Figure 3. Sensor's response of a sensor exposed to increasing concentrations of an analyte. Elaboration from [2], published by ACS Publishing, 2019.

Commercial sensors are mostly based on metal oxides, porous silicon, and polymers, on which water vapor molecules adsorption drastically changes the electrical properties, such as the resistivity and capacity, of the device [4]. Capacitive humidity sensors detect humidity due to a change of capacitance between two detection electrodes. In these devices, the capacitance increases with RH because of a change in the dielectric constant of the sensing material. These sensors exhibit a low power consumption, but have to be calibrated often, according to changes in the sensor permittivity. The resistance or the impedance of the resistive-type sensor decreases as the relative humidity increases for *n*-type semiconductors. Ions or electrons, or both, are the conduction carriers for resistive-type humidity sensors [1,5]. Generally, both resistive and capacitive sensors are cheap and respond over a wide humidity range with good repeatability. However, their response is dependent on the temperature and is influenced by the presence of some other chemical species (i.e., presents low selectivity) [1]. The common construction of the resistive-type ceramic humidity sensors consists of a ceramic substrate with screen-printed interdigitated noble metal electrodes coated with the humidity sensitive materials and a heater to ensure that there is a working temperature in the active region [1]. Currently, this configuration is also the most common for capacitive-type humidity sensors [1].

Nowadays, the capacitive-based sensors are dominating the humidity sensor market, with nearly 75% of the sales [6].

Semiconductor metal oxides (SMOs) are sensitive towards different gases. Their working principle is based on the variation of their electrical properties: on the surface of the grains of an *n*-type SMO gas sensor, oxygen molecules can adsorb. These adsorbed oxygen molecules then attract electrons from the conduction band and trap them at the surface as ions, leading to band bending. Therefore, an electron-depleted layer is formed (also known as a space-charge layer). The space-charge region is more resistive than the bulk of the SMO because of electron depletion. It is now widely accepted that below 150 °C, oxygen is adsorbed in ionic form (ionosorbed) as O_2^- and it dissociates as O^- in the temperature range of 150 °C to 400 °C. Above 400 °C, O^{2-} ions are formed [7–11]. In the *p*-type semiconductors, holes are responsible for the conduction and the signal is the opposite compared to *n*-type materials: after exposure to oxygen, a hole-accumulation layer is formed, and the space-charge region is more conductive than the bulk. This causes a drop in the resistance and impedance of the sensitive material.

Optical-type humidity sensors based on change of the optical properties (such as reflectance, evanescent wave, etc.) and acoustic-type humidity sensors based on the variation in the frequency of acoustic resonance both have good sensitivity and acceptable response times [12]. However, the latter two types of sensors cannot be used as flexible sensors due to the complicated measurement systems and non-flexible active materials [12].

The active materials are the most important part in high-performance flexible sensors and should be both flexible and electrically conductive. To this aim, recently, carbon films have attracted great attention for their potential applications as humidity sensors because of their large sensing area and high chemical inertness [4,12]. Due to their rich surface chemistry and structure designability, carbon materials have become interesting in humidity sensing. In addition, carbon nanomaterials, including carbon nanotubes (CNTs), graphene, carbon black, and carbon nanofibers, are among the most commonly used active materials for the fabrication of high-performance flexible sensors. Specifically, CNTs and graphene can be assembled into one-dimensional fibers, two-dimensional films, and three-dimensional architectures, allowing an easy design of flexible sensors for many practical applications [12]. Other advantages of these materials are their ability to work at room temperature (RT), the possibility to functionalize them for chemical specificity, and their low thermal mass that allows rapid heating with low power consumption [13]. Possible drawbacks are their low selectivity, poor reproducibility, tendency to poisoning, and possible long-term drift [13].

Moreover, low-cost carbon materials, including carbon black and carbon nanofibers, can be utilized as sensing materials integrated with fabrics [12]. Besides carbon nanomaterials, other carbon powders derived from bio-materials through pyrolysis have also been proposed as sensing materials, such as silk [14] and cotton [15]. Biomass is also a qualified carbon source, is available at a high quality and huge amount, and is considered an environmental-friendly renewable resource [16]. Furthermore, different biochars, which are residues of biomass pyrolysis, are now available from pilot plants producing biogas and energy [17,18]. In recent years, biochar applications have been conducted in many fields [19], even though the main application of this material remains field amendment in agriculture [20]. Moreover, in recent years, biochar has been extensively studied as a substitution for more expensive materials, such as carbon nanotubes, graphene, and others [21].

This paper reviews different experimental activities dealing with carbonaceous materials used as capacitive or resistive humidity sensing materials with a special focus on biochar. It discusses their advantages and underlines some limits and drawbacks, suggesting useful strategies for minimizing them.

2. Sensing Materials

2.1. Carbon Nanotubes (CNTs)

In the last two decades, various resistive-type humidity sensors based on polymers (polyimide) [22], sulfonated polyimides (SPIs) [23], poly(2-acrylamido-2-methylpropane sulfonic acid) [24], poly(2-acrylamido-2-methylpropane sulfonate) [25], and poly(4-vinylpyridine)/poly(glycidyl methacrylate [26]) and/or carbon nanotubes [27] have been investigated. Polymer-based resistance sensors are limited in their ability to detect low levels of humidity, mostly because of their high initial resistance value: the minimum detectable RH concentration is around 42% for a polyimide-based resistive humidity sensor [22] and is equal to 30% for a sulfonated polyimide [23]. Thus, more sensitive CNTs-based sensors were developed [5].

CNTs are excellent sensor candidates due to their mechanical and electrical properties, as well as their ability to be functionalized and their easy integration into electronic circuits [28]. CNTs can be either single-walled (SWCNTs) or multi-walled (MWCNTs). MWCNTs are made of multiple concentric layers of SWCNTs. In defect-free tubes, the bonds between carbon atoms in sidewalls are hybridized sp^2 and noncovalent van der Waals forces or π stacking dominate the intermolecular interactions [28]. MWCNTs contain both holes and electrons and at room temperature, present a metallic behavior because of the overlapping of conduction and valence bands with electrons as majority carriers [27,29,30]. MWCNTs can also behave as semiconductors with the energy overlap changing in function of the chirality and thus, the interaction between the different MWCNTs walls [31,32].

In Varghese et al. [27], MWCNTs were grown by the pyrolysis of ferrocene and xylene under an Ar/10% H_2 atmosphere in a two-stage reactor. Ferrocene acted as a Fe catalyst and xylene as a carbon source. The liquid was first pre-heated at 175 °C and subsequently sent to the reactor at 750 °C. MWCNTs were deposited on quartz substrates with interdigitated electrodes to keep the impedance of the sensor low, while providing a maximum surface area for contact with the gas atmosphere. MWCNTs showed a considerable response to humidity with a response time of 2–3 min when increasing RH values, while the recovery times were in the order of few hours [29]. The results revealed a charge transfer between the water molecules and the MWCNTs, typical of a dominant chemisorption process since physisorption does not involve any charge transfer [29]. For *p*-type MWCNTs, the adsorbed water molecules extract holes to the valence band, leading to an increase in the resistance of the film. The proposed sensors showed a cut-off (i.e., sensors start to respond) from 10 RH% and 20 RH%, respectively, for the capacitive and resistive sensors [29].

In Lee et al. [5], poly (acrylic acid) (PAA) was used to disperse the MWCNTs, as PAA is highly hygroscopic and sensitive to low RH levels. To disperse high amounts of MWCNTs in PAA, poly (4-styrenesulfonic acid) (PSS) was used as a surfactant as PSS wraps the MWCNTs' hydrophobic outer shells with noncovalent polymer chains [5]. The PSS and MWCNTs were first mixed together in a mortar to let the MWCNTs wrap with the PSS surfactant. The PSS/MWCNT mixtures were then wetted with distilled water and ground for 10 min to prevent aggregation of the entangled MWCNTs. The mixtures were diluted in distilled water to disperse the nanotubes, prior to sonication for 15 min in an ultrasonic bath, followed by sonication for other 15 min with an ultrasonic probe. Subsequently, PAA powder was dissolved in the MWCNT mixtures at 85 °C by stirring for 1 day. Finally, the sensor was manufactured by depositing a 20 µL drop of this solution on a polyimide film with gold interdigitated electrodes. The MWCNT/PAA films contained up to 33 wt% of MWCNTs [5]. The humidity sensor exploits the volume's change of the MWCNT/PAA film due to humidity adsorption and desorption, which causes an increase or a decrease in the distance between neighboring MWCNTs. Thus, the electrical resistance of the film is dependent on volume changes caused by the adsorption and desorption of water vapor. Films with more space to absorb humidity show a higher degree of swelling, especially with a higher content of PAA. The film made with the ratio of MWCNTs/PAA equal to 1:4 exhibited good sensitivity to humidity: its resistance variation from 30 RH% to 90 RH% was equal to 930 Ω. This result was due to the high concentrations of MWCNTs in the polymer matrices, which led to limited changes in resistance with increasing RH levels. However, the response to humidity was highly linear and the sensitivity was acceptable [5]. The response and recovery times were equal to 670 s and 380 s, respectively (for RH levels increasing from 50% to 90% and decreasing from 90% to 50%).

Polyimide (PI)/MWCNTs composite films were prepared by in-situ polymerization at room temperature in Tang et al. [32]: the precursors (polyamic acid, PA) were obtained by stirring a mixture of pyromellitic dianhydride (PMDA), 4,4-oxydianilline (ODA), and surfactant-dispersed MWCNTs in N,N-dimethylacetamide (DMAc) for 4 h. A series of PA/MWCNTs nanocomposites (with theoretical contents of MWCNTs in the PI of 0.5, 1.0, 2.0, and 3.0 wt% and labeled respectively as PIC05, PIC10, PIC20, and PIC30) were synthesized with the same PA solid content of 18%. The precursors were then coated on glass substrates and kept in an oven at 60 °C for 12 h to let the solvent evaporate. Subsequently, step curing at temperatures of 100, 150, 200, 250, and 300 °C for 1 h was used to obtain PI/MWCNTs films. The sensitivity of PIC10, PIC20, and PIC30 at 30 °C was 0.00138/RH%, 0.00178/RH%, and 0.00146/RH%, respectively. The sensitivity of the PIC30 film is lower compared to that of the PIC20 film, and the sensitivity of composites is higher just around the percolation threshold. In PIC films, the percolation threshold is around 1 wt%. A higher content of MWCNTs make the carbon nanotubes tightly connected to each other and create physical contacts in the network. Therefore, the response due to polymer swelling is smaller compared with the sample near the percolation threshold. Finally, the resistance of the PIC30 film decreased when the temperature increased (Figure 4); thus, temperature compensation should be made when this sensor is used at different temperatures. This negative temperature dependence indicates that the major conductive mechanism of PIC30 was due to the inter-tube effect: higher temperatures provided more energy for the motion of electrons lowering the barriers between MWCNTs, resulting in a decrease of the film resistance.

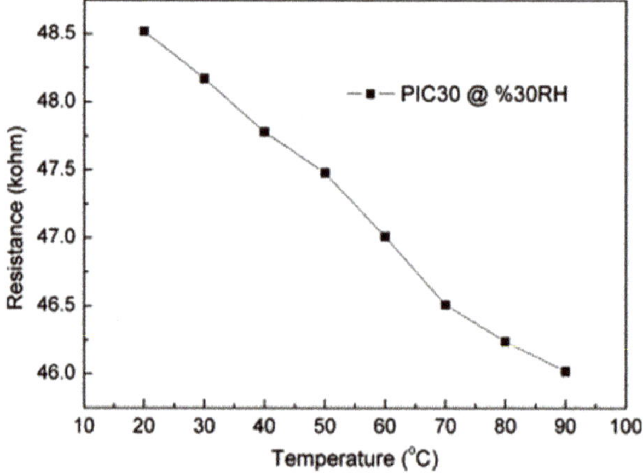

Figure 4. Resistance changes in function of temperature for PIC30 sensor. Reproduced with permission from [32], published by Elsevier, 2011.

In Yoo et al. [33], pristine and radio frequency oxygen plasma-treated MWCNTs (p-MWCNTs) were mixed with PI and spin-coated on a Si_3N_4 membrane. Their resistance values continuously rose with the increasing RH levels, in the relative humidity range 10%–90%, while the initial resistance value decreased with the increasing CNTs content. The measured sensitivities are reported in Table 1.

Table 1. Sensitivity of MWCNTs/PI composites. Reproduced with permission from [33], published by Elsevier, 2010.

CNT Concentration (wt%/PI)	p-MWCNTs/PI Sensitivity (1/RH%)	MWCNTs/PI Sensitivity (1/RH%)
0.1	0.00127	0.00092
0.2	0.00149	0.00103
0.3	0.00305	0.00182
0.4	0.00466	0.00218

The p-MWCNTs/PI sensors showed better linearity and sensitivity than the non-treated ones and a scheme of the gas sensing mechanism is depicted in Figure 5.

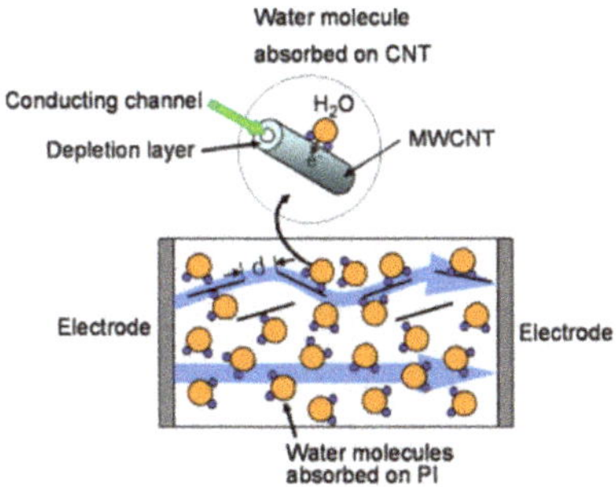

Figure 5. Scheme of the conduction mechanism of *p*-MWCNT/PI composite devices. Reproduced with permission from [33], published by Elsevier, 2010.

The authors explained this result by stating that the surface of the plasma-treated CNTs has a higher presence of defects with respect to pristine ones. They proposed electrical conduction as a bulk phenomenon in contrast to surface adsorption. H_2O molecules in PI possess a polarizability that is around 80% with respect to the condition of free water. As a result, water species are not significantly involved in hydrogen bonding and are situated primarily in free volumes in the polyimide network. Consequently, the high conductivity water in the PI network increases the conductivity of the PI-H_2O system.

Water molecules can be either chemically bound (with the oxygen of the ether linkage and with the four carbonyl groups at low RH levels) in the PI network or can condense in microvoids at higher humidity amounts [34]. As previously described, only some of the water molecules in PI composites films can bind to the CNTs through the hydrogen atom [33]. When the MWCNTs concentration is much lower than the percolation threshold, the distance between the nanotubes is large and CNTs are almost isolated. Thus, the resistance of the sensor is governed by the resistance of the water adsorbed on PI and the sensor's response is nonlinear. On the contrary, when the MWCNTs concentration is close to the percolation threshold, tunneling through a potential barrier across the MWCNTs is effective. In the *p*-MWCNTs/PI sensors, the intertube's resistance increases due to the adsorption of water molecules, while the PI resistance decreases, leading to an almost flat sensor's response in the investigated relative humidity range. If the *p*-MWCNTs fraction is above the percolation threshold, the number of CNT-CNT connections rises and more current paths through intertubes are effective. Then, the adsorption of water molecules on CNTs becomes important. The amplified sensor's resistance is due to the transfer of electrons from H_2O molecules to carbon nanotubes and/or the intertube's distance variation when PIs begin to swell. The electron transfer caused by the adsorption of water molecules will shift the valence band of the CNTs away from the Fermi level, reducing the hole concentrations and increasing the depletion layer thickness. Thus, the resistance of the MWCNTs and of the intertube tunneling barrier will increase [33].

Resistive-type humidity sensors made from composite films of hydroxyethyl cellulose (HEC) and MWCNTs (purity > 99.5%; 7, 8 and 9 wt%) were investigated in Ma et al. [35]. The sensors were prepared by dissolving 0.3 g of HEC powder with MWCNTs in 3.0 g of distilled water at 80 °C for 5 min. Then, the solution was centrifuged for 10 min and defoamed for 1 min. The final mixture was pasted between two electrodes with wires onto an alumina plate. The ceramic plate was finally dried at 130 °C for 1 h to remove the distilled water. The sensor's response ($\Delta R/R_0$) varied in an exponential form. When the sensor's response was expressed on a logarithmic scale, the sensitivity of the studied sensors was 0.0866/RH%, 0.0937/RH%, and 0.10069/RH%, respectively, for the 7 wt%, 8 wt%, and 9 wt% MWCNTs films. The HEC/MWCNTs sensor with 7 wt% MWCNTs showed the best performance from a repeatability and stability point of view, in the dry/wet cyclic tests. However, the HEC/MWCNTs sensor with 9 wt% MWCNTs presented the highest sensitivity.

In the study of Pan et al. [36], MWCNTs were first functionalized by chemical treatment: 150 mg of MWCNTs was ultrasonicated for 4 h in a mixture of concentrated H_2SO_4/HNO_3 at a ratio of 3:1. The treated MWCNTs were then filtered over a 0.45 µm pore size membrane and the filter cake was washed with deionized water. Subsequently, the remaining solid was suspended in solution of H_2O_2 (20 v/v%, 150 mL) in an ultrasonic bath for 2 h. Finally, the product was filtered and dried overnight in a vacuum oven at 50 °C. The oxidized MWCNTs (100 mg) were dispersed in N-N-dimethylformamide (DMF, 8.55 mL) under ultrasonication for a few minutes until a stable suspension was formed. Then, polyvinylpyrrolidone (PVP, 900 mg) was added into the suspension with stirring and the mixture (MWCNTs/PVP weight ratio equal to 1:9) was kept under stirring for another 24 h. The MWCNTs/PVP sensors were prepared on quartz slides with interdigitated electrodes and the films were dried at room temperature for 12 h and at 60 °C for 12 h. Finally, the MWCNTs/PVP films were heat-treated at 350 °C for 1 h. The response was defined as $I/I_{11\%}$, where I was the current of the sensor at different RH values and $I_{11\%}$ was the current of the sensor at 11 RH%. The films started to respond to humidity from 33 RH% (Figure 6). The response and recovery times between 11 RH% and 94 RH% were about 15 and 1.8 s, respectively [36].

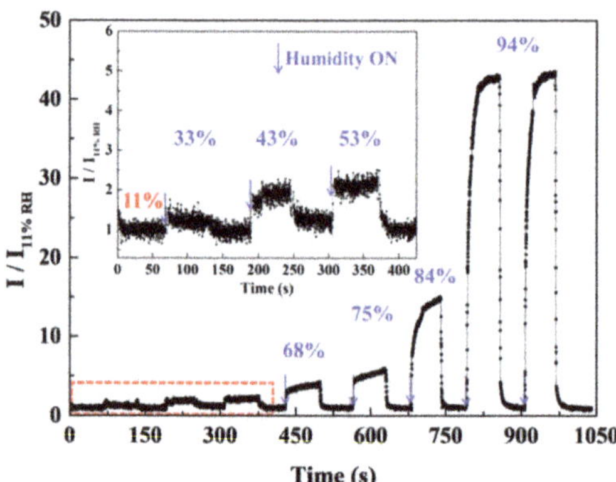

Figure 6. Dynamic response curve for the MWCNTs/PVP film sensor heat-treated at 350 °C at different RH% values. Reproduced with permission from [36], published by Wiley Online Library, 2016.

The interfacial heterojunctions between MWCNTs and PVP strongly influence the sensitivity and response/recovery times of the sensors. At low relative humidity values, few carriers (H^+ and H_3O^+) were injected into the heterojunctions due to the adsorption of water molecules. The ions carrier injected into *n*-type PVP neutralized the intrinsic electrons of PVP and weakened the barrier of *n–p* heterojunctions, decreasing the resistance of heterojunctions. Therefore, the MWCNTs/PVP sensor exhibited a good sensitivity at low RH. At high relative humidity values, the swelling of PVP was strongly limited because of the stability of the MWCNTs net. This allowed the MWCNTs/PVP sensors to exhibit excellent repeatability (Figure 6, the two successive measurements at 94 RH%). In addition, with decreasing MWCNTs content, the number of interfacial heterojunctions and sensitivity decreased, and longer response/recovery times compared with the 10% MWCNTs/PVP sensor were determined [36].

The results of a wearable textile-based humidity sensor utilizing high strength (~750 MPa) and ultra-tough SWCNTs/PVA filaments made by a wet-spinning process were reported in [37] (Figure 7). The diameter of an SWCNT/PVA filament under wet conditions was two times that under dry conditions. Moreover, the electrical resistance of a fiber sensor stitched onto a hydrophobic textile significantly increased by more than 220 times after water was sprayed. Textile-based humidity sensors using a weight ratio of 1:5 between SWCNT and PVA filaments showed a great sensor response. In fact, the electrical resistance increased by more than 24 times under 100% of RH with respect to 2.4 times when the molar ratio was 1:1. Moreover, the response time was rather short (40 s). The effect of operating temperature was also investigated, and a significant decrease in the sensor response of around 50% was measured when the operating temperature increased from 25 to 75 °C. The authors have demonstrated that the textile sensor based on the SWCNT/PVA filament can be utilized to monitor human sweating and water leakage on a high hydrophobic textile, with a contact angle of 115.5°. The as-fabricated sensor also displayed an excellent reversibility under high relative humidity values and wet conditions.

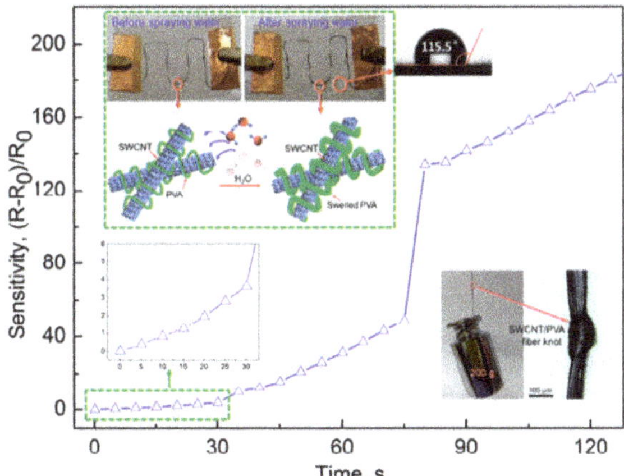

Figure 7. Single-walled carbon nanotube (SWCNT)/Poly (Vinyl Alcohol) (PVA) filament sensor response. Reproduced with permission from [37], published by ACS Publishing, 2017.

In [38], carbon nanotube (CNT)-yarn humidity sensors were manufactured using an MnO_2-coated CNT yarn and the sensor performances were compared with those of a pure CNT yarn. The results of this work demonstrated that an increase in humidity causes a decrease in the hole density of *p*-type nanotubes, resulting in an increase in the resistance of the sensors. An Fe film acting as a catalyst for CNTS growth was first deposited by electron beam evaporation on 330-μm-thick *p*-type silicon wafers and then annealed in a vertical cylinder atmospheric-pressure chamber. After purging the tube with He for 10 min, the temperature in the chamber rose to 780 °C in 15 min. The CNTs were grown at 780 °C by adding acetylene gas to the flow for 5 min. Subsequently, acid treatment was carried out in solutions of H_2SO_4/HNO_3 (3:1, 50 mL) for 1 h at room temperature. The aim of the acid treatment of the CNT during functionalization is to introduce hydrophilic carboxylic functional groups to the sidewalls, in order to to improve the performance of the CNT humidity sensor. After functionalization,

MnO_2 nanoparticles were directly electrodeposited on the CNTs in a fresh aqueous solution of 0.02 M of $MnSO_4 \cdot H_2O$ and 0.2 M Na_2SO_4 at pH ~5.6.

MnO_2 was electrodeposited at a potential of 0.9 V for 40 s, and this resulted in a 150 ± 30 nm thick MnO_2 on CNT yarn. Then, a p-n heterojunction was formed at the interface between the CNT and MnO_2 nanoparticles, after the electrodeposition of MnO_2 on the CNT. This produced band bending in the depletion layers. Because of the relative position of the CNT and of the conduction band edges of MnO_2, the electrons transferred to the conduction band of MnO_2 can be injected to the CNT. Eventually, this process leads to a decrease in the concentration of holes in the CNTs, giving rise to an increase in the electrical resistance of the CNT/MnO_2 composite. In this case, MnO_2 provides active sites for the adsorption of H_2O molecules and constitutes an excellent medium for electron transfer during the sensing process. In this way, the sensitivity of the CNT/MnO_2 composite is significantly improved. A comparison of the two sensors' performances towards humidity is depicted in Figure 8.

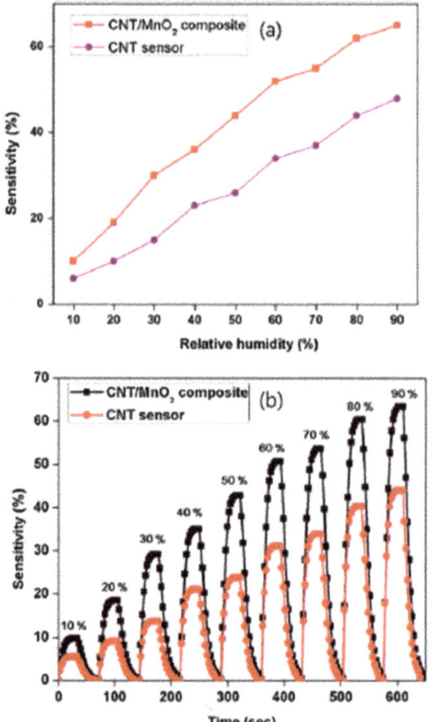

Figure 8. Sensitivities of humidity sensors as a function of relative humidity. (**a**) and as a function of time; (**b**) between 10 and 90% RH. Reproduced with permission from [38], published by Elsevier, 2015.

2.2. Graphene

In the work of Huang et al. [39], sodium (5 g) and sugar (fructose or sucrose, 5 g) first reacted with ethanol as a solvent (50 mL) in a sealed Teflon autoclave at 220 °C for 72 h. The addition of sugar was aimed at creating defects in the graphene sheets. The precursor was then rapidly pyrolyzed at 600 °C for 2 h and washed with deionized water and methanol, prior to filtering and drying. One drop of a graphene methanol solution was subsequently deposited onto an alumina substrate with gold interdigitated electrodes and dried for 5 h to evaporate the solvent. Finally, the device was annealed in vacuum at 180 °C for 2 h to improve the contact. In the absence of any sugar, the response changed slightly from 3 to 30 RH%; however, for the sample with sucrose, the response increased sharply from 0.27 to 3.33 in the same RH range. In addition, the cut-off was around 6 RH% when sugar was used during the synthesis process (Figure 9). XPS, XRD, and Raman spectroscopy results demonstrated that oxygenated groups increased from 17.7% in the sample with no sugar to 24.7% in the sample obtained with sucrose.

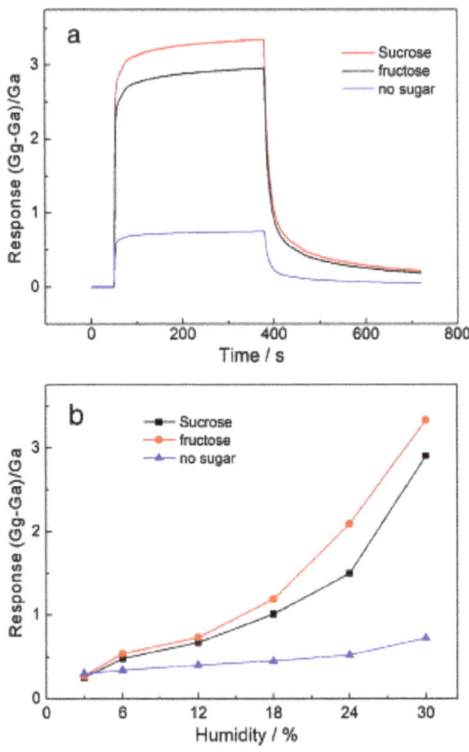

Figure 9. Dynamic response under 30% RH: (**a**) response in terms of conductance's changes in the range 3–30% RH; (**b**) for different graphene-based sensors. Reproduced with permission from [39], published by Elsevier, 2012.

Unlike graphene and reduced graphene oxide (rGO), graphene oxide (GO), because of its oxygen functional groups, is strongly hydrophilic and proton conductive [40,41]. This makes GO an ideal candidate for humidity detection. The Van der Waals interaction between H_2O and graphene is weak (0.044 eV) while H_2O forms hydrogen bonds with the epoxy (0.201 eV) and hydroxyl groups (0.259 eV) [42,43]. Guo et al. [42] used GO produced by Hummers' method from natural graphite (Aldrich, <150 μm) to prepare sensing films on poly(ethylene terephthalate) (PET) substrates by the spin-coating technique. After drying at 60 °C for 1 h, the samples were exposed for 10 s to two beams which were split from the UV laser to reduce and produce patterned hierarchical nanostructures. The content of oxygen atoms in pristine GO was 46.5%, with only 32% of carbon atoms not bonded to oxygen. After reduction with 0.15 W laser treatment, the C–C percentage increased to 68% and C–O percentage decreased to 23%, indicating the loss of oxygen groups from the surface. An increase of the laser power led to a further reduction of the GO film. After two-beam-laser interference (TBLI) reduction with the laser power of 0.15 W, the resistance of the devices dropped by more than two orders of magnitude when RH increased in the 11%–95% range. Moreover, the sensors' response showed good linearity and limited hysteresis (6%) under 60 RH% (Figure 10).

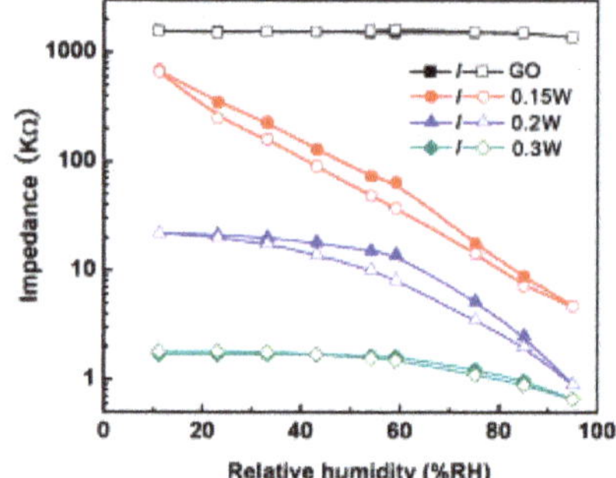

Figure 10. Impedance changes of GO-based sensors. Reproduced with permission from [42], published by Elsevier, 2012.

When the laser power was increased to 0.2 W and 0.3 W, the range of the resistance change became smaller and the resistance versus RH% was no longer linear over a wide humidity range (Figure 10). This result was due to the relatively higher conductivity and to the damaging of hierarchical nanostructures compared with the device fabricated with the 0.15 W laser. For the sensor realized with the laser power of 0.15 W, the response time was very fast (2 s) when increasing from 11 to 95 RH%. Then, when decreasing the RH level from 95% to 11%, the recovery time was higher than 100 s. For the device fabricated by the laser power of 0.2 W, the response time was about 3 s, and the recovery time was about 10 s. Finally, when the laser power increased to 0.3 W, the response time increased until 50 s, while the recovery time was about 3 s [42].

GO was obtained from natural graphite powder by an oxidation reaction according to a modified Hummers' method in Yu et al. [44]. GO ethanol solution (50 mL) with a concentration of 1 mg/mL was sealed in a 100 mL autoclave and then heated to 180 °C for 12 h. After that, the autoclave was left to cool down naturally to room temperature. The prepared ethanol intermediates were removed from the autoclave by a slow and progressive solvent exchange with water, prior to drying with a freeze-dryer and then, at 120 °C for 2 h in a vacuum oven. Subsequently, the sample was annealed at 450 °C in H_2/Ar (5/95, v/v) for 6 h. Finally, the sample was treated in a UV ozone system for 15 min to obtain the final 3D graphene foam (3DGF). When the RH level rose, the obtained channel currents of the sensor decreased continuously. The response and recovery times were determined for the sensor when the RH level was changed from 0 to 85 RH% and from 85 RH% to 0 RH% and were approximately equal to 89 ms and 189 ms, respectively.

Graphene oxide was prepared from natural graphite (Alfa Aesar, 99.999% purity, 200 mesh) using a modified Hummers' method [45]: graphite and H_2SO_4 were first mixed in a flask and $KMnO_4$ was added slowly over 1 h. After 2 h stirring, the solution was kept over an ice water bath. Subsequently, the mixture was stirred vigorously for 18 h and deionized water was added. Then, the solution was stirred for 10 minutes over an ice-water bath and H_2O_2 (30 wt% aqueous solution) was added. Finally, the mixture was stirred for 2 h. The resulting mixture was precipitated and filtered to obtain the graphite oxide powder (GO). The GO was then exfoliated into GO nanosheets in water by bath sonication for 1 h. Subsequently, the GO nanosheets were dispersed in N-methyl-2-pyrrolidone to limit aggregation. Hydrazine monohydrate (N_2H_4) was then added dropwise to the GO solution to a final concentration of 4 mM. The GO was finally reduced in solution by heating at 100 °C for 24 h. Large-area graphene electrodes were grown on a Cu foil (10×10 cm^2) through CVD. The Cu foil was inserted into a quartz tube and annealed at 1000 °C under an H_2 atmosphere for 1 h. Then, 5 sccm of CH_4 were introduced to trigger graphene growth under a continuous H_2 flow of 10 sccm. After 30 min, the CH_4

flow was stopped and the tube was cooled down to room temperature under H_2 flow. The graphene obtained on the Cu foil was then transferred onto a polydimethylsiloxane (PDMS) substrate using standard photolithography and etching processes [45]. The PDMS solution with a base prepolymer and a crosslinking agent (in a weight ratio of 10:1) was poured onto a hexamethyldisilazane-treated glass, and the sample was then put into a chamber to remove bubbles in the PDMS layer. After degassing, the glass with the PDMS layer was spin-coated and cured at 120 °C for 1 h, and then converted into a hydrophilic surface with O_2 plasma. When the relative humidity was increased from 20% to 90%, the capacitance increased from 0.15 to 4.27 pF because the adsorbed water molecules increased the GO capacitance (Figure 11). Two distinct regimes were evidenced in the curves. For RH levels below 60%, water molecules were adsorbed onto the GO surface through double hydrogen bonding. In this regime, the protons' hopping between adjacent hydroxyl groups increased the leak conductivity in the film, enhancing the capacitance of the GO film. As the RH rose above 60%, a larger number of water molecules were adsorbed onto the GO surface and penetrated the GO films. These water molecules made the hydrolysis of carboxyl, epoxy, and hydroxyl groups on the GO surface easier. These ionic species sharply enhanced the ionic conductivity and thus the sensor's capacitance increased exponentially [46].

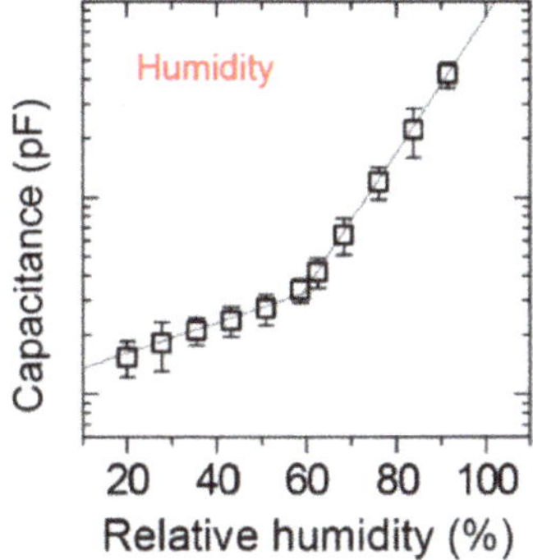

Figure 11. Impedance change of GO-based sensors. Reproduced with permission from [46], published by Wiley Online Library, 2016.

In Borini et Al. [47], GO humidity and temperature sensors with PEN (polyethylene naphthlate) as supporting material and silver interdigitated electrodes were realized by drop casting and spray coating methods. The manufactured sensors had different thicknesses (15 nm, 25 nm, and 1 μm). Sensors obtained by spray coating, with a GO layer roughly 15 nm thick, showed ultrafast response and recovery time, almost 30 ms, due to the high graphene permeability and to its 2D structure. Thi result allows classifying the sensor among the fastest ever produced before. The speed of the sensors was studied as a function of the film thickness, response time 20–30 ms and recovery times 90 ms for 25 nm thick films—30 ms for 15 nm thick films were determined. The response of a 25 nm thick film is lower than that of the 15 nm one. The measurements were done in the R.H. range from 30% to 80% and temperature range 10–40 °C.

GO shows great potential in the sensor technology. For example, it can be exploited, beyond RH and T, to monitor the littlest variation of moisture during a vocal expression.

In [48], a layer of graphene was deposited on the surface of a chip made of p-doped silicon substrates with a 300 nm thick SiO_2 layer by using a CVD graphene wet transfer technique after deposition on a copper foil. The graphene layer was finally patterned using a photoresist mask and O_2 plasma etching. The sensitivity of the sensor in the investigated RH range was 0.31 1/RH%, with very fast response and recovery times (respectively equal to 0.4 and 0.6 s).

A flexible humidity sensor based on SnO_2/reduced graphene oxide (RGO) nanocomposite film was proposed in [49]. The humidity sensor was fabricated on a flexible polyimide (PI) substrate with microelectrodes by one-step hydrothermal synthesis. Compared with traditional humidity sensors, the as-prepared sensor demonstrated an ultrahigh sensitivity together with rapid response and recovery characteristics. The electrodes were fabricated by metal sputtering. In the microfabrication of the sensor, a 20 μm thick Cu/Ni layer was firstly deposited on the PI substrate (75 μm thick) with a sputtering system. Subsequently, photoresist (PR) was applied to make a pair of interdigital electrodes (IDEs) pattern with the lithography technique, and subsequently, the redundant Cu/Ni was etched out to form micro-IDEs. A sensing film of an SnO_2/RGO hybrid composite was finally realized by a hydrothermal treatment with an $SnCl_4$ solution in the presence of graphene oxide. Firstly, 2 mL of GO (0.5 mg·mL^{-1}) and 24 mg of $SnCl_4 \cdot 5H_2O$ were added into 20 mL of deionized water by sonication for 10 min and stirring for 1 h. The solution was then transferred into a 40 mL Teflon-lined, stainless-steel autoclave and heated at 180 °C for 12 h, during which GO was converted into conductive rGO under hydrothermal reduction. After the autoclave was cooled down, the as-prepared products were centrifuged for 10 min and subsequently washed with deionized water. Finally, the resulting SnO_2/RGO dispersion was drop-casted onto the flexible substrate, followed by vacuum-drying in an oven at 50 °C for 2 h. The sensor showed a great variation in capacitance from 246.53 pF to 138267 pF over the humidity range from 11% to 97 RH%. The corresponding capacitance changed by approximately 550-folds of magnitude within the entire humidity range of 11–97 RH%.

A highly sensitive humidity sensor made of silver inter-digital electrodes and a graphene (G)/methyl-red (M-R) composite layer deposited on a low cost transparent polyethyleneterephthalate (PET) substrate through the inkjet printing technique was obtained in [50]. In order to achieve a high sensitivity and a wide sensing range, the methyl-red composite thin film layer was deposited over the silver interdigital electrodes through electrohydrodynamic (EHD) and its thickness was ~300 nm. The graphite powder (0.05 g) was first dispersed in NMP (10 mL) solvent and the solution was then sonicated in an ultrasonic bath for 30 min at room temperature. After bath sonication of the ink, large un-exfoliated graphite flakes were separated by vacuum filtration. Finally, the ink was centrifuged for 30 min and the supernatant was separated from sediment. 10 wt% methyl-red was prepared in dimethylformamide (DMF) by bath sonication for 2 h at 30°C. This temperature was chosen because below 15 °C, the methyl red in DMF forms a gel. The graphene and methyl-red inks were mixed with an optimum 2:1 ratio. The mixed ink was then placed on a bath sonicator at ambient conditions for 1 h to make a uniform dispersion solution of graphene flakes and methyl-red. The sensor electrical resistance changed from 11 MΩ to 0.4 MΩ towards the relative humidity content from 5% to 95%. The proposed humidity sensor showed 96.36% resistive and 2869500% capacitive sensitivity against humidity. The response and recovery time of the sensor was respectively equal to 0.25 s and 0.35 s. Furthermore, it had negligible cross sensitivity from other constituents in air because of M-R addition in the graphene.

Pang et al. [51] realized a porous material, 1.5 μm thick, by means of chemical vapor deposition, growing graphene flakes on a "model" nickel foam. The metallic skeleton was then dissolved with HCl. For performance improvement of Graphene Oxide sensors in moisture monitoring, the GO network was modified with poly (3,4-ethylenedioxythiophene) poly(styrenesulfonate) (PEDOT: PSS) and Ag colloids (AC). In this way, the sensor showed a response time of 31 s and a recovery time of 72 s. The relative resistance change $(R-R_0)/R_0$ reached a value of 1.10% at 12 RH% and increased up to 4.97% at 97 RH%. While the water molecules progressively accumulated in the graphene network, the resistance variation showed improvement. The field of applications of this technology ranged from healthcare to the detection of skin moisture and clinical respiration monitoring.

In Yun et al. [52], reduced graphene oxide and MoS_2 hybrid composites were synthesized by the hydrothermal method and drop-cast on an SiO_2 layer. The amount of GO was defined by the following molar ratio: $GO/(NH_4)_2MoS_4$ = 5:1, 3:2, 1:5. These sensors were labelled as MS-GO1, MS-GO2, and MS-GO3. A p-n junction was formed between rGO and MoS_2 due to van der Waals bonding and was responsible for short response and recovery times and a high selectivity, as well as a greater sensitivity and linearity for humidity sensing in comparison with pristine *p*- or *n*-type sensors. The responses of rGO, MS-GO1, MS-GO2, MS-GO3, and MoS_2 sensors were equal to 3%, 15%, 21%, 14%, and 49%, respectively, at 50 RH% at room temperature (Figure 12). The response times of GO, MSG-O1, MS-GO2, MS-GO3, and MoS_2 were 193.7, 59, 30, 116, and 17 s, and the recovery times of each sample were 418, 343, 253, 282, and 474 s, respectively. Although pristine MoS_2 showed the highest response value, its recovery time was higher with respect to all other investigated samples, so MS-GO2 demonstrated the best overall sensing performances.

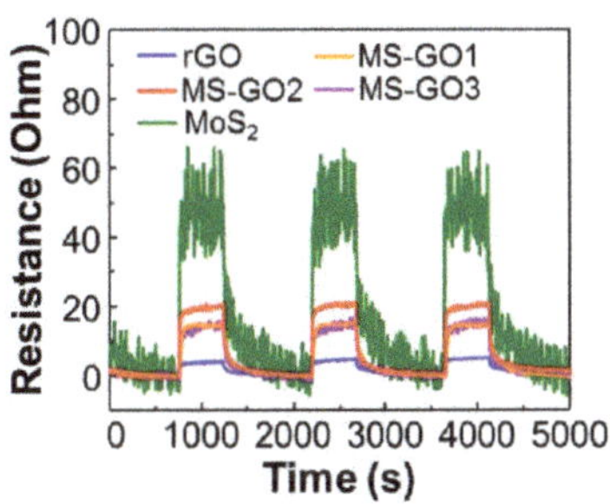

Figure 12. Sensing curves of rGO and MS-GO composites and of MoS2 when RH switched from 0% to 50%. Reproduced with permission from [52], published by Elsevier, 2018.

Finally, Park et al. [53] prepared sensors based on rGo/MoS_2 at a molar ratio of 1: 1, 1: 5, and 1: 10 by simple ultrasonication. These samples were labelled RGMS 1, RGMS 5, and RGMS 10, respectively. RGMS 1; 5; 10 showed a sensor response of 10.5, 872.7, and 86.6% to 50 RH%, respectively. rGo/MoS_2 in the proportion 1:5 showed the highest response to water vapor among the composites and a signal 220 times higher than that of pure rGo. In addition, the base resistance increased with the increasing molar ratio of MoS_2. The responses of RGMS 5 to 50 ppm H_2, 50 ppm CH_3COCH_3, 10 ppm NO_2, and 1000 ppm NH_3 at 27 °C and 1 V were 3.2%, 17.8%, 29.6%, and 47.9%, respectively.

2.3. Carbon Fibers

Carbon nanofiber-based gas sensors are much less popular with respect to those made with CNTs and graphene nanosheets [54].

Commercial carbon nanofibers, produced at temperatures above 1100°C from natural gas and sulfur in a floating nickel catalyst reactor, with a diameter between 20 and 80 nm and lengths of more than 30 μm, were first dispersed in 2-propanol and then, deposited over interdigitated silver electrodes on a PI substrate by means of a spray technique. The formed film was homogenous, with a thickness of hundreds of nanometers, except for the edges, which were thicker. Two kinds of nanofibers were tested: one with a graphitization of about 70% (GANF) and one with about 100% (GANFG) [55,56]. A linear increase of the resistance was observed for the GANF and GANFG sensors with the enhancing water vapor concentration in the range 5–100 RH%. However, a certain drift in the baseline was also observed, especially for relatively high humidity concentrations. Finally, GANFG displayed a better response to radiation (visible light and UV rays) and both types of nanofibers presented a better and faster response and were more repeatable to visible light than UV. It must be underlined that the effect of radiation cannot be ignored as an interference for sensing applications.

2.4. Carbon Films Produced by Physical Vapor Deposition (PVD)

Films of nanostructured carbon (n-C) were grown from a graphite stick under methane and under vacuum by a hot filament physical vapor deposition (HFPVD) method on Si (100) substrates with an SiO_2 buffer layer [4,57,58]. Four gold electrodes were then sputtered onto the n-C film. FESEM micrographs of n-C films showed the presence of nanoparticles (with an average size of about 80 nm), while for the sample grown in vacuum, vertically and well-organized nanohoneycomb structures were observed (Figure 13). The thickness of the two samples was 550 and 210 nm, respectively. The resistance of both n-C films changed almost linearly with RH (Figure 13). When RH rose from 11% to 95%, the resistances of the carbon nanosheets and nanohoneycombs-based sensors increased by 225% and 112%, respectively (Figure 14).

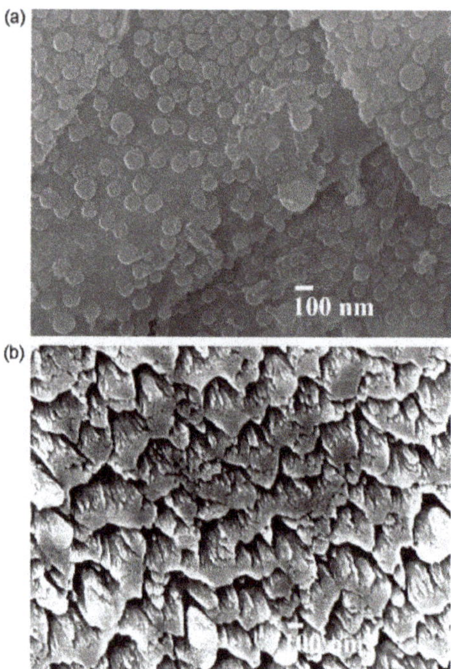

Figure 13. FESEM images of n-C films (thickness 550 nm) grown under CH_4 (57.5 Pa). Nanoparticles with 80 nm size on the surfaces of carbon nanosheets can be observed (**a**). n-C films with 210 nm thickness grown in vacuum, nanohoneycomb structures are visible in (**b**). Reproduced with permission from [4], published by Elsevier, 2013.

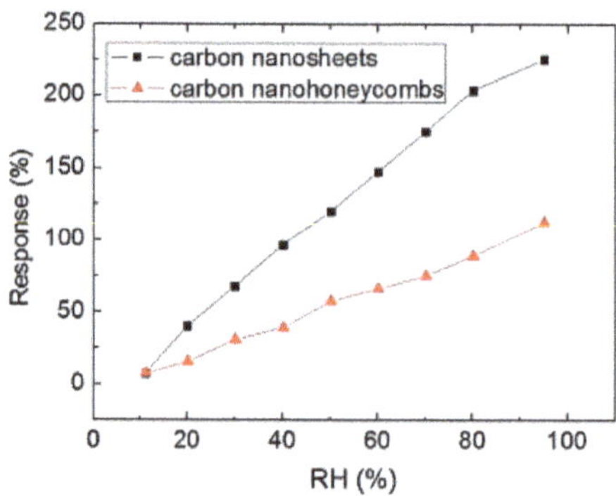

Figure 14. Humidity sensing properties of carbon nanosheets and nanohoneycombs-based sensors. Reproduced with permission from [4], published by Elsevier, 2013.

The maximum hysteresis was about 3.57% and 6.83% under 50 RH% for the carbon nanosheets and nanohoneycombs-based films, respectively. The response time when the relative humidity level changed from 11 to 40 RH% was about 30 s, while the recovery time from 40 to 11 RH% was about 90 s for the carbon nanosheets film.

Amorphous carbon (a-C) capacitive films were deposited on n-silicon from a cold pressed graphite powder target by direct current magnetron sputtering at room temperature in [57]. The results showed that when the RH level changed from 11% to 95%, the junction capacitance showed an increase of ~200% from ~1000 pF at RH = 11% to ~3200 pF at RH = 95%. When the RH amount was enhanced from 33 to 95%, the capacitance of the junction increased by 68% at a 1 kHz frequency. The response and recovery times were about 3 and 4 minutes, respectively.

In [58], hydrogenated amorphous carbon films (a-C:H) were made over printed circuit boards by physical deposition in vapor phase evaporation (PAVD) and plasma pulsed nitriding from graphite rods irradiated with an electron-beam. The aim of the latter treatment was to dilute the sample, to increase its resistance, and to modify the microstructure of the surface by interaction with the plasma. The sensors showed a relative impedance response equal to 0.99 when the RH value changed from 30% to 90%, when alimented with an AC tension of 1 V at 1 kHz.

N-doped carbon spheres (N-CSs)-poly (vinyl alcohol) (PVA) composites were deposited onto an epoxy resin/fiber glass board with gold interdigitated electrodes in [59]. First, acetylene gas (300 mL/min) was bubbled through acetonitrile heated at 80°C as a nitrogen source to produce N-CSs and then, was flowed in an argon atmosphere at 900°C for 10 minutes through a quartz tube [60]. The obtained powder was collected and purified by Soxhlet extraction from toluene for about 48 h, before being dried under vacuum for 48 h at room temperature [60]. The N-doped carbon spheres had a diameter of 181 ± 13 nm. The composite sensors were prepared by dispersing N-CSs in PVA matrix (molecular weight of 1.3×10^5 amu). The dispersion was prepared by adding 6 mg·mL^{-1} hexadecyltrimethylammonium bromide (CTAB) and 4 mg·mL^{-1} of N-CSs to water [59]. This solution was ultrasonicated in an ultrasonic bath at room temperature for 30 min, followed by ultrasonication for 60 min at 0 °C. This dispersion was then left for four days at 10 °C, to let the excess surfactant precipitate in the form of hydrated crystals. After four days, 50% of the total volume of the supernatant free of hydrated crystals was removed and mixed with 6 mg·mL^{-1} of PVA in water. The proportion was chosen to obtain a final N-CSs weight content in PVA of 87%. Finally, sensors were prepared from the N-CSs/PVA—CTAB by dropping a 20, 30, and 35 µL dispersion on top of two types of gold interdigitated electrodes (IDEs) with a gap of 0.1 and 0.3 mm. In the RH range from 9% to 97%, the sensitivity of the sensor with a 0.1 mm gap was 17.1 1/RH%, while the sensitivity of the sensor with a 0.3 mm gap was 8.7 1/RH%. The response and recovery times were equal to 19 s and 178 s, respectively, for the IDEs with a 0.1 mm gap. They were then equal to 8 s and 142 s, respectively, when the IDEs had a gap of 0.3 mm.

An interesting self-powered humidity sensor is described in [61], where an electrical potential of tens of millivolts can be induced by the adsorption of water molecules on a piece of porous carbon film (PCF) with two sides functionalized with different functional groups deposited onto an alumina substrate. PCF was functionalized by simply exposing one half of the PCF to air plasma for one minute while covering the other half with a polyethylene film. This treatment allocates the plasma treated region a higher content of oxygen-containing functional groups, mainly –COOH, resulting in an improved hydrophilicity. The results showed that the water molecules adsorbed on the porous carbon film facilitate the release of protons from –COOH groups. Then, these protons can be freely transported through a bridge formed from adsorbed water molecules in porous carbon, while the –COO$^-$ groups on the carbon surface remain negative. The produced voltage is due to two consecutive processes: proton release and proton transport. Both steps are restricted by the amount of adsorbed water molecules. Ab initio molecular dynamics modeling revealed that a small amount of water molecules could facilitate the proton release from the –COOH groups, forming an effective water bridge for proton transport. These results explain why the voltage remains at zero at low humidity values, starts to rise at 94 RH%, and reaches a maximum when the humidity is close to saturation. Thus, a possible application for these sensors could be as dew point sensors.

2.5. Carbon Black and Biochar

Carbon black-filled polymers have been studied with the aim of developing sensitive and cheap sensors. Carbon black (CB) impairs electrical conductivity to the film which, at high RH levels, swells and increases the resistance of the film. In [62], films of CB and the $Fe[(Htrz)_2(trz)]BF_4$ complex (where *Htrz* = 1,2,4-1H-triazole and *trz* is the deprotonated triazole ligand) were screen-printed. The nanoparticles were realized by a reverse micelle method and had an average diameter of 45 nm. They were dispersed in a commercial carbon black-epoxy ink (ElectroScience Laboratory, ESL RS12113) and screen-printed on an alumina substrate. While sensors based on carbon-black only tend to reach saturation at 50 RH% levels, this result was not observed with the sensors based on carbon-black and Fe(II) compound hybrids. In addition, the latter showed a six-time higher sensitivity compared to the pristine ones.

In [63], carbon quantum dots (CQDs) films were obtained from two identical graphite rods (99.99%, Alfa Aesar, 13 cm in length and 0.6 cm in diameter) vertically inserted in an ultrapure water solution 3 cm above the liquid surface. Both rods were respectively the cathode and the anode of an electrolytic cell and were separated by 7.5 cm. Then, a static potential of 50 V was imposed on the two electrodes by a direct current power supply for 10 days under thorough stirring. Finally, the large graphite particles were removed from the CQDs solution by high speed centrifugation (22,000 rpm). The concentration of the CQDs in aqueous solution was determined to be about 1.2 mg·mL^{-1}. Finally, the CQDs films were manufactured by dropping 2 mL of CQDs aqueous solution on a piece of glass sheet (1 cm × 2 cm), which was then dried at 25 °C. The films were 3 μm in thickness.

In the same work, the authors also used 500 mg of candle soot, which was mixed with 100 mL of 5 M HNO_3. As a result, candle soot nanoparticles were functionalized with oxygen-containing groups by refluxing the fresh candle soot in nitric acid at 100 °C for 24 h. After cooling down to room temperature, the fluorescent carbon nanoparticles were collected by centrifugation (10,000 rpm for 10 min). The films based on candle soot were prepared in the same way as the CQDs.

High-resolution TEM observations evidenced that the resulting CQDs had a uniform particle size distribution ranging from 4 nm to 8 nm.

At low RH levels, only a few high-energy electrons can accomplish the leap migration process between neighboring CQDs. Thus, the conductivity of the solid CQDs films in a dry environment is very low. However, at high RH values, the oxygen-contained groups on the surface of the neighboring CQDs can form hydrogen bonds with water molecules, significantly favoring the migration of electrons and then, increasing the conductivity. Because electron migration between adjacent CQDs through hydrogen bonds is much easier than that through the jumping mode, the conductivity of the CQDs film sharply increases with the humidity of the surrounding environment.

The CQDs films proved to be highly sensitive to water vapor and showed a linear dependence of the conductivity with the relative humidity (Figure 15). The oxygen-containing groups on the surface were considered to be of paramount importance for humidity detection by CQDs. The conductivity of the CQDs film in a dry environment (below 7 RH%) was close to zero, whereas when the CQDs film was placed under 95 RH%, it showed a conductivity increase up to ~35 $\Omega^{-1} \cdot m^{-1}$. In addition, the response and recovery times were measured between 7 and 43 RH% at room temperature and were equal to ~25 s and ~60 s, respectively. Beside the fast response time, the CQDs-based humidity sensor also showed an excellent stability, even for a long-term test [63].

A polyvinylpyrrolidone-grafted carbon black (CB-PVP) composite was prepared by a free radical polymerization reaction [64]. Hydrophilic polyvinylpyrrolidone (PVP) was grafted onto the CB particles (BP-2000 CB from Cabot Limited Corporation, USA) to obtain a water-dispersible conductive material. BP-2000 CB material was selected as the conductive material because of its high specific surface area and small particle size (15–20 nm in diameter). The grafting of PVP onto CB was performed as follows: 1.3 g of CB, 5.4 g of N-vinypyrrolidone (NVP), 0.03 g of 2,2′-azobisisobutyronitrile (AIBN), and 18 mL of tetrahydrofuran (THF) were added to a 100 mL flask and heated at 60 °C for 6 h. Then, the reaction mixture was centrifuged several times to remove the ungrafted polymer and AIBN. The obtained solids were further purified by Soxhlet extraction with THF for 48 h and dried under vacuum at 40 °C. The content of PVP in CB-PVP particles was about 5.3 wt%.

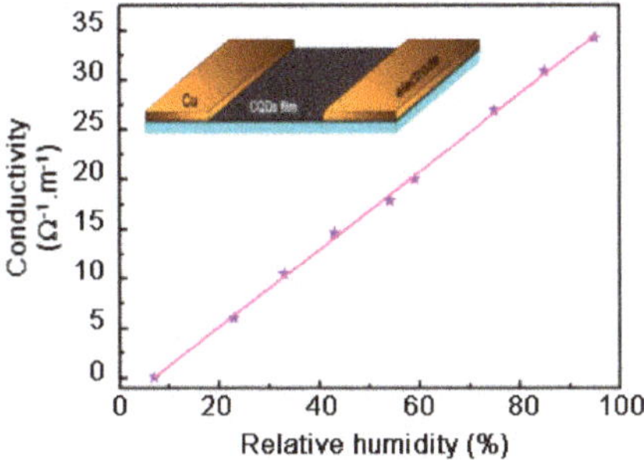

Figure 15. The conductivity of the CQDs film as a function of the relative humidity at room temperature. Reproduced with permission from [63], published by Elsevier, 2013.

For the sensors' preparation, an appropriate amount of CB-PVP and PVA was added into a conical flask with distilled water and then heated at 75 °C and kept under stirring for 2 h to form a stable suspension. After cooling to room temperature, the suspensions (with 3, 5, 6, 7, and 9 wt% of CB-PVP in PVA) were cast onto clean ceramic substrates (10 mm × 8 mm × 0.8 mm) with a screen-printed interdigitated array of Ag–Pd electrodes. The thickness of the film was about 30 µm. A six-hour thermal annealing treatment at 100 °C was performed to enhance the stability of CB-PVP/PVA composite films. The resistance values of composite films were small under low RH levels and increased notably at a certain RH. When the RH value is low, the PVA film adsorbs a small quantity of water, and the space between the CB particles is small enough to form conductive channels. With the increase of RH, the polymers swell and the space between CB particles increases slowly, as does the resistance of composite films. When the RH level is above 85%, rapid swelling of the composite film leads to a sharp decrease of CB concentration, so the resistances of composite membranes increase sharply (Figure 16).

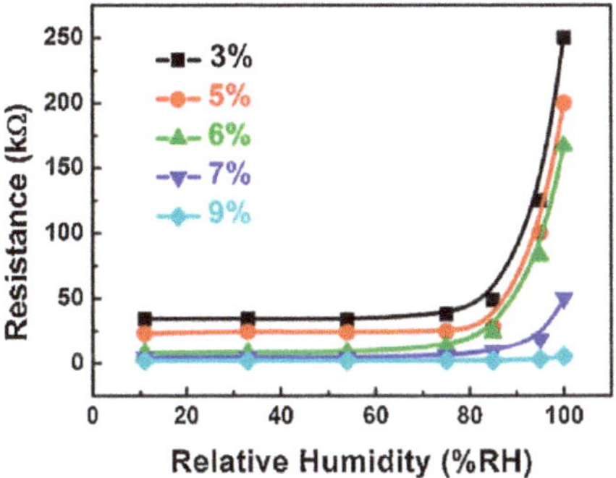

Figure 16. The resistances of CB-PVP/PVA composite films with different contents of CB as a function of RH%. Reproduced with permission from [64], published by Elsevier, 2014.

To the best of our knowledge, only a few studies are dealing with the use of biochar materials for humidity monitoring at room temperature [21,65,66].

In Afify et al. [65], pyrolyzed bamboo screen-printed thick-films were used as novel humidity sensors. Split bamboo culms were cut into small pieces (~1–5 mm) and cleaned with distilled water prior to drying in an oven at 105 ± 5 °C for 48 h. The dried bamboo pieces were then pyrolyzed in a quartz reactor at 800 °C for 1 h under an inert atmosphere of argon gas. The carbonized bamboo (CB) pieces were first manually ground in an agate mortar with an agate pestle, and then by attrition milling in distilled water for 1 h. Subsequently, the powder was dispersed in an organic solvent (ethylene glycol monobutyal ether, Emflow), which provides the appropriate rheological properties to the paste. Once screen-printed onto alumina substrates with platinum interdigitated electrodes, the sensors were dried in air at room temperature and heat-treated at 300 °C in air for 1 h, to remove organic residues from the solvent. CB sensors showed a significant response towards RH at room temperature with a cut-off RH at about 10%. The sensors' resistance started from about 937 kOhm at 0% RH and decreased to 89 kOhm at 95% RH for CB, while the maximum hysteresis was 28.9% at about 33 RH%. The response and the recovery times were quite fast (about 2 min), with recovery times always being shorter with respect to response times. Short response times may be due to the porosity of the film, which gives great accessibility for water molecules' adsorption. Fast recovery times indicate that the process of physisorption was probably the major reason for binding water molecules to the sensing materials. However, the adhesion onto alumina substrates of these sensing films was minimal and they could be easily damaged when testing. Thus, their use in commercial products was precluded.

In Ziegler et al. [21], SWP700 (pyrolyzed mixed softwood pellets) and OSR700 (oil seed rape) commercial biochars (from UK Biochar Research Centre) were used as humidity sensitive materials. The sensors, screen-printed onto alumina substrates with platinum interdigitated electrodes, exhibited a *p*-type behavior at low humidity values (between 5–25 RH% for SWP700 and in the range 5–40 RH% for OSR sensor). Specifically, for the OSR700 film, the initial impedance under dry air was equal to 163.9 kΩ, and the final impedance under 99 RH% decreased to 9 kΩ, leading to a sensor response equal to 94.5%, together with a maximum hysteresis of 52% under 59 of RH% (Figure 17). For the SWP 700 film, due to the highest surface area and porosity compared to the previous biochar, the impedance dropped from 9.5 MΩ in a dry condition to 222 kΩ in a humid environment, with a sensor response of 97.6%. This material exhibited a lower hysteresis, with a maximum value for this parameter of 23% under 55 RH%. Using SWP700, the biochar most sensitive towards humidity, sensors with different amounts of PVP were produced (10 and 20 wt% with respect to biochar). The SWP700 sensor with 10 wt% PVP seemed to represent a satisfying compromise between sensitivity towards water vapor and adhesion of the film onto the substrate (Figure 17).

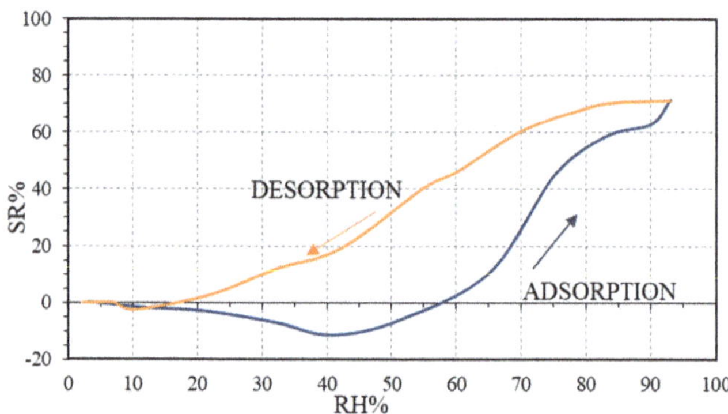

Figure 17. Sensor response (SR%) towards relative humidity for SWP700 biochar with 10 wt% of PVP. Reproduced with permission from [21], published by MDPI, 2017.

No interferences were detected towards 0.5 ppm of ozone, 100 ppm of methane, and 500 and 600 ppm (under dry and humid (40 RH%) air) of carbon dioxide. Only a slight increase in impedance of 2.6% was evidenced under 50 ppm of ammonia at room temperature.

Finally, in Jagdale et al. [66], waste brewed coffee powder (WBCP) was first washed with water, centrifuged, and then filtered. Afterward, the WBCP powder was then dried in an oven at 90°C for 10 h. The pyrolysis of the material was performed at 700 °C for 1 h in nitrogen atmosphere (120 mL/min) with 30 min dwells, respectively, at 250 °C and 400 °C. After the pyrolysis step, the material was manually ground, leading to coffee ground biochar, before being screen-printed onto alumina substrates with platinum interdigitated electrodes. The sensors behaved as *n*-type semiconductors (Figure 18), with the conductivity increasing with the humidity level: the sensor response SR% started at around 20 RH% and reached 51% under humid atmospheres (98% of RH). The impedance decreased from 25.2 MΩ in dry conditions to 12.3 MΩ, in humid conditions. The high initial impedance value was probably due to the limited thickness of the sensing film. The maximum hysteresis was equal to 15% under 74 RH%, confirming that the kinetics of desorption is comparable with respect to the adsorption one. The response and recovery times under 50% of RH were determined and were respectively equal to 4.5 and 1 min. Additionally, negligible interferences were detected towards carbon dioxide 500 ppm, ozone 200 ppb, nitrogen dioxide 200 ppb, and ammonia 50 ppm with this sensing material.

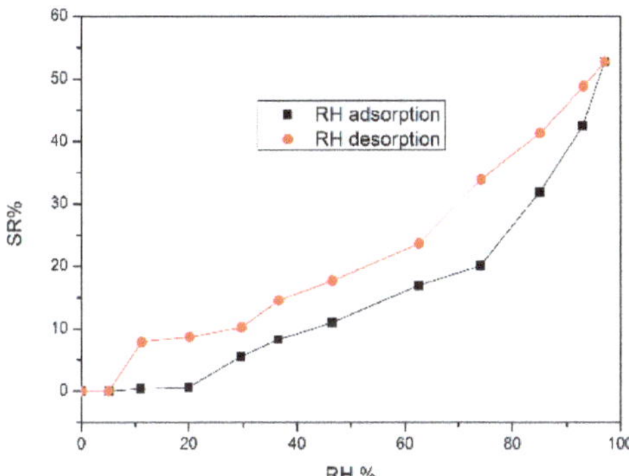

Figure 18. Sensor response (SR%) towards relative humidity for a coffee ground biochar sensor during adsorption and desorption cycles. Reproduced with permission from [66], published by MDPI, 2019.

The main features of the different sensors described in this paper are listed in Table 2.

Table 2. Main features of humidity sensors based on carbonaceous materials. n.d.–not determined.

Sample	Sensor Response	Response Time	Recovery Time	Ref.
MWCNTs in a PAA matrix in a ratio 1:4	930 Ω increase of the resistance when RH changed from 30% to 90%	680 s when RH increased from 50% to 90%	380 s when RH decreased from 90% to 50%	[5]
MWCNTs CVD grown on quartz substrates	~267% increase of resistance under 100 RH%	2–3 min	Few hours	[29]
3 wt% MWCNTs in a PI matrix	~10% increase of resistance when RH changed from 30% to 90%	< 5 s when RH increased from 30% to 90%	n.d.	[32]
0.4 wt% MWCNTs in a PI matrix	~40% increase of resistance when RH changed from 10% to 90%	n.d.	n.d.	[33]
9 wt% MWCNTs in a HEC matrix	~300% increase of resistance change ratio when RH changed from 23% to 80%	~1500 s when RH increased from 0% to 90%	n.d.	[35]
MWCNTs in a PVP matrix in a ratio 1:9	4000% current ratio at 94 RH%	15 s	1.8 s	[36]
SWCNTs/PVA filaments in a molar ratio 1:5	Resistance increase of 24 times	40 s	n.d.	[37]
MnO$_2$-coated CNT yarn	Resistance variation of 65% under 90% of RH	20 s	30 s	[38]
Defect graphene on alumina substrate with gold electrodes	Relative conductance ratio of 3.33 when RH changed from 3% to 30%	n.d.	~150 s when RH% decreased from 30% to 3%	[39]
Spin-coated GO films on PET substrates + laser treatment, 0.2 W	~20 kΩ increase of resistance when RH changed from 11% to 95%	3 s when RH% increased from 11% to 95%	10 s when RH% decreased from 95% to 11%	[42]
GO foam	~37% decrease of the current when RH increased from 0% to 85.9%	89 ms when RH% increased from 0% to 85.9%	189 ms when RH% decreased from 85.9% to 0%	[44]
rGO on PDMS substrate	~2750% increase of the capacitance when RH increased from 20% to 90%	n.d.	n.d.	[46]
GO on polyethylene naphthlate by by drop casting and spray coating methods	The sensor response was normalized	< 100 ms	< 100 ms	[47]
Single layer of graphene on SiO$_2$ by CVD	Relative resistance variation of 1.2% in the range 8–85 RH%	0.6 s	0.4 s	[48]
Tin dioxide/reduced graphene oxide (RGO) nanocomposite film	Capacitance from 246.53 pF at 11% of RH to 138,267 pF at 97%	102 s	Several s	[49]
Graphene/methyl-red composite	SR % 96.36 in terms of reistance and 2869500% in terms of capacitance	0.251 s	0.35 s	[50]
GO modified with poly (3,4-ethylenedioxythiophene) poly(styrenesulfonate) (PEDOT: PSS) and Ag colloids	4.97% under 97 RH%	31 s	72 s	[51]

Table 2. *Cont.*

Sample	Sensor Response	Response Time	Recovery Time	Ref.
Reduced graphene oxide and MoS2 hybrid composites were synthesized by hydrothermal method and drop-cast	21% under 50 RH%	30 s	253 s	[52]
rGo/MoS$_2$ composites by sonication	872.7% under 50 RH%	6.3 s	30.8 s	[53]
C nanofibers sprayed on a PI substrate	~16% increase of relative resistance ratio when RH increased from 5% to 100%	n.d.	n.d.	[56]
C nanosheets produced by physical vapor deposition	Resistance increase of 225% under 95 RH%	30 s when RH% increased from 11% to 40%	90 s when RH% decreased from 40% to 11%	[4]
Amorphous carbon film by DC magnetron sputtering	~200% increase of capacitance when RH increased from 11% to 95%	3 minutes when RH% increased from 33% to 95%	4 minutes when RH% decreased from 95% to 33%	[57]
Hydrogenated amorphous carbon (a-C:H) film	Resistance decrease of 97.3% under 80 RH%	n.d.	n.d.	[58]
N-doped carbon spheres (N-CSs- PVA) by drop coating (0.1 mm gap IDEs)	Conductance increased by 4 order of magnitude in th eRH range 9–97%	19 s	178 s	[59]
Screen-printed commercial composite ink (ESL RS12113) made of epoxy resin and carbon powder	Resistance increase of 4.8% under 80 RH%	n.d.	n.d.	[62]
Carbon quantum dots film made by electrochemical ablation of graphite	Resistivity decrease of 48% under 90 RH%	25 s when RH% increases from 7% to 43%	60 s when RH% decreases from 43% to 7%	[63]
Pyrolyzed bamboo	Resistance decrease of 91% under 95% RH	2 min	2 min	[65]
Pyrolyzed mixed softwood pellets	Impedance decrease of 97.7% under 97.5% RH	1 min	1 min	[21]
Oil seed rape	Impedance decrease of 94.5% under 99% RH	50 s	70 s	[21]
Coffee ground biochar	Impedance decrease of 51% under 98% RH	4.5 min	1 min	[66]

3. Interaction with Water Molecules

Water molecules can adsorb on the surface of carbon films due to a weak binding of water hydrogens to surface carbon atoms [4]. In addition, the sensing performances are strongly influenced by the presence of defects in carbon particles, which could create favorable adsorption sites for water molecules [4]. At low RH levels, a few water vapor molecules chemically adsorb on the carbon grain surfaces. Pati et al. [67] suggested a possible charge transfer between the adsorbate and the carbon film. On carbon nanotubes, H_2O molecule adsorption yields a charge transfer to CNTs of 0.033–0.035 electron per water molecule, as demonstrated by theoretical studies and experimental results [4]. These charge transfer effects determine an increase in the resistance of the CNTs-based gas sensors [68–71]. In fact, this probably leads to electrons transfer to the valence band and to an increase of the impedance as the conductivity of p-type semiconductors is determined by holes scattering through the material. Typically, carbon-based materials contain numerous oxygen derivatives on their surface. These functional groups can improve the carbon surface hydrophilicity and increase water vapor molecules adsorption [72]. When these adsorbed water molecules donate electrons to the valence band of the carbon material, the number of holes decreases and the separation in energy between the Fermi level and valence band increases [73,74], decreasing the conductivity of the p-type semiconductor. The higher the RH level, the more water molecules are adsorbed, and more electrons are transferred, lowering the holes' concentration and leading to an increase of the resistance value of samples by raising RH values [4]. An increase in sensitivity towards H_2O was also observed upon the functionalization of pristine CNTs with different molecules [75].

On the contrary, for higher RH levels, when more water molecules are adsorbed on the surface, a liquid-like multilayer of hydrogen-bonded water molecules will form at room temperature [76]. In addition, these physisorbed water molecules can condense into pores with a size in the range of 1–250 nm. Since the formation of clusters of H_2O and the hydration of H^+ into H_3O^+ are thermodynamically favored in liquid water, H^+ are the dominant charge carriers in the water adsorbed in the mesopores. As the amount of H^+ enhances when increasing the moisture content, H^+ can move freely in liquid water, leading to a decrease of grain surface impedance with heightening RH values [76].

According to Ref. [77], at high humidity amounts, the conduction is well-described by the Grotthuss mechanism. It might be that the proton mobility correlates with H-bond cleavage rather than with its formation, since it enhances when decreasing the H bond content in water, i.e., at higher temperatures and pressures. The proton mobility is an incoherent proton hopping phenomenon. The actual proton motion is much faster compared to the solvent reorganization. The rate-limiting step involves the cleavage of a single H having an activation energy of about 2.6 kCal/mol. During the elementary transfer step itself, the donor and acceptor H_2O with their solvation shells present equivalent structures. Since the rate-limiting step involves the cleavage of a H bond, but not one in the first solvation shell, H-bond cleavage in the second solvation shell is the rate-limiting step. This leads to isomerization of the $H_9O_4^+$ cation into $H_5O_2^+$ The proof of this phenomenon is more likely to come from computational chemistry than from any single experiment.

When increasing the concentration of the reducing gas to which the nanotubes are exposed, the Fermi level may shift and the CNTs could turn from p-type into n-type materials: for instance, in [71], the response of the composite was p-type at low concentrations of NH_3 under low RH values and switched to n-type in higher humidity conditions. This behavior was attributed to a hole compensation effect by water molecules [71].

Zhang et al. showed that the resistance reduction of the single-walled carbon nanotube (SWCNTs) networks is caused by the SWCNTs, while the resistance increase is due to the inter-tube junctions [63]. Then, the overall resistive humidity response of the SWCNT network is the result of a competition between the SWCNT resistance changes and intertube junction resistances' variations. In fact, intertube junctions play a crucial role in limiting the SWCNT network conductance due to the suppression of intertube carrier hopping caused by water molecules. Covalent modifications to the SWCNTs were shown to be able to raise the humidity detection dynamic ranges by increasing the resistance of the SWCNTs film. These modifications also enhanced the sensitivities of as-grown SWCNT networks to humidity by increasing their hydrophilicity.

4. Conclusions and Perspectives

Flexible sensors with a high sensitivity, excellent flexibility, acceptable stretchability, and good stability, can be mounted on the human body or clothing to provide the long-term detection of human activities and physiological information [12]. Thus, future research aims at developing more sensitive, selective, and stable humidity sensors able to withstand harsh environments in a wide range of temperatures and in dynamic operating modes [1]. Though important progress in flexible sensors based on carbonaceous materials has been achieved, some key issues, including the preparation processes of carbon materials, the fabrication processes, and the performance of flexible sensors, still need to be further investigated for practical applications [12]. Mass and cost-effective production of carbon materials with a high quality is a fundamental prerequisite for these targeted applications. CNT powders and GO sheets have been embedded in mass production, but the manufacturing processes can also introduce many defects, significantly affecting their humidity sensing properties. A monolayer and few-layers of graphene with a high quality are mandatory for fabricating high-performance devices. However, the current high-cost preparation methods limit their use in view of industrial applications. In addition, the properties of macroscopic assemblies are clearly lower than those of individual CNTs or graphene [12].

Moreover, bio-materials (cotton fibers) with macro-scale architectures (woven structures) are available. After a simple carbonization treatment (pyrolysis), they can be used as active materials of high-performance flexible sensors [14,15]. As shown, from the pyrolysis of bio-materials, it is possible to obtain additionally encouraging carbon sensing materials for fabricating high-performance flexible sensors.

Gas sensors are mainly produced in two ways: solution casting and CVD deposition. Solution casting processes include simple casting, dropping-drying, spin-coating, and dip-coating. The suspensions for film casting on certain substrates can be simply produced by direct dispersion of the nanocarbon components in solvents. These simple and cost-effective methods can be easily scaled-up for industrial applications [1]. In CVD methods, nanocarbon (CNTs and graphene sheets) layers can be directly grown on substrates. Solution casting methods are more flexible when preparing composite materials, while thin-films manufactured by CVD are more uniform and homogeneous [1]. It is expected that composites of nanocarbons and nanoparticles of polymers and carbonaceous materials should be the main topic of this research direction, due to the possibility to obtain easily flexible films.

Finally, the integration of flexible sensors with energy conversion and storage devices is required in view of the practical applications of wearable electronics. However, these self-powered sensors still require external power for operating the measurement systems and reading the signals [12]. Therefore, the integration of self-powered components, energy storage devices (such as supercapacitors or batteries), and flexible sensors is of paramount importance for manufacturing wearable systems [12].

Author Contributions: D.Z. and J.-M.T. defined the structure of the manuscript and J.-M.T. wrote it. All the authors discussed the content of the paper and contributed to its revision.

Funding: This research received no external funding.

Conflicts of Interest: The authors declare no conflict of interest.

References

1. Blank, T.A.; Eksperiandova, L.P.; Belikov, K.N. Recent trends of ceramic humidity sensors development: A review. *Sens. Actuators B Chem.* **2016**, *228*, 416–442. [CrossRef]
2. Schroeder, V.; Savagatrup, S.; He, M.; Lin, S.; Swager, T.M. Carbon nanotube chemical sensors. *Chem. Rev.* **2019**, *119*, 599–663. [CrossRef]

3. Staerz, A.; Berthold, C.; Russ, T.; Wicker, S.; Weimar, U.; Barsan, N. The oxidizing effect of humidity on WO_3 based sensors. *Sens. Actuators B Chem.* **2016**, *237*, 54–58. [CrossRef]

4. Chu, J.; Peng, X.; Feng, P.; Sheng, V.; Zhang, J. Study of humidity sensors based on nanostructured carbon films produced by physical vapor deposition. *Sens. Actuators B Chem.* **2013**, *178*, 508–513. [CrossRef]

5. Lee, J.; Cho, D.; Jeong, Y. A resistive-type sensor based on flexible multi-walled carbon nanotubes and polyacrylic acid composite films. *Solid-State Electron.* **2013**, *87*, 80–84. [CrossRef]

6. Yeo, T.L.; Sun, T.; Grattan, K.T.V. Fibre-optic sensor technologies for humidity and moisture measurement. *Sens. Actuators A Phys.* **2018**, *144*, 280–295. [CrossRef]

7. Korotcenkov, G.; Cho, B.K. Metal oxide composites in conductometric gas sensors: Achievements and challenges. *Sens. Actuators B Chem.* **2017**, *244*, 182–210. [CrossRef]

8. Bârsan, N.; Weimar, U. Conduction model of metal oxide gas sensors. *J. Electroceram.* **2001**, *7*, 143–167. [CrossRef]

9. Gőpel, W.; Schierbaum, K.D. SnO_2 sensors: Current status and future prospects. *Sens. Actuators B Chem.* **1995**, *26–27*, 1–12. [CrossRef]

10. Xu, C.; Tamaki, J.; Miura, N.; Yamazoe, N. Grain size effects on gas sensitivity of porous SnO_2-based elements. *Sens. Actuators B Chem.* **1991**, *3*, 147–155. [CrossRef]

11. Bârsan, N.; Schweizer-Berberich, M.; Göpel, W. Fundamental and practical aspects in the design of nanoscaled SnO_2 gas sensors: A status report. *Fresenius J. Anal. Chem.* **1999**, *365*, 287–304. [CrossRef]

12. Jian, M.; Wang, C.; Wang, Q.; Wang, H.; Xia, K.; Yin, Z.; Zhang, M.; Liang, X.; Zhang, Y. Advanced carbon materials for flexible and wearable sensors. *Sci. China Mater.* **2017**, *60*, 1026–1062. [CrossRef]

13. Korotcenkov, G.; Brinzari, V.; Ham, M.-H. Materials acceptable for gas sensor design: Advantages and limitations. *Key Eng. Mater.* **2018**, *780*, 80–89. [CrossRef]

14. Wang, C.; Li, X.; Gao, E.; Jian, M.; Xia, K.; Wang, Q.; Xu, Z.; Ren, T.; Zhang, Y. Carbonized Silk Fabric for Ultrastretchable, Highly Sensitive, and Wearable Strain Sensors. *Adv. Mater.* **2016**, *28*, 6640–6664. [CrossRef]

15. Zhang, M.; Wang, C.; Wang, H.; Jian, M.; Hao, X.; Zhang, Y. Carbonized cotton fabric for high-performance wearable strain sensors. *Adv. Funct. Mater.* **2017**, *27*, 1604795. [CrossRef]

16. Nanda, S.; Dalai, A.K.; Berruti, F.; Kozinski, J.A. Biochar as an exceptional bio resource for energy, agronomy, carbon sequestration, activated carbon and specialty materials. *Waste Biomass Valor.* **2015**, *7*, 201–235. [CrossRef]

17. Available online: https://www.biochar.ac.uk/cms/i/user/standard_materials/21_SWP%20700-web.pdf (accessed on 19 February 2017).

18. Available online: https://www.biochar.ac.uk/cms/i/user/standard_materials/19_OSR_700-web.pdf (accessed on 19 February 2017).

19. Ahmad, S.; Khushnood, R.A.; Jagdale, P.; Tulliani, J.-M.; Ferro, G.A. High performance self-consolidating cementitious composites by using micro carbonized bamboo particles. *Mater. Des.* **2015**, *76*, 223–229. [CrossRef]

20. Chan, K.Y.; Van Zwieten, V.; Meszaros, I.; Downie, A.; Joseph, S. Agronomic values of green waste biochar as a soil amendment. *Aust. J. Soil Res.* **2007**, *45*, 629–634. [CrossRef]

21. Ziegler, D.; Palmero, P.; Giorcelli, M.; Tagliaferro, A.; Tulliani, J.-M. Biochars as innovative humidity sensing materials. *Chemosens* **2017**, *5*, 35. [CrossRef]

22. Packirisamy, M.; Stiharu, I.; Li, X.; Rinaldi, G. A polyimide based resistive humidity sensor. *Sens. Rev.* **2005**, *25*, 271–276. [CrossRef]

23. Ueda, M.; Nakamura, K.; Tanaka, K.; Kita, H.; Okamoto, K. Water-resistant humidity sensors based on sulfonated polyimides. *Sens. Actuators B Chem.* **2007**, *127*, 463–470. [CrossRef]

24. Sakai, Y.; Matsuguchi, M.; Yonesato, N. Humidity sensor based on alkali salts of poly (2-acrylamido-2-methylpropane sulfonic acid). *Electrochim. Acta* **2001**, *46*, 1509–1514. [CrossRef]

25. Su, P.-G.; Uen, C.-L. A resistive-type humidity sensor using composite films prepared from poly (2-acrylamido-2-methylpropane sulfonate) and dispersed organic silicon sol. *Talanta* **2005**, *66*, 1247–1253. [CrossRef]

26. Chen, Y.S.; Li, Y.; Yang, M.J. A fast response resistive thin film humidity sensor based on poly (4-vinylpyridine) and poly (glycidyl methacrylate). *J. Appl. Polym. Sci.* **2007**, *105*, 3470–3475. [CrossRef]

27. Varghese, O.K.; Kichambre, P.D.; Gong, D.; Ong, K.G.; Dickey, E.C.; Grimes, C.A. Gas sensing characteristics of multi-wall carbon nanotubes. *Sens. Actuators B Chem.* **2001**, *81*, 32–41. [CrossRef]

28. Fennell, J.F., Jr.; Liu, S.F.; Azzarelli, J.M.; Weis, J.G.; Rochat, S.; Mirica, K.A.; Ravnsbæk, J.B.; Swager, T.M. Nanowire chemical/biological sensors: Status and a roadmap for the future. *Angew. Chem. Int. Ed.* **2016**, *55*, 1266–1281. [CrossRef]

29. Charlier, J.-C.; Issi, J.-P. Electronic structure and quantum transport in carbon nanotubes. *Appl. Phys. A Mater. Sci. Eng.* **1998**, *67*, 79–87. [CrossRef]

30. Langer, L.; Stockman, L.; Heremans, J.P.; Bayot, V.; Olk, C.H.; Vanhaesendonck, C.; Bruynseraede, Y.; Issi, J.P. Electrical-resistance of a carbon nanotube bundle. *J. Mater. Res.* **1994**, *9*, 927–932. [CrossRef]

31. Kaneto, K.; Tsuruta, M.; Sakai, G.; Cho, W.Y.; Ando, Y. Electrical conductivities of multi-wall carbon nano tubes. *Synth. Met.* **1999**, *103*, 2543–2546. [CrossRef]

32. Tang, Q.Y.; Chan, Y.C.; Zhang, K. Fast response resistive humidity sensitivity of polyimide/multiwall carbon nanotube composite films. *Sens. Actuators B Chem.* **2011**, *152*, 99–106. [CrossRef]

33. Yoo, K.P.; Lima, L.-T.; Min, N.-K.; Lee, M.J.; Lee, C.J.; Park, C.-W. Novel resistive-type humidity sensor based on multiwall carbon nanotube/polyimide composite films. *Sens. Actuators B Chem.* **2010**, *145*, 120–125. [CrossRef]

34. Melcher, J.; Daben, Y.; Ark, G. Dielectric effects of moisture in polyimide. *IEEE Trans. Electr. Insul.* **2002**, *24*, 31–38. [CrossRef]

35. Ma, X.; Ning, H.; Hu, N.; Liu, Y.; Zhang, J.; Xu, C.; Wu, L. Highly sensitive humidity sensors made from composites of HEC filled by carbon nanofillers. *Mater. Technol.* **2015**, *30*, 134–139. [CrossRef]

36. Pan, X.; Xue, Q.; Zhang, J.; Guo, Q.; Jin, Y.; Lu, W.; Li, X.; Ling, C. Effective enhancement of humidity sensing characteristics of novel thermally treated MWCNTs/Polyvinylpyrrolidone film caused by interfacial effect. *Adv. Mater. Interfaces* **2016**, *3*, 1–7. [CrossRef]

37. Zhou, G.; Byun, J.H.; Oh, Y.; Jung, B.-M.; Cha, H.-J.; Seong, D.G.; Um, M.K.; Hyun, S.; Chou, T.W. High sensitive wearable textile-based humidity sensor made of high-strength, single-walled carbon nanotube (SWCNT)/Poly(Vinyl Alcohol) (PVA) filaments. *ACS Appl. Mater. Interfaces* **2017**, *9*, 4788–4797. [CrossRef]

38. Jung, D.; Kim, J.; Lee, G.S. Enhanced humidity-sensing response of metal oxide coated carbon nanotube. *Sens. Actuators A Phys.* **2015**, *223*, 11–17. [CrossRef]

39. Huang, Q.; Zeng, D.; Tian, S.; Xie, C. Synthesis of defect graphene and its application for room temperature humidity sensing. *Mater. Lett.* **2012**, *83*, 76–79. [CrossRef]

40. Karim, M.R.; Hatakeyama, K.; Matsui, T.; Takehira, H.; Taniguchi, T.; Koinuma, M.; Matsumoto, Y.; Akutagawa, T.; Nakamura, T.; Noro, S.; et al. Graphene oxide nanosheet with high proton conductivity. *J. Am. Chem. Soc.* **2013**, *135*, 8097–8100. [CrossRef]

41. Hatakeyama, K.; Karim, M.R.; Ogata, C.; Tateishi, H.; Funatsu, A.; Taniguchi, T.; Koinuma, M.; Hayami, S.; Matsumoto, Y. Proton conductivities of graphene oxide nanosheets: Single, multilayer, and modified nanosheets. *Angew. Chem. Int.* **2014**, *53*, 6997–7000. [CrossRef]

42. Guo, L.; Jiang, H.-B.; Shao, R.-Q.; Zhang, Y.-L.; Xie, S.-Q.; Wang, J.-N.; Li, X.-B.; Jiang, F.; Chen, Q.-D.; Zhang, T.; et al. Two-beam-laser interference mediated reduction, patterning and nanostructuring of graphene oxide for the production of a flexible humidity sensing device. *Carbon* **2012**, *50*, 1667–1673. [CrossRef]

43. Toda, K.; Furue, R.; Hayami, S. Recent progress in applications of graphene oxide for gas sensing: A review. *Anal. Chim. Acta* **2015**, *878*, 43–53. [CrossRef]

44. Yu, Y.; Zhang, Y.; Jin, L.; Chen, L.; Li, Y.; Li, Q.; Cao, M.; Che, Y.; Yang, J.; Yao, J. A fast response-recovery 3d graphene foam humidity sensor for user interaction. *Sensors* **2018**, *18*, 4337. [CrossRef]

45. Marcano, D.C.; Kosynkin, D.V.; Berlin, J.M.; Sinitskii, A.; Sun, Z.; Slesarev, A.; Alemany, L.B.; Lu, W.; Tour, J.M. Improved synthesis of graphene oxide. *ACS Nano* **2010**, *24*, 4806–4814. [CrossRef]

46. Ho, D.H.; Sun, Q.; Kim, S.Q.; Han, J.T.; Kim, D.H.; Cho, J.H. Stretchable and multimodal all graphene electronic skin. *Adv. Mater.* **2016**, *28*, 2601–2608. [CrossRef]

47. Borini, S.; White, R.; Astly, M.; White, R.; Wei, D.; Haque, S.; Spigone, E.; Harris, N. Ultrafast Graphene Oxide Humidity Sensors. *ACS Nano* **2013**, *12*, 11166–11173. [CrossRef]

48. Smith, A.D.; Elgammal, K.; Niklaus, F.; Delin, A.; Fischer, A.C.; Vaziri, S.; Forsberg, F.; Råsander, M.; Hugosson, H.; Bergqvist, L.; et al. Resistive graphene humidity sensors with rapid and direct electrical readout. *Nanoscale* **2015**, *7*, 19099–19109. [CrossRef]

49. Zhang, D.; Chang, H.; Li, P.; Liu, R.; Xue, Q. Fabrication and characterization of an ultrasensitive humidity sensor based on metal oxide/graphene hybrid nanocomposite. *Sens. Actuators B Chem.* **2016**, *225*, 233–240. [CrossRef]

50. Ali, S.; Hassan, A.; Hassan, G.; Bae, J.; Lee, C.H. All-printed humidity sensor based on graphene/methyl-red composite with high sensitivity. *Carbon* **2016**, *105*, 23–32. [CrossRef]

51. Pang, Y.; Jian, J.; Tu, T.; Yang, Z.; Ling, J.; Li, Y.; Wang, X.; Qiao, Y.; Tian, Y.Y.; Ren, T.L. Biosensors and Bioelectronics Wearable humidity sensor based on porous graphene network for respiration monitoring. *Biosens. Bioelectron.* **2018**, *116*, 123–129. [CrossRef]

52. Park, S.Y.; Lee, J.E.; Kim, Y.H.; Kim, J.J.; Shim, Y.-S.; Kim, S.Y.; Lee., M.H.; Jang, H.W. Room temperature humidity sensors based on rGO/MoS2 hybrid composites synthesized by hydrothermal method. *Sens. Actuators B Chem.* **2018**, *258*, 775–782. [CrossRef]

53. Park, S.Y.; Kim, Y.H.; Lee, S.Y.; Sohn, W.; Lee, J.E.; Kim, D.H.; Shim, Y.-S.; Kwon, K.C.; Choi, K.S.; Yoo, H.J.; et al. Highly selective and sensitive chemoresistive humidity sensors based on rGO/MoS2 van der Waals composites. *J. Mater. Chem. A* **2018**, *6*, 5016–5024. [CrossRef]

54. Xiao, Z.; Kong, L.B.; Ruan, S.; Li, X.; Yu, S.; Li, X.; Jiang, Y.; Yao, Z.; Ye, S.; Wang, C.; et al. Recent development in nanocarbon materials for gas sensor applications. *Sens. Actuators B Chem.* **2018**, *274*, 235–267. [CrossRef]

55. Weisenberger, M.; Martin-Gullon, I.; Vera-Agullo, J.; Varela-Rizo, H.; Merino, C.; Andrews, R.; Qian, D.; Rantell, T. The effect of graphitization temperature on the structure of helical-ribbon carbon nanofibers. *Carbon* **2009**, *47*, 2211–2218. [CrossRef]

56. Monereo, O.; Claramunt, S.; de Marigorta, M.M.; Boix, M.; Leghrib, R.; Prades, J.D.; Cornet, A.; Merino, P.; Merino, C.; Cirera, A. Flexible sensor based on carbon nanofibers with multifunctional sensing features. *Talanta* **2013**, *107*, 239–247. [CrossRef]

57. Chen, H.-J.; Xue, Q.-Z.; Ma, M.; Zhou, X.-Y. Capacitive humidity sensor based on amorphous carbon film/n-Si heterojunctions. *Sens. Actuators B Chem.* **2010**, *150*, 487–489. [CrossRef]

58. Epeloa, J.; Repetto, C.E.; Gómez, B.J.; Nachez, L.; Dobry, A. Resistivity humidity sensors based on hydrogenated amorphous carbon films. *Mater. Res. Express* **2019**, *6*, 1–9. [CrossRef]

59. Cunha, B.B.; Greenshields, M.W.C.C.; Mamo, M.A.; Coville, N.J.; Hümmelgen, I.A. A surfactant dispersed N-doped carbon sphere-poly (vinyl alcohol) composite as relative humidity sensor. *J. Mater. Sci. Mater. Electron.* **2015**, *26*, 4198–4201. [CrossRef]

60. Mamo, M.A.; Machado, W.S.; Coville, N.J.; Hümmelgen, I.A. A comparative study on hydrostatic pressure response of sensors based on N-doped, B-doped and undoped carbon-sphere poly(vinyl alcohol) composites. *J. Mater. Sci. Mater. Electron.* **2012**, *23*, 1332–1337. [CrossRef]

61. Liu, K.; Yang, P.; Li, S.; Li, J.; Ding, T.; Xue, G.; Chen, Q.; Feng, G.; Zhou, J. Induced Potential in Porous Carbon Films through Water Vapor Absorption. *Angew. Chem. Int. Ed.* **2016**, *55*, 8003–8007. [CrossRef]

62. Llobet, E.; Barberà-Brunet, R.; Etrillard, C.; Létard, J.F.; Debéda, H. Humidity Sensing Properties of Screen-printed Carbon-black an Fe(II) Spin Crossover Compound Hybrid Films. *Procedia Eng.* **2014**, *87*, 132–135. [CrossRef]

63. Zhang, X.; Ming, H.; Liu, R.; Han, X.; Kang, Z.; Liu, Y.; Zhang, Y. Highly sensitive humidity sensing properties of carbon quantum dots films. *Mater. Res. Bull.* **2013**, *48*, 790–794. [CrossRef]

64. Jiang, K.; Fei, T.; Jiang, F.; Wang, G.; Zhang, T. A dew sensor based on modified carbon black and polyvinyl alcohol composites. *Sens. Actuators B Chem.* **2014**, *192*, 658–663. [CrossRef]

65. Afify, A.S.; Ahmad, S.; Khushnood, R.A.; Jagdale, P.; Tulliani, J.-M. Elaboration and characterization of novel humidity sensor based on micro-carbonized bamboo particles. *Sens. Actuators B Chem.* **2017**, *239*, 1251–1256. [CrossRef]

66. Jagdale, P.; Ziegler, D.; Rovere, M.; Tulliani, J.-M.; Tagliaferro, A. Waste Coffee Ground Biochar: A Material for Humidity Sensors. *Sensors* **2019**, *19*, 801. [CrossRef]

67. Pati, R.; Zhang, Y.; Nayak, S.K.; Ajayan, P.M. Effect of H2O adsorption on electron transport in a carbon nanotube. *Appl. Phys. Lett.* **2002**, *81*, 2638–2640. [CrossRef]

68. Cao, C.L.; Hu, C.G.; Fang, L.; Wang, S.X.; Tian, Y.S.; Pan, C.Y. Humidity sensor based on multi-walled carbon nanotube thin films. *J. Nanomater.* **2011**, *5*, 1–5. [CrossRef]

69. Steven, E.; Saleh, W.R.; Lebedev, V.; Acquah, S.F.A.; Laukhin, V.; Alamo, R.G.; Brooks, J.S. Carbon nanotubes on a spider silk scaffold. *Nat. Commun.* **2013**, *4*, 1–8. [CrossRef]

70. Rigoni, F.; Drera, G.; Pagliara, S.; Goldoni, A.; Sangaletti, L. High sensitivity, moisture selective, ammonia gas sensors based on single-walled carbon nanotubes functionalized with indium tin oxide nanoparticles. *Carbon* **2014**, *80*, 356–363. [CrossRef]

71. Pokhrel, S.; Nagaraja, K.S. Electrical and humidity sensing properties of chromium (III) oxide-tungsten (VI) oxide composites. *Sens. Actuators B Chem.* **2003**, *92*, 144–150. [CrossRef]
72. Kong, J.; Franklin, N.R.; Zhou, C.W.; Chapline, M.G.; Peng, S.; Cho, K.; Dai, H.J. Nanotube molecular wires as chemical sensors. *Sci.* **2000**, *287*, 622–625. [CrossRef]
73. Zahab, A.; Spina, L.; Poncharal, P.; Marliere, C. Water-vapor effect on the electrical conductivity of a single-walled carbon nanotube mat. *Phys. Rev. B* **2000**, *62*, 10000–10003. [CrossRef]
74. Liu, L.; Ye, X.; Wu, K.; Zhou, Z.; Lee, D.; Cui, T. Humidity sensitivity of carbon nanotube and poly (dimethyldiallylammonium chloride) composite films. *IEEE Sens. J.* **2009**, *9*, 1308–1314. [CrossRef]
75. Traversa, E. Ceramic sensors for humidity detection. *Sens. Actuators B Chem.* **1995**, *23*, 135–156. [CrossRef]
76. Malard, L.M.; Pimenta, M.A.; Dresselhaus, G.; Dresselhaus, M.S. Raman spectroscopy in graphene. *Phys. Rep.* **2009**, *473*, 51–87. [CrossRef]
77. Agmon, N. The Grotthuss mechanism. *Chem. Phys. Lett.* **1995**, *244*, 456–462. [CrossRef]

 *micromachines*

Review

Rheological Issues in Carbon-Based Inks for Additive Manufacturing

Charlie O' Mahony *, Ehtsham Ul Haq, Christophe Silien and Syed A. M. Tofail *

Department of Physics, and Bernal Institute, University of Limerick, National Technological Park, Limerick V94 T9PX, Ireland; Ehtsham.U.Haq@ul.ie (E.U.H.); Christophe.Silien@ul.ie (C.S.)
* Correspondence: Charlie.OMahony@ul.ie (C.O.M.); Tofail.Syed@ul.ie (S.A.M.T.)

Received: 17 December 2018; Accepted: 27 January 2019; Published: 29 January 2019

Abstract: As the industry and commercial market move towards the optimization of printing and additive manufacturing, it becomes important to understand how to obtain the most from the materials while maintaining the ability to print complex geometries effectively. Combining such a manufacturing method with advanced carbon materials, such as Graphene, Carbon Nanotubes, and Carbon fibers, with their mechanical and conductive properties, delivers a cutting-edge combination of low-cost conductive products. Through the process of printing the effectiveness of these properties decreases. Thorough optimization is required to determine the idealized ink functional and flow properties to ensure maximum printability and functionalities offered by carbon nanoforms. The optimization of these properties then is limited by the printability. By determining the physical properties of printability and flow properties of the inks, calculated compromises can be made for the ink design. In this review we have discussed the connection between the rheology of carbon-based inks and the methodologies for maintaining the maximum pristine carbon material properties.

Keywords: carbon Inks; rheology; additive manufacturing; graphene; carbon nanotubes; printing

---

1. Introduction

Carbon (C) is a many allotrope material that can exist in various forms. These forms vary from diamond, an ultra-hard optically isotropic (directionally transparent or opaque) to graphite, a soft grey material [1]. This high variability in properties gives rise to a similarly high amount of utility. Carbon allotropes such as Graphene and Carbon nanotubes pose interesting properties in surface areas, tensile strength, low density, stretchability, thermal conductivity, current density, gas impermeability, and overall electrical properties [2–7]. An outline of the important properties and potential uses is seen in Figure 1.

Conventionally, carbon is solid in nature, arising from strong covalent bonds, which makes printing it alone impossible, with exceptions at extreme temperature and pressure. To utilize these materials in printing methods, the particles must be suspended in a fluid and used as a vehicle for printing. This mixture of solid and liquid forms is known as a colloidal system.

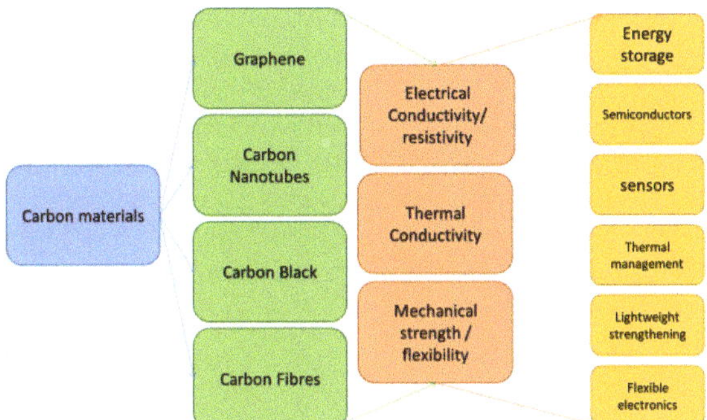

Figure 1. Outline of various forms of carbon (C) used for inks in additive manufacturing and conventional printing techniques, along with their beneficial properties and potential uses.

Colloidal systems are heterogeneous solutions with a dispersed phase uniformly distributed throughout the second dispersion phase. This system presents interesting rheological properties, which bears importance due to the use of these colloidal inks in additive manufacturing and conventional 2D printing processes. Printability determines the finished print quality in terms of porosity, surface finish, and resolution of geometry. The determination of ideal printability requires tailored rheological properties while maintaining the intended properties of the novel carbon particles [4–12].

Rheology is the study of the deformation and flow of matter, which holds a great amount of relevance to ink printing. Designing the flow of inks is of great importance, as the ease at which the ink flows and stiffens when shaped into the intentional design is key. Efforts to determine the idealized flow properties must be investigated to ensure maximum printability. A representation of this compromise of properties for printability is seen in Figure 2. Since these carbon-based inks are colloidal, a clear understanding of the flow nature of the suspended particles in additive the manufacturing and conventional printing methods will dictate the overall printability of the carbon allotropes.

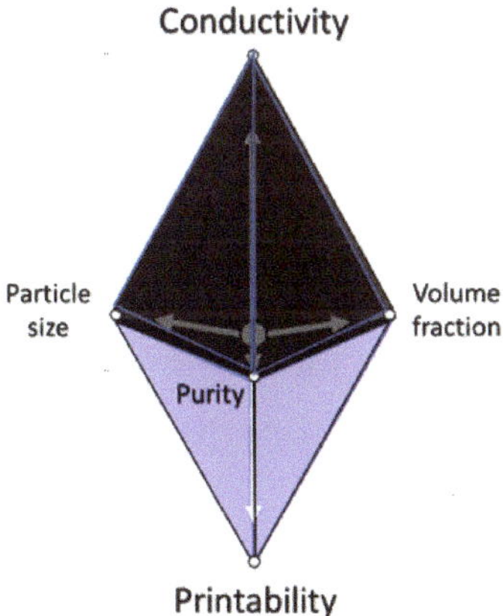

Figure 2. Graphical representation of the general relationship of the conductivity trade-off with printability, in terms of tailoring conductive inks. The variables in the base triangle increase (particle size (μm), volume fraction (%), and purity (%)) the conductivity increases. Conversely, the printability increases with decreases in these variables.

This review paper intends to bridge the gap between the rheological importances to printing inks while maintaining the intended properties of the printed product. It is clear that there is no one size fits all when weighing the rheological properties versus the desired properties. One of these properties is conductivity, as both graphene and carbon nanotubes present usability in semiconductors and electronic circuits, as examples. A large conductivity range opens the inks to a plethora of purposes, which require important controls. Previous works investigating this have been done by Lucja Dybowska-Sarapuk et al. [6], where the various carbon-based inks were compared rheologically. M.I. Maksud et al. [7] investigated the printability of carbon nanotubes, though understanding how to optimize the printability of idealized material properties in carbon-based inks is still absent in the literature. A review paper of relevance was conducted by Derby [8], in which the rheological aspects of 3D printing ceramics were investigated.

2. Background

The process capabilities of printing go hand in hand with the material's rheological and mechanical properties. Printing performance, or the "printability" of ink, is defined by many physical parameters and fluidic properties, such as density (ρ), viscosity (η) and surface tension (σ). When considering the carbon-based inks used in 3D printing, these properties must be accounted for to ensure high printability and resolution [9].

2.1. Rheology

Rheology is the study of flow and the structural properties of viscous materials. To investigate these flow properties, rotational shear is applied, causing the sample to flow. There are two methods to investigate flow properties: Linear (oscillatory) and Non-linear (Steady-Shear) [10].

2.1.1. Linear Rheology

Small amplitude oscillatory shear (linear rheology) is widely used in viscoelastic material characterization. It is a non-destructive method. Due to the small amplitude the deformation does not exceed the linear viscoelastic region of the material. This is achieved through cyclically varying stress and strain in a sinusoidal fashion. The material responds elastically to deformation, rather than plastic deformation within this region. Contributions from the viscous and elastic responses in the material are measured, this gives both complex modulus G^* (comprised of loss and storage) and the phase angle δ. During this test, the sample is under harmonic strain causing harmonic stress. The harmonic strain can be represented by:

$$\gamma = \gamma_0 \sin(\omega t) \tag{1}$$

where γ is the strain, γ_0 is the strain amplitude, ω is the angular frequency of oscillation and t is the time. The stress varies with the same angular frequency, ω, amplitude of σ_0. However, it is out of phase with the strain by an angle, δ. The linear response of the material in terms of stress can be written as:

$$\sigma = \sigma_0 \sin(\omega t + \delta) \tag{2}$$

This equation is only valid at low strain amplitudes. At larger strains, however, a non-linear response is observed in the sample. The strength of small amplitude shear oscillation is that the stress response gives quantifiable material values, in storage (G') and loss (G'') moduli. These moduli are the ratios of stress and strain amplitudes, storage being the real (elastic response, in-phase stress) and loss being the imaginary (viscous response, out-of-phase stress). Equation (2) can be further described as [12]

$$\sigma = G'(\omega) \sin \omega t + G''(\omega) \cos \omega t \tag{3}$$

Maxwell created a model based on the linear region of viscoelastic behavior, comparing it to that of a spring and a dashpot in series, the total strain consisting of the spring strain (ε_1) and the dashpot strain (ε_2). With the strain the same in both elements, we are given the equations:

$$\varepsilon_1 = \tfrac{1}{E}\sigma \quad \dot{\varepsilon_2} = \tfrac{1}{\eta}\sigma \quad \varepsilon = \varepsilon_1 + \varepsilon_2 \tag{4}$$

Differentiating the left and right equations above with respect to time and inputting the left and middle into the right we get:

$$\dot{\varepsilon} = \frac{1}{E}\dot{\sigma} + \frac{1}{\eta}\sigma \tag{5}$$

We put this in the standard form with the stress on the left and strain on the right, giving Maxwell's equation:

$$\sigma + \frac{\eta}{E}\dot{\sigma} = \eta\dot{\varepsilon} \tag{6}$$

η is the viscosity, E is Young's modulus [12].

2.1.2. Non-Linear Rheology

Steady-state shear experiments consist of a continuous stress sweep. Because the large deformities created during the test promote deviation for the linear relationship between the shear stress and shear rate. These deformities make this method non-linear and destructive towards the sample. This test is commonly applied to complex interfaces like emulsions, foams, biological fluids, polymers and colloidal particles, and dispersion of vesicles etc., which tend to have a non-linear response to applied deformations even if they are relatively small. This form of experimentation gives the fluid type (Newtonian, shear thinning, etc.). The properties related to this are:

$$Shear\ rate = \frac{d\gamma}{dt}, \ Shear\ stress, \ \sigma, \ viscosity, \ \eta$$

The viscosity is the proportionality constant between the shear rate and the shear stress.

2.2. Rheological Connection to Additive Manufacturing

In this review, the importance of these rheological properties will be discussed via their impact in 3D printing techniques. A factor of practicality to 3D printing is the non-Newtonian fluid type called "shear-thinning". Shear-thinning is a phenomenon in which the viscosity of the fluid decreases with increasing shear stress. Shear thinning can be time-dependent, this behavior is called thixotropic. This is characterized by the fluidification of the material under shear stress and stiffening at rest. It is a reversible property of the material [13]. Thixotropic materials show shear rate dependent rheological properties. Benefits arise from this property in inkjet printing, as the fluid has high viscosity under standard conditions but low viscosity when passing through the print head. This avoids clogging and fluidizes through the nozzle. Once the drop is detached from the nozzle, the viscosity increases again, suppressing satellite drop formations. Solidification, when deposited, is also improved [14].

Shear thinning suppresses satellite droplets, as shown by Hoath et al. and confirmed with work from Morrison and Harlen [8,9]. Satellite drops are partial, unintended droplets between the drop stream formed from surface tension. These shear thinning effects on satellite droplets formation have been studied by Hoath et al. in aqueous PEDOT:PSS. These satellites reduce printability performance, in terms of resolution [15].

2.3. Factors Affecting Rheology

2.3.1. Temperature

Viscosity is highly sensitive to temperature, making it a very important factor in processing conditions and end result quality. An increase in temperature creates thermal motion of the molecules, resulting in a displacement that overcomes the intermolecular interactions. With increasing temperature, viscosity decreases, as predicted by the Arrhenius equation:

$$\eta = \eta_\infty e^{\frac{Ea}{RT}} \tag{7}$$

where η is the viscosity, η_∞ is the pre-exponential factor, Ea is the activation energy for flow, R is the universal gas constant, and T is the absolute temperature in Kelvin and the Boltzmann constant R [9,10]. This equation assumes that there are no physical/chemical changes being induced by the applied heat energy.

2.3.2. Pressure

Viscosity is largely dependent on the free volume of the system. Since the free volume of a system is influenced by pressure, the viscosity is pressure dependent. The pressure reduces free volume and as a result, it reduces molecular mobility. This, however, becomes noticeable only at high pressures. The rise in pressure increases both the T_g and T_m, which also reflects an increase of viscosity [16]. Viscosity generally increases with increasing pressure; the general correlation is given by:

$$\eta = A_0' e^{B_0' p} \tag{8}$$

where A_0' and B_0' are constant and p is the pressure.

2.3.3. pH

From one investigation by Alias et al., the dependence of pH on rheology was observed to be that friction increases with increasing pH. Highly acidic graphene suspensions showed lower friction/viscosity. This is associated with higher pH increasing agglomeration and reduced dispersions. As seen in Figure 3. In low pH levels, graphene oxide (GO) sees better dispersion in distilled water, reducing the friction coefficient.

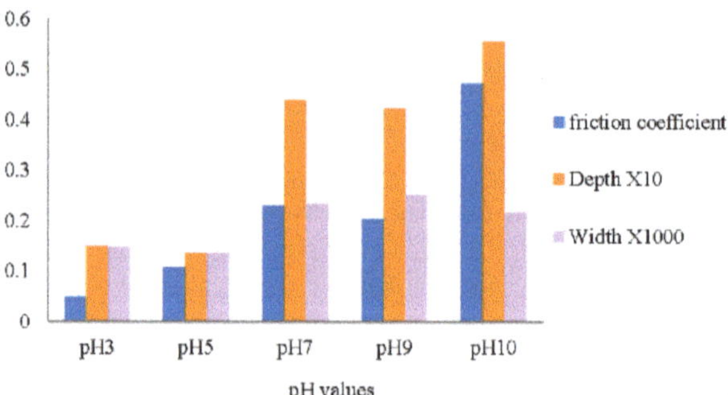

Figure 3. Dependence of the flow behavior on pH level of graphene oxide (GO) in water lubrication, from this graph we observe the trend of higher friction being created. Taken from Aias et al. open access © University Malaysia Pahang Publishing, Malaysia [17].

2.3.4. Topography & Shape of the Suspension

The viscosity of the colloidal systems is also dependent on the topography and shape of the suspended particles. From Reinhardt et al., a correlation between the shape of a suspended particle (flakes suspensions as in graphene-based inks) and its viscosity are determined, shown below in Figure 4 [18]. This shows that the shear-thinning behavior in the shear rate $\dot{\gamma} < 1~\text{s}^{-1}$ is intensified for flake suspensions, giving similar viscosity within the shear-thinning region, while generally having higher viscosity compared to the sphere-shaped suspensions, as seen in Figure 4. This would likely be associated with the high particle aspect ratio involved with flakes. Particle-particle effects increase due to the orientating of the flakes at higher shear rates, this creates a second Newtonian plateau in the colloid. This specific shear thinning region becomes important when considering manufacturing parameters, as the shear rate is design specific. Although, this applies to the zero-shear rate, this does not describe the effects of the shear rates above the linear viscoelastic region. Steady shear causes de-agglomeration and alignment behaviors of large agglomerates causing strong shear thinning [19].

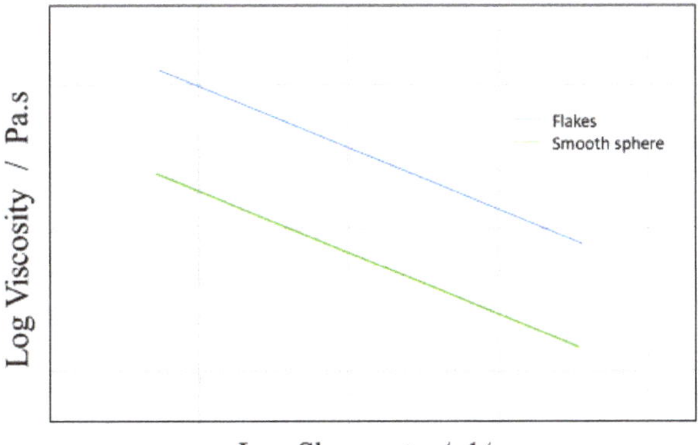

Figure 4. General relationship of viscosity and shear rate, for flake suspension versus smooth sphere suspension (Sketch representation of the general trend concluded from Ran Niu et al. [19]).

2.3.5. Surface Tension in Printing

Surface tension is an important property for the additive manufacturing printing process, as it can affect the printability and formation on the substrate. Higher surface tension shrinks more rapidly and has a far shorter tail, which leads to fewer satellite droplets. Higher surface tension in the nozzle obtains higher printability and resolution with fewer defects from satellite drops. Leakage and liquid accumulation at the nozzle are also problems associated with low surface tension, or viscosity for that matter [13,14]. The effects of surface tension on ink droplet formation are depicted in Figure 5.

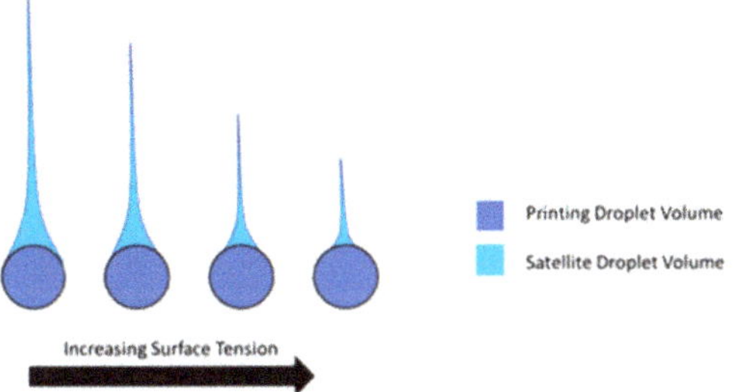

Figure 5. Comparison of droplet volume to satellite volumes with increasing surface tension. As droplets are calculated by their diameter in additive manufacturing, the satellite droplets and tail are not accounted for and cause error. For effective printing, the closer the droplet can realize a spherical shape, the less the error.

On the other side, high surface tension leads to problems on the substrate, as surface wettability decreases. This can lead to the agglomeration of the printed droplets. The high contact angle formed by the drops due to high surface tension causes the droplets to combine together, either decreasing resolution or causing an error in the build [20]. On the other hand, high wettability increases the image resolution due to lower ink spreading. As concluded by Vafaei et al., as wettability decreases, continuous lines become more difficult to print. It is also noted that the cross-sectional area and volume of printed lines increases with decreasing wettability [20].

The surface tension of the substrate also plays an important role during ink-jet printing. The surface must have higher surface tension than the ink, along with attractive forces to allow for transfer, which in turn gives good adhesion of the print to the substrate. Wettability of the droplets to the surface can determine the feature size (resolution) and cross-section. The values of surface tensions and solubility for various solvents of interest are given in Table 1.

Table 1. Surface tensions and solubility parameters of solvents for GO, reduced graphene oxide (rGO). Adapted from Konios et al. [21].

Solvent	Surface Tension (mN/m)	GO Solubility (μg/mL)	rGO Solubility (μg/mL)
De-ionized water	72.8	6.6	4.74
Acetone	25.2	0.8	0.9
Methanol	22.7	0.16	0.52
Ethanol	22.1	0.25	0.91
2-propanol	21.66	1.82	1.2
Ethylene glycol	47.7	5.5	4.9
Tetrahydrofuran (THF)	26.4	2.15	1.44
N,N-dimethyformamide (DMF)	37.1	1.96	1.73
N-methyl-2-pyrrolidone (NMP)	40.1	8.7	9.4
n-Hexane	18.43	0.1	0.61
Dichloromethane (DCM)	26.5	0.21	1.16
Chloroform	27.5	1.3	4.6
Toluene	28.4	1.57	4.14
Chlorobenzene (CB)	33.6	1.62	3.4
o-Dichlorobenzene (o-DCB)	36.7	1.91	8.94
1-Chloronaphthalene (CN)	41.8	1.8	8.1
Acetylaceton	31.2	1.5	1.02
Diethyl ether	17	0.72	0.4

2.4. Relation to Inkjet Printing

The importance of characterizing rheological properties is well known in additive manufacturing. Carefully measured rheological parameters are paramount for simulations, which are becoming an integral part of the additive manufacturing process recently. In additive manufacturing, the ink is subjected to shear flow over a wide range of shear rates, which is described through steady-state jetting. The knowledge of rheological properties is essential for process design and optimization. To estimate the value of Force F exerted on a fluid, for a real jet discharge from a small orifice, of area A and uniform velocity V, the following equation is used:

$$\frac{F}{2\pi\rho v^2} \approx \frac{1}{2\pi}\frac{AV^2}{v^2} \tag{9}$$

where ρ and v are the density and dimensional consistency constant. The mass flux through this orifice too can be defined:

$$\rho AV = \frac{F}{V} \tag{10}$$

The dimensionless parameter $\frac{F}{2\pi\rho v^2}$ in Equation (9) is known as the Reynolds number, this is generally depicted as:

$$Re = \frac{\rho VL}{\eta} \tag{11}$$

where V, L, and η is the velocity of the fluid with respect to the object (m/s), characteristic linear dimension (m) and the dynamic viscosity of the fluid (Pa·s), respectively [22]. Low values for the Reynolds number signify high viscosity, high values signify low viscosity, which usually leads to satellite droplet formation. For an idealized printability though, this number alone does not tell the whole story.

Another dimensionless value of interest is the Weber number, which is the characteristic number describing droplet formation ability. Two forces form the basis of the Weber number, one being the fluid-mechanical force and the other being surface tension. When a liquid flows through a second fluid phase, either a gas or a liquid, then the fluid-mechanical force F_A causes the drops to deform and ultimately disperse:

$$F_A = \frac{1}{2} C_w \frac{\pi}{4} L^2 \rho v^2 \tag{12}$$

C_w, L, ρ and v is the Drag coefficient, Characteristics length, Density and Flow rate respectively. Surface tension involves a cohesion force F_k, opposes the increase in surface area, which is caused by the falling deformation. The droplet is held together by:

$$F_k = \pi L \sigma \tag{13}$$

The Weber number is the ratio of these forces and hold the following relation: [22]

$$W_e = \frac{8F_A}{C_w F_k} = \frac{\rho v^2 L}{\sigma} \tag{14}$$

A combination of Reynolds and Weber numbers provide an understanding of fluid drop formation, the Ohnesorge number, defined as:

$$Oh = \frac{\sqrt{We}}{Re} = \frac{\eta}{\sqrt{\rho L \sigma}} \tag{15}$$

where η and σ are the dynamic viscosity and surface tension of the fluid, respectively [22]. This value is normally utilized as its inverse $Z = 1/Oh$, where $1 < Z < 10$ are the limits to stable drop formation [23]. This idealized region for printability is seen in Figure 6.

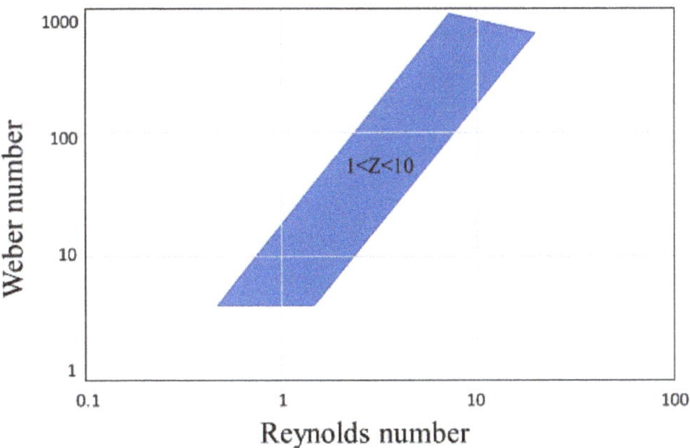

Figure 6. Graphical representation of the idealized region for stable printing, where $1 < Z < 10$, the inverse of the Ohnesorge number. The depicted blue area is the Goldilocks zone for 3D printing.

These equations describe few of the defining properties of printability for an ink like dynamic viscosity, density, characteristic linear dimension (Nozzle diameter), and surface tension. The accuracy of the calculations from the above equations is dependent on the printing parameters such as the waveform and temperature, which also have an effect on the printability [24]. The values are not ubiquitous, Lee et al. have shown nozzle clogging causing nonjetting even in viable Z values. The observed printability at a Z range of $2.5 < Z < 26$ for Newtonian fluids, however, they were unable to jet colloidal ZnO suspensions for the identical range [4].

The Carbon-based colloidal suspensions present issues in terms of printability and their rheological properties. The largest problem associated with printing carbon-based inks is clogging of the nozzle and print head. Clogging occurs as a result of the suspended carbon particles agglomerating, preventing an even flow through the print head. Special care must be taken for designing carbon-based inks, to ensure consistent flow. Another rheological issue is satellite or unintentional trailing droplets, arising from low viscosity, which compromises the print accuracy and precision. The additions of surfactants in the ink to improve particle distribution further exasperate the issue since they lead to larger satellite drops [25]. Viscosity and density ratios also play a major role in droplet formation. Low-density ratio leads to larger satellite droplet formations, similarly, lower ink viscosity has been reported to give way for easier satellite droplet formation [26]. On the other hand, low viscosity is a necessity for printability, therefore compromises have to be made, and all important parameters like conductivity, mechanical strength, flexibility, particle size, flow properties must be carefully optimized (as shown in Figure 2) during the ink design process to suit a specific printer. Speed and precision of print rely heavily on the ink viscosity, with pinching speed of droplets being proportional to surface tension and inversely proportional to viscosity [27]. Typical viscosities for ink-jet printing are between 5 and 20 mPa·s and surface tensions of 25 to 35 mN/m. Though much like Z-numbers, this is not a ubiquitous rule for 3D printing [28]. The zero shear viscosity of various printable carbon inks is shown in Table 2.

Table 2. Examples from the literature of 3D printable carbon nanoform suspensions (zero shear viscosity).

Carbon Form	Ink Method	Zero Shear Viscosity/Pa·s	Reference
Graphene	Organic solvent	~18.5	[29]
	Organic solvent + dispersant	~6	-
	Water 0.5%	0.478 mPa·s	[30]
	Water 1%	~ 0.52 mPa·s	-
	Water 1.5%	~ 0.56 mPa·s	-
Graphene Oxide	LC viscoelastic gel 2 mg/mL	~3.8	[31]
	LC viscoelastic gel 9 mg/mL	~100	-
	PMMA matrix 0.05% GO	~80	[32]
	PMMA matrix 1.2% GO	~20,000	-
Carbon Nanotubes	epoxy resin 0.3% treated CNT	20	[33]
	PIB 1.7% MWCNT	~1	[34]
	PIB 3.0% MWCNT	~7	-
	PIB 6.0% MWCNT	~900	-
	PDMS 1% MWCNT	~10	[35]
	PDMS 4% MWCNT	30–40	-
Carbon Black	Poly-acrylate, 5.3% spherical	~0.6	[36]
	Poly-acrylate, 11% spherical	~6000	-

At a given shear rate through the nozzle of radius (R) and length L, the pressure required to extrude a shear thinning liquid, such as carbon-based suspensions, can be calculated by:

$$\Delta p = \frac{8\eta QL}{\pi R^4} \tag{16}$$

where Q is the volumetric flow rate and η is the viscosity [37]. This equation presents the importance of the radius and length of the nozzle in ink design. The radius of the nozzle is a major factor for printing resolution and droplet formation.

Methods for producing droplets from the printing nozzle can be split into three methods. One is continuous inkjet printing, where a continuous ink stream is broken into droplets of uniform size and spacing [38]. The nozzle is held at a potential relative to ground that transfers charge to the drops. Deflector plates are utilized to steer droplets, as due to the continuous flow of drops, unwanted drops are deflected into a gutter to be recycled back.

Another method of producing droplets is drop-on-demand technologies, allowing the printhead to produce singular droplets. This process is normally driven by the application of voltage pulses to a piezoelectric actuator, creating pressure through its mechanical motions. Optimization of voltage is required dependent on material, nozzle dimensions and environment [39]. Ejected columns of liquid are pinched off to form a drop. The volume ranges are from 1 pL–1 nL, with a diameter range of 10–100 μm [40].

Modeling plays a role in optimizing these nozzles, with issues such as clogging. Computer-based simulation allows for rapid experimentation and parameter variation in an aim to optimize and predict the most effective system. Barati et al. [41] presented a model for reducing clogging through transient simulation. Looking at the wall-fluid adhesion mechanisms and interactions, clogging could be simulated, and nozzle design can be varied so as to reduce clogging. Simulations are then followed up by a validation experiment, verifying the simulations results [42].

3. Carbon Based Inks

3.1. Carbon Based Inks—A Colloidal Suspension

Carbon conventionally is a hard solid in nature, which stems from the covalent bonding between carbon atoms. Direct printing of carbon alone would be impossible, with the exception of extreme combinations of temperature and pressure. Hence, to utilize carbon forms in additive manufacturing, the particles must be assisted by a liquid, acting as a vehicle to make printing possible. This combination of solid particles in a liquid is known as a colloidal system.

3.1.1. Colloidal Systems

A colloid is a heterogeneous solution, with a dispersed phase uniformly distributed throughout the second medium, the dispersion phase. When the dispersed phase is smaller than 1nm in diameter, the system assumes the properties of a true solution. Conversely, when this dispersion is larger than 1000 nm, the separation is large enough that it is considered a suspension. Suspensions containing much larger solid particles or high solid content can form sedimentation [43].

3.1.2. Einstein Viscosity & the Krieger–Dougherty Equation

From the energy dissipation calculation of suspensions, Einstein derived that a dilute suspension of rigid spherical particles behaves as a Newtonian fluid with relative viscosity (η_r) represented by the following equation:

$$\eta_r = 1 + 2.5\Phi \tag{17}$$

where Φ is the volume fraction of the dispersed phase. This assumes that all particles are separated by such a distance so that there is no interaction between them.

$$\Phi = n4\pi\frac{a^3}{3} \tag{18}$$

where n is the number density of particles and a is the particle radius [44]. When looking for higher particle concentrations, Krieger and Dougherty proposed a semi-empirical equation for the concentration dependence of the viscosity:

$$\eta_r = \left(1 - \frac{\Phi}{\Phi_{max}}\right)^{-2.5\Phi_{max}} \tag{19}$$

where Φ_{max} is the maximum packing fraction or the volume fraction at which the zero shear viscosity diverges. When particle packing density is low, this reduces to the Einstein relation. Approaching the maximum volume fraction Φ_{max} the particle packing density is such that the dispersion flow is impossible and $\eta_r \rightarrow \infty$ [45]. This increase in viscosity is attributed with the increase in particle concentration in a suspension. The restriction in the relative motion of the particles results in an increase of the particle collision in the suspension, subsequently leading to an increase in the frictional forces. The theoretical models explaining colloidal systems assume hard sphere systems that only interact through hydrodynamics, where the distribution of particles is highly sensitive to the shape, size, and surface charge of the particles in suspension [44]. However, real colloidal suspensions lack the hard sphere shape, which further affects aggregation and flocculation at higher concentrations of particles and significantly alters the system's flow properties. This promotes non-Newtonian rheological behavior.

Adding particles in the ink does not simply increase the viscosity of the liquid as a result of the hydrodynamic disturbance of the flow, it can also be a cause for a deviation from the Newtonian behavior, including shear rate dependent viscosity, elasticity, and time-dependent rheological behavior. Colloidal dispersions at low to moderate volume fraction exhibit shear-thinning behavior analogous to low viscous liquids. On the other hand at high concentrations they behave like solids, which require higher stress to start the flow. The rheological behavior of a suspension is strongly dependent on the nature of colloidal interactions attraction. Depending on whether the colloidal interactions are attractive or repulsive, the particles can form different structures, which determine the rheological behavior of the material [46].

3.2. Graphene

Graphene is a transparent two-dimensional sheet of carbon atoms arranged in hexagons. Graphene is the single layer equivalent of graphite. Much interest and research have been conducted on graphene due to its exceptional electrical and mechanical properties. Because of these properties, there is a growing interest in additive manufacturing and printing, opening possibilities to light, strong, and conductive builds.

Jakus et al. [47] demonstrated 3D printed biocompatible scaffolds from graphene inks, which show in vivo compatibility over at least 30 days. This opens up biomedical usages of graphene printed inks, which can be applied to in vitro and in vivo tissue regenerative engineering applications. The potential biocompatibility of Graphene opens completely new realms of applicability for graphene due to its combination of conductivity and printability. Though, there are other reports in the literature that conflict on this biocompatibility, with factors such as surface functionalization (which reduces toxicity and is in most cases a must for functional use), size and shape all playing effective roles in possible toxicity. [48] Zhang et al. and Shen et al. [22,23] discussed the potential for graphene-based inks in drug deliverance, gene therapy, cancer therapy, tissue engineering, biosensing, and bioimaging. Recently, Graphene 3D Labs have produced conductive graphene composite filaments for Fused Filament Fabrication in the commercial market.

3.3. Graphene Oxide

Graphene oxide is a functionalized Graphene, created through the oxidation of graphite. The oxidation expands layer separation in the graphite and makes the layers hydrophilic, allowing for dispersion in water. Sonication exfoliates the graphite further, creating single and few-layer GO. Importantly to the functional use of GO, the lower the oxygen content, the more conductive. GO forms non-covalent networks with optimum rheological properties with respect to printing. Shear thinning behavior of the colloidal suspensions along with the relatively high storage modulus (G') gives strong printability and self-supporting structures. Reduced graphene oxide (rGO) is the restoration of pristine graphene properties to GO through the reduction, removal of Oxide. This is of value due to the ease at which GO can be handled and suspended in water, but the need to recuperate conductivity properties in the end product [29,30].

Due to the highly anisotropic nature of graphene sheets, with the thickness of the sheet being in single atomic layers and the lateral being in the micrometers scale, the properties of the graphene oxide are dependent on how it is assembled. Careful control of the assembly of the flakes then is of necessity, as agglomerations, bends and crimples will affect the properties of the final product. For additive manufacturing, the directional dispersion of the flakes after printing is crucial to the effective properties. Kim et al. [49] studied the surface activity of GO in a suspension and have shown that graphene aligns with gas bubbles. GO inks have also been used in electronics, with lithium-ion battery electrodes fabricated through the use of high viscosity GO-based electrode inks by Fu et al. [50].

Using 2D and 3D graphene printed inks, both planar and volumetric structures can potentially be made with this material. Graphene retains mechanical flexibility, high electrical conductivity, and stability to thermal and chemical effects after deposition [51]. Due to Graphene's mechanical properties, its use in composites for printing mechanical reinforcement is well documented [2,30,31].

Interest is also around the flexible nature of graphene prints. The combination of conductivity and its flexible nature open new paradigms of consumer electronic capabilities. This conductive and flexible nature was investigated by Secor et al. for inkjet printing. They developed a graphene/EC powder that was produced at room temperature and was capable of stable jetting of features, boasting excellent printability and geometrical shaping [52]. On the toxicity of Graphene and Graphene Oxide, the available data is still insufficient for conclusive answers [53].

Table 2 presents the conductivity of reduced graphene oxide-based inks, from Fernandez-Merino et al. [54].

The higher concentration of rGO is key to higher conductivity. The conductivity of 100% rGO far exceeds the conductivity at lower concentrations, rGO/SDS closest, with similarly high rGO concentrations in wt%. Uddin et al. [55], studied the impact of surfactant on conductivity, as shown in Table 3. Covalent dispersion techniques also are investigated, from work by Kuila et al. [56] in Table 4.

Table 3. This is a table presenting the Conductivity properties of various tested reduced graphene oxide-based inks. Reproduced with permission from M.J.Fernández-Merino et al., Carbon; published by Elsevier, 2012 [54].

Film	rGO (wt%)	Conductivity $(S \cdot m^{-1})$	Specific Capacitance $(F \cdot g^{-1})$
rGO	100	7548	38
rGO/PBA	36	13.31	1
rGO/DOC	47	0.06	1
rGO/TDOC	36	2.18	3
rGO/PSS	41	10.51	114
rGO/SDBS	29	0.87	7
rGO/SDS	87	4679	46
rGO/CHAPS	36	0.92	2
rGO/DBDM	11	0.01	3
rGO/P-123	38	5.53	12
rGO/Brij 700	10	1.08	6
rGO/Tween 80	13	0.41	95

Table 4. Comparison of materials of absorbed surfactant and electrical conductivity. Reproduced with permission from Md. Elias Uddin et al., Journal of Alloys and Compounds; published by Elsevier, 2013 [55].

Sample	Adsorbed Surfactant (%)	Conductivity $(S \cdot m^{-1})$
GO	34.34	0.002
CR-G	-	4760
SDBS-0.25-G	7.02	108
SDBS-0.5-G	6.13	106
SDBS-1-G	9.31	97
SDS-0.25-G	17.41	94
SDBS-0.5-G	17.93	93
SDBS-1-G	21.62	95
TRX-0.25-G	9.63	98
TRX-0.5-G	9.63	92
TRX-1-G	9.37	89

3.4. Carbon Nanotubes

Carbon nanotubes (CNT) are similar to Graphene with great interest and research being conducted due to its extraordinary, electrical, optoelectronic, and biosensing capabilities [25–27]. There is a wealth of literature on the mechanical, electrical, and thermal properties of carbon nanotubes [45,46]. Owing to these properties, a whole plethora of applications where carbon nanotubes could be used. Additive CNTs are often combined with the polymers to form strong, electrically conductive composites [5] or in water suspensions for nanoelectronics and sensors [57].

Inkjet printing of carbon nanotubes has been demonstrated, notably by Kordas et al., to create conductive patterns, [58] using carboxylated Multi-Wall Carbon Nanotubes (MWCNTs). The notable advantage of CNTs over other conventional conductive inks is the lack of carbon nanotubes to require curing. Electrically conductive CNT-based inks have been designed for dip coating and screen-printing methods by Shin et al. [59]. These samples proved to be highly flexible, bendable, and stretchable while maintaining electrical connectivity and very little change in resistance. Single-Walled Carbon Nanotubes (SWCNTs) have been printed with inkjet printers as a thin film, the flexible electrode on cloth by Chen et al. [60]. Control over geometry and pattern, in this case, showed promise for wearable energy storage, as a printable electrochemical capacitor.

3.5. Carbon Black

Carbon black is a finely particulate paracrystalline carbon produced by the incomplete combustion of heavy petroleum products or vegetable oil. [61] Carbon black has been used as part of compounds due to its additional mechanical strength, conductivity, black pigmentation, and absorption of ultraviolet light [44,45]. Talarico et al. have designed carbon black based electrochemical sensors by using a screen printing process. Printing energy storage devices and supercapacitors have also been fabricated from carbon black inks [46,47]. Inkjet printed carbon black composites have also been used as a catalyst layer in fuel cells due to their high conductivity and corrosion resistance, presented by Taylor et al. [62]. One of the advantages of carbon black is that it has a history of being used in lithographic inks and black inkjet printers, the printing process is well established.

3.6. Carbon Fiber

Carbon fiber is a well-established and go-to engineering material due to its high strength mechanical properties and light-weight. Using carbon fibers in additive manufacturing is still less established though. However, the strength of carbon fiber comes from the length of the fibers, however, the additive manufacturing process requires a small enough length of the fibers, however, the additive manufacturing process requires a small enough length to fit through the nozzle of the printer; therefore, compromises have to be made on the printing speed.

Tekinalp et al. achieved highly orientated carbon fiber-polymers (0.2–0.4 mm) through Fused Filament Fabrication (FFF). Additive manufacturing here controls the orientation and allows for good dispersions, but also presents higher porosity, which is detrimental to the improvement seen in the orientation [63].

Because of the relation of carbon fibers strength with length and orientation, printing methods with continuous carbon fibers have been investigated. A methodology for in-nozzle impregnation has been demonstrated by Matsuzaki et al. wherein, the carbon fiber is fed through the nozzle with polylactic acid (PLA) in one continuous line [64]. Through the printing of carbon fiber with PLA, Tian et al. tested and optimized conditions necessary for continuous fiber printing [65]. Table 5 summarizes the relationship of conductivity with print thickness for various type of ink compositions.

Table 5. Comparison of covalent dispersion techniques dispersibility and electrical conductivity. Reproduced with permission from Tapas Kuila et al., Progress in Materials Science; published by Elsevier, 2012 [56].

Modification Techniques	Modifying Agent	Dispersing Medium	Dispersibility (mg/mL)	Electrical Conductivity ($S \cdot m^{-1}$)
Nucleophilic Substitution	Alkyl amine/amino acid	$CHCl_3$, THF, toluene, DCM	-	-
	4-Aminobenzene sulfonic acid	Water	0.2	-
	4,4′-Diaminodiphenyl ether	Xylene, methanol	0.1	-
	POA	THF	0.2	-
	Allylamine	Water, DMF	1.55	-
	APTS	Water, ethanol, DMF, DMSO	0.5	-
	IL-NH$_2$	Water, DMF, DMSO	0.5	-
	PLL	Water	0.5	-
	Dopamine	Water	0.05	-
	Polyglycerol	Water	3	-
	Poly(norepinephrine)	Water, methanol, acetone, DMF, NMP, THF	0.1	-
Electrophilic Substitution	ANS	Water	3	145
	4-Bromo aniline	DMF	0.02	-
	Sulfanilic acid	Water	2	1250
	NMP	Ethanol, DMF, NMP, PC, THF	0.2–1.4	21,600
Condensation Reaction	Organic isocyanate	DMF, NMP, DMSO, HMPA	1 (DMF)	-
	Organic diisocyanate	DMF	-	1.9×10^4
	ODA	THF, CCl$_4$, 1,2-dichloroethane	0.5 (THF)	-
	TMEDA	THF	0.2	-
	PEG-NH$_2$	Water	1	-
	CS	Water	2	-
	TPAPAM	THF	-	-
	β-CD	Water, acetone, DMF	1 (DMF)	-
	α-CD, β-CD, γ-CD	Water, ethanol, DMF, DMSO	>2.5	-
	PVA	Water, DMSO	-	-
	TPP-NH$_2$	DMF	-	-
	Adenine, cystine, nicotamide, OVA	Water	0.1	-
Addition Reaction	POA	THF	0.2	-
	Polyacetylene	Ortho dichlorobenzene (O-DCB)	0.1	-
	Aryne	DMF, O-DCB	0.4	-
	Cyclopropanated malonate	Toluene, O-DCB, DMF, DCM	0.5	-

Looking at all samples, a comparison of the methodology of a few ink compositions was made, observing the conductivity found in Table 5.

4. Problems Associated with Printing Carbon Based Inks

4.1. Agglomeration

One of the major concerns for carbon-based inks is its reaction with the liquid medium in which it is suspended. Due to the hydrophilic nature of aromatic carbon forms, water, the first choice for ink suspension, however, water-based inks suffer agglomeration. Water-based inks are ideally suited due to their environmentally friendly nature, the ease at which they can be stored and handled [66]. Agglomeration is a challenge both for additive manufacturing due to limitations of the nozzle area and in graphene, the restacking of sheets to form graphite, which possesses inferior properties [67]. One method for avoiding the hydrophilic nature of graphene is through functionalization of the graphene sheets, an example of such is graphene oxide, as discussed earlier on [26,68]. The oxidation of graphene results in a reduction in the conductivity, which is undesirable.

Similarly, for CNTs, three methods are taken to counter this problem. One is the functionalization of the side walls of the CNT. This is generally done with Carboxylation, the addition of hydrophilic carboxyl (−COOH) groups to the carbon nanotube walls. Carboxylation can prove to be counterproductive as it decreases the conductivity and hence, effectiveness [58].

Sonication is another method commonly used for CNTs and graphene-based inks. Sonification is the irradiation of a liquid sample with ultrasonic (>20 kHz) waves. These high-frequency sound waves propagate in the liquid, resulting in high-pressure and low-pressure cycles, creating agitation in the medium [69]. Another common solution to avoid agglomeration is the addition of dispersants in the solvent to avoid agglomeration. The use of polymers and surfactants have been utilized to this end, by coating the CNTs, Van der Waal forces can be suppressed [70]. The presence of both hydrophilic heads and hydrophobic tails in the dispersants are known to disperse the CNTs and colloids in the inks with a huge reduction in agglomeration [71].

On the other hand, the problem of hydrophobicity altogether by utilizing organic solvents in the inks in place of water. Organic solvents can avoid the conglomeration of the CNTs effectively without functionalization and there is no compromise on the conductivity. The organic solvent molecules are attracted to the surface of the CNT due to its hydrophobic nature, which prevents the Van der Waal attraction of the CNTs [63,64].

Though organic solvents present their own set of problems, as they pose a hazard to the environment and health. Careful cartridge design and disposal are of importance since most of the organic solvents can potentially be highly corrosive. These hazards need to be properly addressed, especially when the potential use is in medical applications. Most of the organic solvents are highly volatile and evaporate faster, despite lower surface tension compared to water. The evaporation rate of the solvents needs to be properly optimized according to the printing method, otherwise, the ink may clog the nozzle and agglomeration could result through a loss of solvent [65,72].

4.2. Maintaining Suspension and Dispersion

The initial aggregation rate for GO flakes can be described as:

$$k_a N_0 \propto \left(\frac{dR_h(t)}{dt} \right)_{t \to 0} \tag{20}$$

where N_0 is the initial particle concentration, R_h hydrodynamic radius. From this the aggregation attachment efficiency α (quantification of particle aggregation kinetics)

$$\alpha = \frac{1}{W} = \frac{k_a}{k_{a,fast}} = \frac{\frac{1}{N_0} \left(\frac{dR_h(t)}{dt} \right)_{t \to 0}}{\frac{1}{(N_0)_{fast}} \left(\frac{dR_h(t)}{dt} \right)_{t \to 0, fast}} \tag{21}$$

"fast" here refers to favorable aggregation conditions [73].

The shelf life of a 3D ink is also important for practical use in industry. The main concern with carbon-based suspensions is the stability, the settling, and agglomeration of particles with time. Methods such as solvents that maintain constant dispersion [74] dispersing using sonication [71] or the addition of copolymers to increase stability [75]. Su et al. conducted testing on colloidal stabilities of high concentration graphene inks, 1 mg/L–3 mg/L suspensions have a constant distribution for at least as an hour, whereas concentrations of graphene above 3 mg/L only have a shelf life of less than one minute [76]. An environmental impact study into GO in water by Chowdhury et al. investigated the stability of GO nanoparticles in various water types, 10 mg/L was shown to be stable in fresh water for almost a month [77]. Comparisons of different dispersion methods for carbon-based inks are shown in Table 6.

Table 6. Various carbon based ink examples from the literature, showing the relationship of ink medium with conductivity and film.

Carbon Form	Ink	Conductivity ($S \cdot m^{-1}$)	Thickness of Prints	Reference
Graphene	Pristine	~40,000	-	[72]
	GO + water	~400	20 prints	[78]
	Few layer GO + water	~875	20 prints	[78]
	G + NMP (Substrate O_2 plasma treated)	~0.08	50 nm	[79]
	G + NMP (Substrate Pristine)	~30	50 nm	[79]
	G + NMP (Substrated HMDS-coated)	~95	50 nm	[79]
	G + Cyrene	37,000	7.8 μm	[80]
Carbon Nanotube	SWNT + water + SDBS (substrate paper)	~550	50 nm	[81]
	MWCNT 12% + PAN + DMF	~100	300 nm	[82]
	MWCNT 89% + PAN + DMF	~333	300 nm	[82]
	MWCNT + aqueous solution	2400 ± 180	10 μm	[59]
Carbon Black	Cold microwave plasma, CO_2 1.7%	256	-	[83]
Silver	Ag microparticles + Organic binder + solvent (Substrate PET/glass)	46,700	Screen printed	[84]

4.3. Health, Safety, and Environmental Concerns

As mentioned previously, the biocompatibility of carbon nanoforms still requires further investigation. A review into the potential insurability of such nanoparticles has been carried out by Mullins et al., in which the minimization of exposure and framework for the transfer of technology is established [85]. The extent or potential of harm from these nanoforms could potentially damage the use of graphene and CNTs in personal electrical and medical devices. Graphene presents a number of potential issues ranging from environmental risks and toxicity, due to the nanoscale, which also reveals the difficulties related to removing and filtering the particles [86]. CNTs can also cause damage due to their scale, with oxidation stress and biocompatibility. Factors that appear to affect this are length, diameter, purity, production method, and functionalization, and that by modifying these factors, CNTs may be safe for human use [87]. The majority of the solvents that are utilized in printable inks technology present environmental health risks, [88] from handling to evaporation, hazards pertain.

Though, accounting for well-established inks and historical influence, printing is an environmentally damaging operation, especially when considering the heavy metals and volatile organic solvents involved [80,81]. Comparing the other conductive inks utilized in additive manufacturing, silver nanoparticles similarly contain potential hazards to the environment [89]. The relative unknown level of environmental risk from Carbon nanoforms is comparatively lower than the established heavy metal and organic solvent hazards [90].

5. Applications of Printable Carbon Inks

5.1. Electronics

Digital circuits have been printed using CNT at sub-3V voltages by Ha et al. onto plastic substrates [91]. Nanowires have been printed as nano-arches using rGO suspended in water by Kim et al. [92]. These nanowires were functionalized in a gas sensor prototype as a 3D transducer.

5.1.1. Transistors

Printed Graphene thin film transistors have been demonstrated to have electron mobility up to ~95 $cm^2V^{-1}s^{-1}$ by Torrisi et al. [79]. The fabrication of field effect transistors through inkjet printing of graphene has numerous examples. [93–95] Carbon nanotubes are also presenting very promising results as thin film transistor, exhibiting properties similar to CMOS devices [93,94]. Showing the potential viability of flexible, transparent electronics, created from additive manufacturing carbon based inks. Paper printable transistors of carbon black and rGO have been developed, presenting the flexibility of carbon transistors developed through additive manufacturing [96]. With graphene being suggested as the long term air to silicon in conventional computing, [97] additive manufactured transistors will allow for rapid testing and design.

5.1.2. Sensors

Sensors have been designed using polymer/carbon black composites, by Loffredo et al. [98]. Highly stretchable sensors based on embedded based on embedded 3D printing of carbon-based resistive ink within an elastomer [99]. Graphene in this ink is used to add conductivity along with its elastic properties to retain the desired elastomers use. GO and FGO based inks have been shown to be designable for sensors directly with standard office inkjet printers while maintaining high electrical conductivity [78]. Glucose biosensors are one such example of carbon-based ink demonstrating the practical electrical properties of GOs from inkjet printing [100].

5.1.3. Electrodes

The first work into GO-based electrode inks for use in lithium-ion battery prototypes using 3D printing has been designed by Fu et al. with optimization of the viscosity and viscoelastic properties [50]. This 3D printed electrode exhibited stable cyclic performance with an LTO anode, with specific capacities of $\approx$160 mAhg^{-1} (LFP) and $\approx$170 mAg^{-1} (LTO). Electrodes made from 3D printed graphene/PLA were demonstrated by Browne et al., these were electrochemically treated for higher conductivity [101]. Carbon nanotube inks present numerous examples of electrode capabilities [102–104].

5.1.4. Supercapacitor

Yao et al. in 2018 printed a record-breaking capacitance with a graphene-based scaffold and pseudocapacitive electrodes of Manganese Oxide (MnO2). This shows promise for the feasibility of practical pseudocapacitive electrodes [105]. This was further improved upon with the "wrapping" of the supercapacitor with CNTs [106]. GO has also been utilized in the design of All-Solid-State, flexible Micro-supercapacitors. Pei et al. [107] demonstrated this using a carbon-based hybrid ink using GO, showing promising potential for lightweight energy storage. A novelty of additive manufacturing allows for full packaging of electrical components during the printing process, supercapacitors of this elk have been designed by Chen et al. for SW-CNTs [108].

5.2. Biological Scaffolding

Lee et al. utilized Multi-Wall CNTs (MWCNT) with PEGDA polymer to print an electroconductive scaffold for nerve regeneration through therapeutic electrical stimulation [109]. Similar lines to this, Ho et al. fabricated a composite scaffold using CNT and polycaprolactone (PCL) with biological compatibilities to cardiac tissue engineering using a CNT based 3D ink [110]. Bone cell growth has been presented using PCL-hydroxyapatite scaffolds filled with CNTs, to stimulate cell growth [111]. Graphene too can be utilized for cell regeneration, with Jakus et al. showing the possible use of graphene-based inks in biological scaffolding [7]. The carbon forms are utilized in each of these composites act to add conductivity and protein absorption to the polymers, promoting faster cell growth. Conductivity is important to stimulate cells with electrical pulses. A summary of various applications of carbon inks suggested in the literature are shown in Table 7.

Table 7. A table presenting the advantages and disadvantages of methods of dispersing carbon-based materials, adapted from Liang et al. under open access © School of Materials Science and Engineering, Southwest Jiaotong University, Chengdu 610031, China [112].

Dispersion Method	Mechanism	Advantage	Disadvantage
Physical methods	Applying physical force to separate agglomerated graphene	Simple operation	Low dispersion rate and possible damage to nanoparticles
Covalent bonding methods	Introducing various active groups by chemical reaction on the surface or edge of the graphene	Making the graphene more workable and operable	Causing damage to the initial structure of the graphene
Noncovalent bonding methods	Modifying the graphene's structure with functionalized molecules through non-covalent interaction	Functionalizing carbon forms, allowing ease of use	Introduces other components and impurities to the carbon forms

Summating the applications of carbon based inks is seen in Table 8 below.

Table 8. Summary table for possible applications for carbon-based inks.

Carbon Form	Ink Method	Application	Reference
Graphene	rGO + water	Nanowire arches	[92]
	Graphene/h-BN + NMP + ethanol	Transistor	[93]
	N-Methylpyrrolidone	Transistor	[79]
	PBT/Graphene composite	Conductive polymer	[113]
	GO + water	Lithium ion battery electrodes	[50]
	Graphene/PLA	Electrodes	[101]
	Graphene/PLA	Energy storage	[114]
	Graphene + Hypromellose, aerogel suspension	Pseudocapacitive Electrodes	[105]
	Graphene + poly-lactide-co-glycolide	Electrical and biomedical scaffolding	-
	PEDOT:PSS	Digital circuit	[91]
Carbon Nanotubes	PBT/CNT composite	Conductive polymer	[113]
	Amine functionalization MWCNT/PEGDA matrix	Nerve regeneration scaffolding	[109]
	CNT + PCL in chloroform	Cardiac tissue scaffolding	[110]
	PCL-hydroxyapatite scaffold + CNT	Stimulate bone cell growth	[111]
Carbon Black	Polymer-carbon black	Chemical sensor	[99]
	Conductive carbon grease (Dimethylpolysiloxane)	Strain sensor	[99]
Carbon Fibers	Active carbon + water	supercapacitor	[115]
	Epoxy	Lightweight cellular composites, controlled alignment	[116]

6. Conclusions

Since its inception in the 1980s, additive manufacturing has become a technology of choice due to its ability for the rapid prototyping (RP) of complex shapes and geometry directly from Computer Aided Design (CAD). Despite huge interest, the technology still suffers some technological barriers that hinder its use in wider applications. Main areas of concern are quality of materials (inks), limitations of equipment, optimization of the manufacturing process and lack of self-correction during the printing process. These limitations needed to be addressed through improvements in the instrument design and optimization of the process.

In this paper, state of the art additive manufacturing of carbon based materials is described. Carbon nanoforms possess huge potential for industrial application through the creation of superior properties in advanced composites. One of the key properties of carbon nanoforms that makes them suitable for many applications is that they offer a range in conductivities of the printed materials, however, it is strongly size dependent. The particle size on the other hand controls the ink rheology and printability. Therefore, due to the nature of the composites, printing process compromises are made to the properties of these idealized carbon forms to make them printable. The paper also discusses the reliance of printability on the rheological and flow properties of the ink.

The rheological characterization and understanding of the flow behavior at various shear rates and material loadings helps to assess ink processability and optimization of the process design. Another reason for understanding material rheology is to simulate and link the flow behavior with the actual printing process which is becoming an integral part of the additive manufacturing process. The paper highlights several issues encountered by graphene and carbon nanotube-based materials, one of the main problems being their poor solubility in water, which leads to problems in terms of rheology and dispersion. While the oxidation of these nanoforms improves it, oxidation has a negative effect on conductivity, a pivotal property of the material for many applications. With considerate design of the material and slurry, carbon based ink can be optimized to produce inks with the desired properties. The combination of high quality inks with versatile design capabilities and additive manufacturing could revolutionize the consumer, medical, and industrial electronics.

Author Contributions: C.M. wrote-original draft, E.U.H. and S.A.M.T. critically reviewed the paper. C.S. contributed in the project management and supervision.

Funding: The authors acknowledge funding from the European Union's Horizon 2020 research and innovation programme, M3DLoC (Additive Manufacturing of 3D Microfluidic MEMS for Lab-on-a-Chip applications) under grant agreement No 760662.

Conflicts of Interest: The authors declare no conflict of interest.

References

1. The Editors of Encyclopaedia Britannica. Carbon. In *Encyclopædia Britannica*; Encyclopædia Britannica, Inc.: Chicago, IL, USA, 2018.
2. Wei, X.; Li, D.; Jiang, W.; Gu, Z.; Wang, X.; Zhang, Z.; Sun, Z. 3D Printable Graphene Composite. *Sci. Rep.* **2015**, *5*, 1–7. [CrossRef] [PubMed]
3. Ebbesen, T.W.; Ajayan, P.M. Large-scale synthesis of carbon nanotubes. *Nature* **1992**, *358*, 220–222. [CrossRef]
4. Lee, A.; Sudau, K.; Ahn, K.H.; Lee, S.J.; Willenbacher, N. Optimization of Experimental Parameters to Suppress Nozzle Clogging in Inkjet Printing. *Ind. Eng. Chem. Res.* **2012**, *51*, 13195–13204. [CrossRef]
5. Postiglione, G.; Natale, G.; Griffini, G.; Levi, M.; Turri, S. Conductive 3D microstructures by direct 3D printing of polymer/carbon nanotube nanocomposites via liquid deposition modeling. *Compos. Part A Appl. Sci. Manuf.* **2015**, *76*, 110–114. [CrossRef]
6. Dybowska-Sarapuk, Ł.; Szalapak, J.; Wróblewski, G.; Wyżkiewicz, I.; Słoma, M.; Jakubowska, M. Rheology of inks for various techniques of printed electronics. In *Advanced Mechatronics Solutions*; Springer: New York, NY, USA, 2016; Volume 393.
7. Maksud, M.I.; Yusof, M.; Embong, Z.; Nodin, M.; Rejab, N.A. Investigation on Printability of Carbon Nanotube (CNTs) Inks By Flexographic onto Various Substrates. *Int. J. Mater. Sci. Eng.* **2014**, *2*, 49–55. [CrossRef]
8. Derby, B. Additive Manufacture of Ceramics Components by Inkjet Printing. *Engineering* **2015**, *1*, 113–123. [CrossRef]
9. Guo, Y.; Patanwala, H.S.; Bognet, B.; Ma, A.W.K. Inkjet and inkjet-based 3D printing: Connecting fluid properties and printing performance. *Rapid Prototyp. J.* **2017**, *23*, 562–576. [CrossRef]
10. Motyka, A.L. An Introduction to Rheology with an Emphasis on Application to Dispersions. *J. Chem. Educ.* **1996**, *73*, 374. [CrossRef]
11. Deshpande, A. Techniques in Oscillatory Shear Rheology. Available online: http://www.physics.iitm.ac.in/~{}compflu/Lect-notes/abhijit.pdf (accessed on 29 January 2019).
12. Marques, S.; Creus, G. Rheological Models: Integral and Differential Representations. In *Computational Viscoelasticity*; Springer: New York, NY, USA, 2012; pp. 11–21.
13. Wallevik, O.H.; Feys, D.; Wallevik, J.E.; Khayat, K.H. Avoiding inaccurate interpretations of rheological measurements for cement-based materials. *Cem. Concr. Res.* **2015**, *78*, 100–109. [CrossRef]
14. Atala, A.; Lanza, R.; Mikos, T.; Nerem, R. *Principles of Regenerative Medicine*; Elsevier Science: Amsterdam, The Netherlands, 2018; ISBN 9780128098936.

15. Hoath, S.D.; Hsiao, W.-K.; Jung, S. Properties of PEDOT: PSS from Oscillating Drop Studies. In *NIP & Digital Fabrication Conference*; The Society for Imaging Science and Technology: Cambridge, MA, USA, 2014; pp. 299–303.

16. Shenoy, A.V. *Rheology of Filled Polymer Systems*; Springer: Dordrecht, The Netherlands, 1999.

17. Alias, A.A.; Kinoshita, H.; Nishina, Y.; Fujii, M. Dependence of ph level on tribological effect of graphene oxide as an additive in water lubrication. *Int. J. Automot. Mech. Eng.* **2016**, *13*, 3150–3156. [CrossRef]

18. Reinhardt, K.; Hofmann, N.; Eberstein, M. The importance of shear thinning, thixotropic and viscoelastic properties of thick film pastes to predict effects on printing performance. In Proceedings of the EMPC 2017 21st European Microelectronics and Packaging Conference (EMPC) & Exhibition, Warsaw, Poland, 10–13 September 2017; pp. 1–7.

19. Niu, R.; Gong, J.; Xu, D.; Tang, T.; Sun, Z.-Y. The Effect of Particle Shape on the Structure and Rheological Properties of Carbon-Based Particle Suspensions. *Chin. J. Polym. Sci.* **2015**, *33*, 1550–1561. [CrossRef]

20. Vafaei, S.; Tuck, C.; Ashcroft, I.; Wildman, R. Surface microstructuring to modify wettability for 3D printing of nano-filled inks. *Chem. Eng. Res. Des.* **2016**, *109*, 414–420. [CrossRef]

21. Konios, D.; Stylianakis, M.M.; Stratakis, E.; Kymakis, E. Dispersion behaviour of graphene oxide and reduced graphene oxide. *J. Colloid Interface Sci.* **2014**, *430*, 108–112. [CrossRef] [PubMed]

22. Batchelor, G.K. *An Introduction to Fluid Dynamics*; Cambridge University Press: Cambridge, UK, 2000; ISBN 9780521663960.

23. Derby, B.; Reis, N. Inkjet Printing of Highly Loaded Particulate Suspensions. *MRS Bull.* **2003**, *28*, 815–818. [CrossRef]

24. Kuscer, D.; Shen, J.Z. Chapter 18—Advanced Direct Forming Processes for the Future. In *Advanced Ceramics for Dentistry*; Shen, J.Z., Kosmač, T., Eds.; Butterworth-Heinemann: Oxford, UK, 2014; pp. 375–390. ISBN 978-0-12-394619-5.

25. Kovalchuk, N.M.; Nowak, E.; Simmons, M.J.H. Kinetics of liquid bridges and formation of satellite droplets: Difference between micellar and bi-layer forming solutions. *Colloids Surfaces A Physicochem. Eng. Asp.* **2017**, *521*, 193–203. [CrossRef]

26. Kim, C.; Bernal, L. Density and viscosity ratio effects in droplet formation. In *38th Aerospace Sciences Meeting and Exhibit*; Aerospace Sciences Meetings; American Institute of Aeronautics and Astronautics: Reston, VA, USA, 2000.

27. Hoath, S.; Martin, G.D.; Hatchings, I.M. Effects of fluid viscosity on drop-on-demand ink-jet break-off. In *NIP & Digital Fabrication Conference*; Society for Imaging Science and Technology: Washington, DC, USA, 2010.

28. Prudenziati, M.; Hormadaly, J. 1—Technologies for printed films. In *Woodhead Publishing Series in Electronic and Optical Materials*; Prudenziati, M., Hormadaly, J.B.T.-P.F., Eds.; Woodhead Publishing: Cambridge, UK, 2012; pp. 3–29. ISBN 978-1-84569-988-8.

29. Dybowska-Sarapuk, L.; Kielbasinski, K.; Arazna, A.; Futera, K.; Skalski, A.; Janczak, D.; Sloma, M.; Jakubowska, M. Efficient Inkjet Printing of Graphene-Based Elements: Influence of Dispersing Agent on Ink Viscosity. *Nanomater* **2018**, *8*, 602. [CrossRef] [PubMed]

30. Udawattha, D.S.; Narayana, M.; Wijayarathne, U.P.L. Predicting the effective viscosity of nanofluids based on the rheology of suspensions of solid particles. *J. King Saud Univ. Sci.* **2017**, in press. [CrossRef]

31. Naficy, S.; Jalili, R.; Aboutalebi, S.H.; Gorkin, R.A., III; Konstantinov, K.; Innis, P.C.; Spinks, G.M.; Poulin, P.; Wallace, G.G. Graphene oxide dispersions: Tuning rheology to enable fabrication. *Mater. Horizons* **2014**, *1*, 326–331. [CrossRef]

32. Vallés, C.; Young, R.J.; Lomax, D.J.; Kinloch, I.A. The rheological behaviour of concentrated dispersions of graphene oxide. *J. Mater. Sci.* **2014**, *49*, 6311–6320. [CrossRef]

33. Ma, A.; Mackley, M.; Chinesta, F. The Microstructure and Rheology of Carbon Nanotube Suspensions. *Int. J. Mater. Form.* **2008**, *1*, 75–81. [CrossRef]

34. Hobbie, E.K.; Fry, D.J. Rheology of concentrated carbon nanotube suspensions. *J. Chem. Phys.* **2007**, *126*, 124907. [CrossRef] [PubMed]

35. Huang, Y.Y.; Ahir, S.V.; Terentjev, E. Dispersion rheology of carbon nanotubes in a polymer matrix. *Phys. Rev. B* **2006**, *73*, 125422. [CrossRef]

36. Barrie, C.L.; Griffiths, P.C.; Abbott, R.J.; Grillo, I.; Kudryashov, E.; Smyth, C. Rheology of aqueous carbon black dispersions. *J. Colloid Interface Sci.* **2004**, *272*, 210–217. [CrossRef] [PubMed]

37. Ajinjeru, C.; Kishore, V.; Liu, P.; Hassen, A.A.; Lindahl, J.; Kunc, V.; Duty, C. Rheological evaluation of high temperature polymers to identify successful extrusion parameters. In Proceedings of the 27th Annual International Solid Freeform Fabrication Symposium, Additive Manufacturing Conference, Austin, TX, USA, 7–9 August 2017; pp. 485–494.

38. Hoath, S.D. *Fundamentals of Inkjet Printing: The Science of Inkjet and Droplets*; Wiley: New York, NY, USA, 2016; ISBN 9783527337859.

39. Taris, L.; Poirier, S.; Vinsonneau, S.; Mesnilgrente, F.; Temple-Boyer, P. Experimental Temperature Compensation on Drop-On-Demand Inkjet Printing. *Micro Nanosyst.* **2010**, *2*, 137–141. [CrossRef]

40. Hutchings, G.D.M.; Hoath, S.D.; Hutchings, I.M. Inkjet printing—The physics of manipulating liquid jets and drops. *J. Phys. Conf. Ser.* **2008**, *105*, 12001.

41. Barati, H.; Wu, M.; Kharicha, A.; Ludwig, A. A transient model for nozzle clogging. *Powder Technol.* **2018**, *329*, 181–198. [CrossRef]

42. Barati, H.; Wu, M.; Kharicha, A.; Ludwig, A. A transient model for nozzle clogging—Part II: Validation and verification. *Powder Technol.* **2017**. [CrossRef]

43. Everett, D.H. *Basic Principles of Colloid Science*; Royal Society of Chemistry: Cambridge, UK, 1988.

44. Barrie, C.L. *Rheology of Carbon Black Dispersions*; Cardiff University: Cardiff, UK, 2004.

45. Willenbacher, N.; Georgieva, K. *1 Rheology of Disperse Systems*; Wiley: New York, NY, USA, 2013; pp. 1–57.

46. Mueller, S.; Llewellin, E.; Mader, H.M.; Mueller, B.S.; Mader, A.H.M. The rheology of suspensions of solid particles. *Proc. R. Soc. A Math. Phys. Eng. Sci.* **2009**, *2010*, 1201–1228. [CrossRef]

47. Jakus, A.E.; Secor, E.B.; Rutz, A.L.; Jordan, S.W.; Hersam, M.C.; Shah, R.N. Three-dimensional printing of high-content graphene scaffolds for electronic and biomedical applications. *ACS Nano* **2015**, *9*, 4636–4648. [CrossRef]

48. Zhang, Y.; Nayak, T.R.; Hong, H.; Caia, W. Graphene: A versatile nanoplatform for biomedical applications. *Nanoscale* **2013**, *4*, 3833–3842. [CrossRef]

49. Kim, F.; Cote, L.J.; Huang, J. Graphene oxide: Surface activity and two-dimensional assembly. *Adv. Mater.* **2010**, *22*, 1954–1958. [CrossRef] [PubMed]

50. Fu, K.; Wang, Y.; Yan, C.; Yao, Y.; Chen, Y.; Dai, J.; Lacey, S.; Wang, Y.; Wan, J.; Li, T.; et al. Graphene Oxide-Based Electrode Inks for 3D-Printed Lithium-Ion Batteries. *Adv. Mater.* **2016**, *28*, 2587–2594. [CrossRef] [PubMed]

51. Jakus, A.E.; Shah, R.N. Creating electronic and biomedical structures and devices. *Mater. Matters* **2016**, *11*, 43–48.

52. Secor, E.B.; Prabhumirashi, P.L.; Puntambekar, K.; Geier, M.L.; Hersam, M.C. Inkjet printing of high conductivity, flexible graphene patterns. *J. Phys. Chem. Lett.* **2013**, *4*, 1347–1351. [CrossRef] [PubMed]

53. Ou, L.; Song, B.; Liang, H.; Liu, J.; Feng, X.; Deng, B.; Sun, T.; Shao, L. Toxicity of graphene-family nanoparticles: A general review of the origins and mechanisms. *Part. Fibre Toxicol.* **2016**, *13*, 57. [CrossRef] [PubMed]

54. Fernández-Merino, M.J.; Paredes, J.I.; Villar-Rodil, S.; Guardia, L.; Solís-Fernández, P.; Salinas-Torres, D.; Cazorla-Amorós, D.; Morallón, E.; Martínez-Alonso, A.; Tascón, J.M.D. Investigating the influence of surfactants on the stabilization of aqueous reduced graphene oxide dispersions and the characteristics of their composite films. *Carbon* **2012**, *50*, 3184–3194. [CrossRef]

55. Uddin, M.E.; Kuila, T.; Nayak, G.C.; Kim, N.H.; Ku, B.-C.; Lee, J.H. Effects of various surfactants on the dispersion stability and electrical conductivity of surface modified graphene. *J. Alloys Compd.* **2013**, *562*, 134–142. [CrossRef]

56. Kuila, T.; Bose, S.; Mishra, A.; Khanra, P.; Kim, N.H.; Lee, J. Chemical Functionalization of graphene and its applications. *Prog. Mater. Sci.* **2012**, *57*, 1061–1105. [CrossRef]

57. Lin, Z.; Le, T.; Song, X.; Yao, Y.; Li, Z.; Moon, K.; Tentzeris, M.M.; Wong, C. Preparation of Water-Based Carbon Nanotube Inks and Application in the Inkjet Printing of Carbon Nanotube Gas Sensors. *J. Electron. Packag.* **2013**, *135*, 011001. [CrossRef]

58. Kordás, K.; Mustonen, T.; Tóth, G.; Jantunen, H.; Lajunen, M.; Soldano, C.; Talapatra, S.; Kar, S.; Vajtai, R.; Ajayan, P.M. Inkjet printing of electrically conductive patterns of carbon nanotubes. *Small* **2006**, *2*, 1021–1025. [CrossRef]

59. Shin, S.R.; Farzad, R.; Tamayol, A.; Manoharan, V.; Mostafalu, P.; Zhang, Y.S.; Akbari, M.; Jung, S.M.; Kim, D.; Comotto, M.; et al. A Bioactive Carbon Nanotube-Based Ink for Printing 2D and 3D Flexible Electronics. *Adv. Mater.* **2016**, *28*, 3280–3289. [CrossRef] [PubMed]

60. Chen, P.; Chen, H.; Qiu, J.; Zhou, C. Inkjet printing of single-walled carbon nanotube/RuO2 nanowire supercapacitors on cloth fabrics and flexible substrates. *Nano Res.* **2010**, *3*, 594–603. [CrossRef]

61. Park, C.; Allaby, M. *Oxford Dictionary of Environment and Conservation*; University Press Oxford: Oxford, UK, 2013.

62. Taylor, A.D.; Kim, E.Y.; Humes, V.P.; Kizuka, J.; Thompson, L. Inkjet Printing of Carbon Supported Platinum 3-D Catalyst Layers for Use in Fuel Cells. *J. Power Sources* **2007**, *171*, 101–106. [CrossRef]

63. Tekinalp, H.L.; Kunc, V.; Velez-Garcia, G.M.; Duty, C.E.; Love, L.J.; Naskar, A.K.; Blue, C.A.; Ozcan, S. Highly oriented carbon fiber-polymer composites via additive manufacturing. *Compos. Sci. Technol.* **2014**, *105*, 144–150. [CrossRef]

64. Matsuzaki, R.; Ueda, M.; Namiki, M.; Jeong, T.K.; Asahara, H.; Horiguchi, K.; Nakamura, T.; Todoroki, A.; Hirano, Y. Three-dimensional printing of continuous-fiber composites by in-nozzle impregnation. *Sci. Rep.* **2016**, *6*, 1–7. [CrossRef] [PubMed]

65. Tian, X.; Liu, T.; Yang, C.; Qingrui, W.; Li, D. Interface and performance of 3D printed continuous carbon fiber reinforced PLA composites. *Compos. Part A* **2016**, *88*, 198–205. [CrossRef]

66. Srinivasan, S.; Praveen, V.K.; Philip, R.; Ajayaghosh, A. Bioinspired superhydrophobic coatings of carbon nanotubes and linear π systems based on the "bottom-up" self-assembly approach. *Angew. Chem. Int. Ed.* **2008**, *47*, 5750–5754. [CrossRef] [PubMed]

67. Konkena, B.; Vasudevan, S. Understanding Aqueous Dispersibility of Graphene Oxide and Reduced Graphene Oxide through pKa Measurements. *J. Phys. Chem. Lett.* **2012**, *3*, 867–872. [CrossRef] [PubMed]

68. Ghasemi-Kahrizsangi, A.; Neshati, J.; Shariatpanahi, H.; Akbarinezhad, E. Improving the UV degradation resistance of epoxy coatings using modified carbon black nanoparticles. *Prog. Org. Coat.* **2015**, *85*, 199–207. [CrossRef]

69. Suslick, K.S. Applications of Ultrasound to Materials Chemistry. *MRS Bull.* **1995**, *20*, 29–34. [CrossRef]

70. Kharisov, B.I.; Kharissova, O.V.; Méndez, U.O. Methods for Dispersion of Carbon Nanotubes in Water and Common Solvents. *MRS Proc.* **2014**, *1700*, 109–114. [CrossRef]

71. Tortorich, R.; Choi, J.-W. Inkjet Printing of Carbon Nanotubes. *Nanomaterials* **2013**, *3*, 453–468. [CrossRef] [PubMed]

72. Woltornist, S.J.; Oyer, A.J.; Carrillo, J.-M.Y.; Dobrynin, A.V.; Adamson, D.H. Conductive Thin Films of Pristine Graphene by Solvent Interface Trapping. *ACS Nano* **2013**, *7*, 7062–7066. [CrossRef] [PubMed]

73. Chen, K.L.; Elimelech, M. Aggregation and deposition kinetics of fullerene (C60) nanoparticles. *Langmuir* **2006**, *22*, 10994–11001. [CrossRef] [PubMed]

74. You, X.; Yang, J.; Feng, Q.; Huang, K.; Zhou, H.; Hu, J.; Dong, S. Three-dimensional graphene-based materials by direct ink writing method for lightweight application. *Int. J. Light. Mater. Manuf.* **2018**, *1*, 96–101. [CrossRef]

75. Popescu, M.T.; Tasis, D.; Papadimitriou, K.D.; Gkermpoura, S.; Galiotis, C.; Tsitsilianis, C. Colloidal stabilization of graphene sheets by ionizable amphiphilic block copolymers in various media. *RSC Adv.* **2015**, *5*, 89447–89460. [CrossRef]

76. Su, Y.; Yang, G.; Lu, K.; Petersen, E.J.; Mao, L. Colloidal properties and stability of aqueous suspensions of few-layer graphene: Importance of graphene concentration. *Environ. Pollut.* **2017**, *220*, 469–477. [CrossRef]

77. Chowdhury, I.; Duch, M.C.; Mansukhani, N.D.; Hersam, M.C.; Bouchard, D. Colloidal properties and stability of graphene oxide nanomaterials in the aquatic environment. *Environ. Sci. Technol.* **2013**, *47*, 6288–6296. [CrossRef]

78. Huang, L.; Huang, Y.; Liang, J.; Wan, X.; Chen, Y. Graphene-based conducting inks for direct inkjet printing of flexible conductive patterns and their applications in electric circuits and chemical sensors. *Nano Res.* **2011**, *4*, 675–684. [CrossRef]

79. Torrisi, F.; Hasan, T.; Wu, W.; Sun, Z.; Lombardo, A.; Kulmala, T.S.; Hsieh, G.W.; Jung, S.; Bonaccorso, F.; Paul, P.J.; et al. Inkjet-printed graphene electronics. *ACS Nano* **2012**, *6*, 2992–3006. [CrossRef]

80. Pan, K.; Fan, Y.; Leng, T.; Li, J.; Xin, Z.; Zhang, J.; Hao, L.; Gallop, J.; Novoselov, K.S.; Hu, Z. Sustainable production of highly conductive multilayer graphene ink for wireless connectivity and IoT applications. *Nat. Commun.* **2018**, *9*, 5197. [CrossRef] [PubMed]

81. Hu, L.; Pasta, M.; La Mantia, F.; Cui, L.; Jeong, S.; Deshazer, H.D.; Choi, J.W.; Han, S.M.; Cui, Y. Stretchable, Porous, and Conductive Energy Textiles. *Nano Lett.* **2010**, *10*, 708–714. [CrossRef] [PubMed]

82. Shim, W.; Kwon, Y.; Jeon, S.-Y.; Yu, W.-R. Optimally conductive networks in randomly dispersed CNT:graphene hybrids. *Sci. Rep.* **2015**, *5*, 16568. [CrossRef] [PubMed]

83. Hof, F.; Kampioti, K.; Huang, K.; Jaillet-Bartholome, C.; Derr, A.; Poulin, P.; Yusof, H.; White, T.; Koziol, K.; Paukner, C.; et al. Conductive inks of graphitic nanoparticles from a sustainable carbon feedstock. *Carbon* **2016**, *111*, 142–149. [CrossRef]

84. Liang, J.; Tong, K.; Pei, Q. A Water-Based Silver-Nanowire Screen-Print Ink for the Fabrication of Stretchable Conductors and Wearable Thin-Film Transistors. *Adv. Mater.* **2016**, *28*, 5986–5996. [CrossRef] [PubMed]

85. Mullins, M.; Murphy, F.; Baublyte, L.; McAlea, E.M.; Tofail, S.A. The insurability of nanomaterial production risk. *Nat Nanotechnol.* **2013**, *8*, 222–224. [CrossRef] [PubMed]

86. Arvidsson, R.; Molander, S.; Sandén, B. Review of Potential Environmental and Health Risks of the Nanomaterial Graphene. *Hum. Ecol. Risk Assess.* **2013**, *19*, 873–887. [CrossRef]

87. Madani, S.Y.; Mandel, A.; Seifalian, A.M. A concise review of carbon nanotube's toxicology. *Nano Rev.* **2013**, *4*. [CrossRef]

88. Roy, W.R. The environmental fate and movement of organic solvents in water, soil, and air. In *Handbook of Solvents Vol.3, s*; Illinois State Geological Survey: Richland, WA, USA, 2019; Volume 16.1, pp. 1149–2000.

89. Yu, S.; Yin, Y.; Liu, J. Silver nanoparticles. *Environ. Sci. Process. Impacts* **2013**, *15*, 78–92. [CrossRef]

90. Saad, A.A.E.E. Environmental Pollution Reduction by Using VOC-Free Water-Based Gravure Inks and Drying them with a New Drying System Based on Dielectric Heating. Ph.D. Thesis, Universität Wuppertal, Wuppertal, Germany, 2007.

91. Ha, M.; Xia, Y.; Green, A.A.; Zhang, W.; Renn, M.J.; Kim, C.H.; Hersam, M.C.; Frisbie, C.D. Printed, Sub-3V Digital Circuits on Inks. *ACS Nano* **2010**, *4*, 4388–4395. [CrossRef]

92. Kim, J.H.; Chang, W.S.; Kim, D.; Yang, J.R.; Han, J.T.; Lee, G.W.; Kim, J.T.; Seol, S.K. 3D printing of reduced graphene oxide nanowires. *Adv. Mater.* **2015**, *27*, 157–161. [CrossRef] [PubMed]

93. Carey, T.; Cacovich, S.; Divitini, G.; Ren, J.; Mansouri, A.; Kim, J.M.; Wang, C.; Ducati, C.; Sordan, R.; Torrisi, F. Fully inkjet-printed two-dimensional material field-effect heterojunctions for wearable and textile electronics. *Nat. Commun.* **2017**, *8*, 1202. [CrossRef] [PubMed]

94. Monne, M.A.; Enuka, E.; Wang, Z.; Chen, M.Y. Inkjet printed graphene-based field-effect transistors on flexible substrate. In Proceedings of the SPIE, San Diego, CA, USA, 25 August 2017; Volume 10349.

95. Xiang, L.; Wang, Z.; Liu, Z.; Weigum, S.E.; Yu, Q.; Chen, M.Y. Inkjet-Printed Flexible Biosensor Based on Graphene Field Effect Transistor. *IEEE Sens. J.* **2016**, *16*, 8359–8364. [CrossRef]

96. Ji, A.; Chen, Y.; Wang, X.; Xu, C. Inkjet printed flexible electronics on paper substrate with reduced graphene oxide/carbon black ink. *J. Mater. Sci.* **2018**, *29*, 1–11. [CrossRef]

97. Friedman, J.S.; Girdhar, A.; Gelfand, R.M.; Memik, G.; Mohseni, H.; Taflove, A.; Wessels, B.W.; Leburton, J.-P.; Sahakian, A.V. Cascaded spintronic logic with low-dimensional carbon. *Nat. Commun.* **2017**, *8*, 15635. [CrossRef] [PubMed]

98. Loffredo, F.; Del Mauro, A.D.G.; Burrasca, G.; La Ferrara, V.; Quercia, L.; Massera, E.; Di Francia, G.; Sala, D. Della Ink-jet printing technique in polymer/carbon black sensing device fabrication. *Sens. Actuators B Chem.* **2009**, *143*, 421–429. [CrossRef]

99. Muth, J.T.; Vogt, D.M.; Truby, R.L.; Mengüç, Y.; Kolesky, D.B.; Wood, R.J.; Lewis, J.A. Embedded 3D printing of strain sensors within highly stretchable elastomers. *Adv. Mater.* **2014**, *26*, 6307–6312. [CrossRef]

100. Wang, T.; Cook, C.C.; Serban, S.; Ali, T.; Drago, G.; Derby, B. Fabrication of glucose biosensors by inkjet printing. *arXiv* **2012**, arXiv:1207.1190v1.

101. Browne, M.P.; Novotný, F.; Sofer, Z.; Pumera, M. 3D Printed Graphene Electrodes' Electrochemical Activation. *ACS Appl. Mater. Interfaces* **2018**, *10*, 40294–40301. [CrossRef]

102. Schlatter, S.; Rosset, S.; Shea, H. Inkjet printing of carbon black electrodes for dielectric elastomer actuators. In Proceedings of the SPIE, San Diego, CA, USA, 25 August 2017; Volume 10163.

103. Kwon, O.-S.; Kim, H.; Ko, H.; Lee, J.; Lee, B.; Jung, C.-H.; Choi, J.-H.; Shin, K. Fabrication and characterization of inkjet-printed carbon nanotube electrode patterns on paper. *Carbon* **2013**, *58*, 116–127. [CrossRef]

104. Tortorich, R.P.; Song, E.; Choi, J.-W. Inkjet-Printed Carbon Nanotube Electrodes with Low Sheet Resistance for Electrochemical Sensor Applications. *J. Electrochem. Soc.* **2014**, *161*, B3044–B3048. [CrossRef]

105. Yao, B.; Chandrasekaran, S.; Zhang, J.; Xiao, W.; Qian, F.; Zhu, C.; Duoss, E.B.; Spadaccini, C.M.; Worsley, M.A.; Li, Y. Efficient 3D Printed Pseudocapacitive Electrodes with Ultrahigh MnO_2 Loading. *Joule* **2018**, in press. [CrossRef]

106. Yu, G.; Hu, L.; Liu, N.; Wang, H.; Vosgueritchian, M.; Yang, Y.; Cui, Y.; Bao, Z. Enhancing the supercapacitor performance of graphene/MnO_2 nanostructured electrodes by conductive wrapping. *Nano Lett.* **2011**, *11*, 4438–4442. [CrossRef] [PubMed]

107. Pei, Z.; Hu, H.; Liang, G.; Ye, C. Carbon-Based Flexible and All-Solid-State Micro-supercapacitors Fabricated by Inkjet Printing with Enhanced Performance. *Nano-Micro Lett.* **2017**, *9*, 19. [CrossRef] [PubMed]

108. Chen, B.; Jiang, Y.; Tang, X.; Pan, Y.; Hu, S. Fully Packaged Carbon Nanotube Supercapacitors by Direct Ink Writing on Flexible Substrates. *ACS Appl. Mater. Interfaces* **2017**, *9*, 28433–28440. [CrossRef] [PubMed]

109. Lee, S.J.; Zhu, W.; Nowicki, M.; Lee, G.; Heo, D.N.; Kim, J.; Zuo, Y.Y.; Zhang, L.G. 3D printing nano conductive multi-walled carbon nanotube scaffolds for nerve regeneration. *J. Neural Eng.* **2018**, *15*, 016018. [CrossRef] [PubMed]

110. Ho, C.M.B.; Mishra, A.; Lin, P.T.P.; Ng, S.H.; Yeong, W.Y.; Kim, Y.J.; Yoon, Y.J. 3D Printed Polycaprolactone Carbon Nanotube Composite Scaffolds for Cardiac Tissue Engineering. *Macromol. Biosci.* **2017**, *17*, 1–9. [CrossRef] [PubMed]

111. Gonçalves, E.M.; Oliveira, F.J.; Silva, R.F.; Neto, M.A.; Fernandes, M.H.; Amaral, M.; Vallet-Regí, M.; Vila, M. Three-dimensional printed PCL-hydroxyapatite scaffolds filled with CNTs for bone cell growth stimulation. *J. Biomed. Mater. Res. Part B Appl. Biomater.* **2016**, *104*, 1210–1219. [CrossRef] [PubMed]

112. Liang, A.; Jiang, X.; Hong, X.; Jiang, Y.; Shao, Z.; Zhu, D. Recent Developments Concerning the Dispersion Methods and Mechanisms of Graphene. *Coatings* **2013**, *8*, 33. [CrossRef]

113. Gnanasekaran, K.; Heijmans, T.; van Bennekom, S.; Woldhuis, H.; Wijnia, S.; de With, G.; Friedrich, H. 3D printing of CNT- and graphene-based conductive polymer nanocomposites by fused deposition modeling. *Appl. Mater. Today* **2017**, *9*, 21–28. [CrossRef]

114. Foster, C.W.; Down, M.P.; Zhang, Y.; Ji, X.; Rowley-Neale, S.J.; Smith, G.C.; Kelly, P.J.; Banks, C.E. 3D Printed Graphene Based Energy Storage Devices. *Sci. Rep.* **2017**, *7*, 42233. [CrossRef]

115. Jost, K.; Stenger, D.; Perez, C.R.; McDonough, J.K.; Lian, K.; Gogotsi, Y.; Dion, G. Knitted and screen printed carbon-fiber supercapacitors for applications in wearable electronics. *Energy Environ. Sci.* **2013**, *6*, 2698–2705. [CrossRef]

116. Compton, B.G.; Lewis, J.A. 3D-Printing of Lightweight Cellular Composites. *Adv. Mater.* **2014**, *26*, 5930–5935. [CrossRef] [PubMed]

 *micromachines*

Review

Transparent Conductive Electrodes Based on Graphene-Related Materials

Yun Sung Woo

Department of Advanced Materials Application, Korea Polytechnics, Seoul 13122, Korea; yswoo@kopo.ac.kr;
Tel.: +82-31-749-4085

Received: 30 November 2018; Accepted: 18 December 2018; Published: 26 December 2018

Abstract: Transparent conducting electrodes (TCEs) are the most important key component in photovoltaic and display technology. In particular, graphene has been considered as a viable substitute for indium tin oxide (ITO) due to its optical transparency, excellent electrical conductivity, and chemical stability. The outstanding mechanical strength of graphene also provides an opportunity to apply it as a flexible electrode in wearable electronic devices. At the early stage of the development, TCE films that were produced only with graphene or graphene oxide (GO) were mainly reported. However, since then, the hybrid structure of graphene or GO mixed with other TCE materials has been investigated to further improve TCE performance by complementing the shortcomings of each material. This review provides a summary of the fabrication technology and the performance of various TCE films prepared with graphene-related materials, including graphene that is grown by chemical vapor deposition (CVD) and GO or reduced GO (rGO) dispersed solution and their composite with other TCE materials, such as carbon nanotubes, metal nanowires, and other conductive organic/inorganic material. Finally, several representative applications of the graphene-based TCE films are introduced, including solar cells, organic light-emitting diodes (OLEDs), and electrochromic devices.

Keywords: transparent conducting electrode; flexible electrode; graphene; optoelectronic device

1. Introduction

Transparent conductive materials have been extensively used as essential components of optoelectronic devices, such as liquid crystal displays, touch panels, organic light-emitting diodes (OLEDs), and solar cells. Furthermore, the development of foldable or wearable displays and photoelectric devices has led to a need for electronic conductors that are transparent as well as stretchable. Owing to its relatively high electrical conductivity and transparency, indium tin oxide (ITO) is considered as a standard transparent electrode material for such devices. However, the high cost of raw materials, poor mechanical flexibility, and relatively high temperature of ITO deposition significantly limit the scope of its practical applications. For this reason, it has been attempted to use various kinds of nanoscale materials, such as carbon nanotubes (CNT), graphene, metal nanowires, metal nanogrids, and thin films as a replacement for ITO in transparent conducting electrodes (TCEs) [1–18]. Of these, CNT-based transparent electrodes showed the TCE performance, with a sheet resistance of 24 $\Omega \cdot Sq^{-1}$ at 83% transmittance [18]. TCEs that are composed of randomly distributed metal nanowire networks have also been reported to have high optical transparency, low sheet resistance, and excellent mechanical flexibility [9,16,17]. However, silver or copper nanowires are easily damaged by moisture and external mechanical impact and their adhesion to the plastic substrate is poor [19–23]. Graphene-based electrodes have been investigated using the liquid suspension of graphene and macro-scale graphene synthesized via chemical vapor deposition (CVD) [12,24–28]. Liquid-based suspensions of graphene have an advantage in coating, as they would enable relatively low-cost methods of spin coating, roll-to-roll processing, and printing to be used. Additionally, CVD graphene has excellent physical properties, namely a sheet resistance of 30 $\Omega\cdot$Sq^{-1} at 90% transmittance, and thus is a promising candidate for TCE technology [12].

Despite the fact that TCEs based on graphene are easy to process, low in cost, and have excellent stability, they have a disadvantage in that the sheet resistance is larger than that of metal-based transparent electrodes that exhibit the same level of transparency [29]. Therefore, in recent years, an attempt has been made to complement the shortcomings and disadvantages of each material by constructing a hybrid structure of metallic nanostructure and graphene or graphene oxide (GO) or reduced GO (rGO) [3,30]. It has been reported that the performance of TCEs is improved by hybridizing organic and inorganic materials, such as ITO and poly(3,4-ethylenedioxythiophene) polystyrene sulfonate (PEDOT:PSS) with graphene [31,32].

This paper will review the fabrication methods of TCEs using graphene or GO (or rGO), which have been studied previously, and the optical, electrical, and mechanical properties with their limited application. Next, recent studies that have attempted to overcome the limitations of TCEs made with graphene or GO (or rGO) by introducing hybrid TCEs containing other materials are summarized. The performances of various hybrid TCEs that are based on graphene are summarized in Table 1. Lastly, various applications in optics and optoelectronics, especially in several newly emerging areas, such as electrochromic devices, are addressed along with their challenges and prospects in these fields.

Table 1. The performance of transparent conducting electrodes (TCEs) based on graphene-related materials. CVD–chemical vapor deposition, DETA–diethylenetriamine, rGO–reduced graphene oxide, CNT–carbon nanotubes, SWNT–single-walled carbon nanotubes, MWNT–multi-walled carbon nanotubes, NW–nanowire, GP–Graphene, PEDOT:PSS–poly(3,4-ethylenedioxythiophene) polystyrene sulfonate, ITO–indium tin oxide.

Material	Details	Deposition/Transfer Techniques	Sheet Resistance ($\Omega\cdot$sq^{-1})	Transmission (%)	Ref.
CVD graphene	HNO$_3$ doping	Dry transfer/thermal release tape	~30 (4-layers)	90	[12]
	Cu catayst	Polymer-free transfer	810 (1-layer) 230 (4-layers)	97.4 (1-layer) 89.4 (4-layers)	[33]
	Cu catalyst, HNO$_3$ doping	Clean-lifting transfer	50 (4-layers)	~90 (4-layers)	[34]
	Cu catalyst	Roll-to-Roll green transfer	97.5	5.2k	[35]
	Ni catalyst	Wet transfer	500	75	[28]
	No catalyst	Direct CVD	370–510	82	[36]
	No catalyst, 400–600 °C	Direct CVD	5.2k	84.6	[37]
	Ni/C films on dielectrics	Transfer-free growth	50	96	[38]
	Cu-Ni alloy	Wet transfer	409	96.7	[39]
	Layer-by-layer, acid-doping	Wet transfer	80 (4-layers)	90 (4-layers)	[40]
	Dual n-doping (NH$_2$-SAMs/DETA)	Wet transfer	86 ± 39	96	[41]
rGO	Theraml reduction of GO	Spin -coating	10^2–10^3	80	[42]
	rGO/POEGMA layer	Dip-coating	23.8k	90	[43]
	Thermal reduction of GO	Filtration	43k	95	[44]
	Thermal reduction of GO	Dip-coating	1.8 ± 0.08k	70.7	[45]
Graphene/ CNT	Graphene growh on SWNT	Wet transfer	300	96.4	[46]
	Graphene flake/SWNT	Filtration	100	80	[47]
	CVD Synthesis	Wet transfer	~600	95.8	[48]
	Thermal reduction of rGO on MWNT	Electrostatic adsorption	151k	93	[48]
	Chemically converted grpahene/SWNT hybrid suspension	Spin-coating	636	92	[49]
	Ultralarge GO/SWNT	Langmuir-Blodgett	180–560	77–86	[50]

Table 1. *Cont.*

Material	Details	Deposition/Transfer Techniques	Sheet Resistance ($\Omega \cdot sq^{-1}$)	Transmission (%)	Ref.
Graphan/ metallic nano- structure	CVD graphene on AgNW	-	22	88	[51]
	AgNW on GP	-	33	94	[52]
	GP on AgNW	-	64 ± 6.1	93.6	[53]
	Roll-to-roll encapsulation	-	8	94	[54]
	Graphene/CuNW -Core/shell structure	-	36	79	[55]
	Graphene/CuNW Embedded structure	-	25	82	[56]
	Ag-mesh/Graphene	-	5.39 (GP on mesh) 4.54 (Mesh on GP)	88.1 (GP on mesh) 89.3 (Mesh on GP)	[57]
Graphene/ organics	Graphene/PEDOT:PSS, hybrid ink	Spray coating	600	80	[58]
	rGO/PEDOT:PSS, hybrid ink	Filtration	2.3k	80	[59]
	PEDOT:PSS supproting layer on CVD graphene	Wet transfer	80 ± 4	84.6	[60]
Graphene/ inorganics	CVD graphene on ITO film	-	76.46	88.25	[61]
	ITO nanoparticle on CVD graphene	-	522.21	85	[62]

2. Fabrication of Graphene-Based Transparent Conducting Electrodes (TCEs)

2.1. Chemical Vapor Deposition (CVD) Graphene-Based TCEs

CVD graphene is usually produced by flowing hydrocarbon gas onto a transition metal catalyst in a high-temperature furnace, which has been regarded as the most promising way to synthesize high-quality large-area graphene [12,63–66]. CVD graphene that is grown on transition metals, such as Ni, Cu, Pt, and Co, can be used as a TCE after transferring it onto the desired transparent substrate by removing the underlying metal [67].

In the first report on CVD graphene, multiple layers of graphene were grown on a Ni substrate via carbon dissolution and segregation using a Ni catalyst by a CVD process [48,68]. However, Ni possesses high carbon solubility, which makes it difficult to control the number of graphene layers. Thus, a mixture of monolayer graphene and multilayer graphene were formed on Ni foil. A breakthrough in CVD graphene has been achieved by developing synthetic ways of producing large-area monolayer graphene on a Cu foil using a roll-to-roll method [12,64,69–74]. Unlike Ni, Cu has a low carbon solubility, which makes it possible to grow monolayer graphene with a grain size of several centimeters on Cu foil using a mixture of methane and hydrogen gas at a high temperature of 1000°C as shown in Figures 1a–c and 2; the sheet resistance was reported to be 125 $\Omega \cdot sq^{-1}$ at 97.4% transparency [12]. It has also been demonstrated that Cu–Ni alloy can be used to produce monolayer and multilayer graphene using CVD with methane gas as precursor, since the carbon solubility can be controlled by adjusting the atomic fraction of Ni in Cu (Figure 1d) [39,75–79]. For example, Chen et al. presented the CVD synthesis of large-area, primarily bilayer, graphene on Cu–Ni foil by the use of a cold-wall reactor with methane and hydrogen as precursors [74]. Additionally, Cho et al. recently reported that extremely thin Cu–Ni alloy film could promote the formation of monolayer graphene, regardless of alloying contents by constraining the total amount of carbon that was absorbed into the film [78]. There have also been various attempts to directly grow graphene on a glass substrate [36–38,80]. Recently, it was reported by Sun et al. that large-area and uniform graphene film could be directly grown on glass substrate using catalyst-free atmospheric CVD (APCVD), with the resulting material presenting a sheet resistance of 370–510 $\Omega \cdot sq^{-1}$ at a transmittance of 82% [36,38]. In comparison, annealing-based capping-metal catalyzed synthesis provides a fairy high quality of graphene. Xiong et al. showed that monolayer graphene that was grown on various dielectric substrates via rapid thermal process of substrate coated with amorphous carbon and Ni thin films exhibited a low sheet resistance of ~50 $\Omega \cdot sq^{-1}$ at 95.8% transparency [37]. However, the processing temperature of 1100 °C was too high and not applicable to glass and plastic substrates.

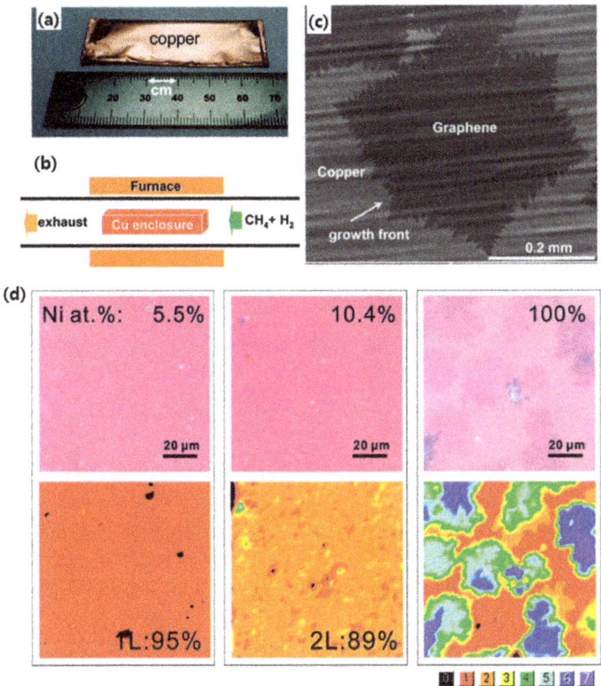

Figure 1. (**a**) Copper foil enclosure prior to insertion in the furnace. (**b**) Schematic of the chemical vapor deposition (CVD) system for graphene on copper. (**c**) SEM image of graphene on copper grown by CVD. Graphene grown at 1035 °C on Cu at an average growth rate of ~6 µm/min. Reproduced with the permission of Reference [65], Copyright 2011. American Chemical Society (**d**) Morphology and layer distribution of various few layers graphene segregated from Cu-Ni alloy at 900 °C after transfer to 300 nm SiO_2/Si substrate. Reproduced with the permission of Reference [76]. Copyright 2011, American Chemical Society.

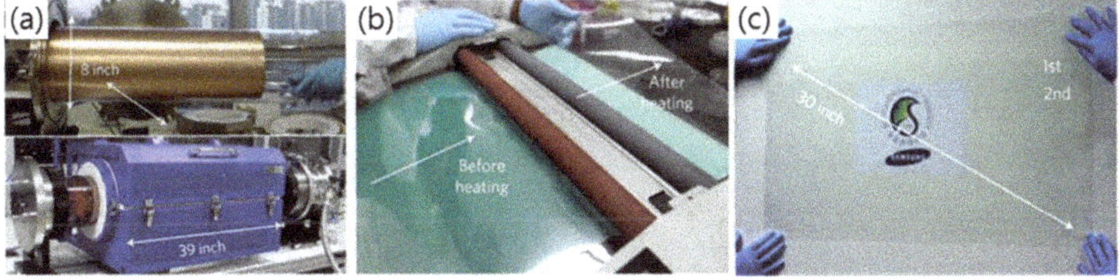

Figure 2. (**a**) Copper foil wrapping around a 7.5-inch quartz tube to be inserted into a 8-inch quartz reactor. The lower image shows the stage in which the copper foil reacts with CH_4 and H_2 gases at high temperatures. (**b**) Roll-to-roll transfer of graphene films from a thermal release tape to a polyethylene terephthalate (PET) film at 120 °C. (**c**) A transparent ultralarge-area graphene film transferred on a 35-inch PET sheet. Reproduced with the permission of Reference [12]. Copyright 2010, Nature Publishing Group.

In order to utilize CVD graphene grown on catalyst metal as a TCE, a transfer step is required to separate graphene film from the catalyst metal and move it to a transparent substrate, such as glass or polyethylene terephthalate (PET) [3,33,52,67,81–85]. Generally, a polymer-support layer is employed to protect the graphene from the external force during the transfer process. After coating the graphene onto the catalyst metal with a protective layer, such as polydimethylsiloxane (PDMS) or poly(methylmethacrylate) (PMMA), a metal catalyst, such as Cu or Ni, is etched away using a chemical etchant, such as HCl, HNO_3, etc. The as-prepared polymer/graphene film after etching the metal catalyst is cleaned with deionized (DI) water and then transferred onto the target substrate. After removing the supporting polymer layer by organic solvent, only some polymer residues are left on the substrate for use as TCEs [67,80,82]. However, one of the disadvantages of this polymer-support transfer method is that polymer residues and defects are generated on graphene [52]. Therefore, various methods have been devised for transferring graphene without a support layer, which are called "polymer-free" methods as described in Figure 3; for example, a thermal release tape-assisted transfer, using a tape with specific adhesives that strongly adheres to substrates at room temperature while losing adhesion at high temperature, and a metal-assisted transfer method, using a metal as a protective layer instead of a polymer [83,85]. In particular, the thermal release tape-assisted transfer method is likely to be suitable for large-scale production because it can inherit the roll-to-roll (R2R) production process for graphene growth [62,63]. For instance, Bae et al. have successfully demonstrated the R2R production and coating of CVD graphene onto flexible substrate using a thermal release tape-assisted transfer method (Figure 2) [12]. Additionally, Lin et al. transferred the graphene to the substrate using "graphite holder", in which monolayer graphene was etched and the etchant was pulled out to be replaced by mixed solvents when the solution was pulled out with the syringe [82]. Although the polymer-free transferred graphene exhibits superior electronic characteristics, the size and shape of the transferred graphene film is limited by the graphite holder. Furthermore, Wang et al. developed a unique "clean-lifting transfer" technique using a controllable electrostatic attraction force to transfer graphene film on various rigid or flexible substrates [81]. In this method, the graphene is attracted to the negatively charged target substrate by the electrostatic force during the transfer process. However, there was a problem with the "polymer-free" transfer method, in that the graphene was deformed easily by external force, such as liquid fluctuation, during metal etching because of the lack of a support layer. Therefore, the development of the "transfer-free" method is crucial for promoting the application of graphene, although the transfer process will still play an important role in the production of graphene devices before the "transfer-free" method becomes mature.

2.2. Graphene Oxide-Based TCEs

Another potential way of producing graphene film on a large scale for TCEs is to deposit GO sheets on the desired substrate and then reduce them using thermal and chemical methods [45,86,87]. Since GO with hydrophilic properties is easily dispersed in water or other organic solvents, it has the advantage that GO solutions can be easily made into a large-area film using a liquid-based, cost-effective method.

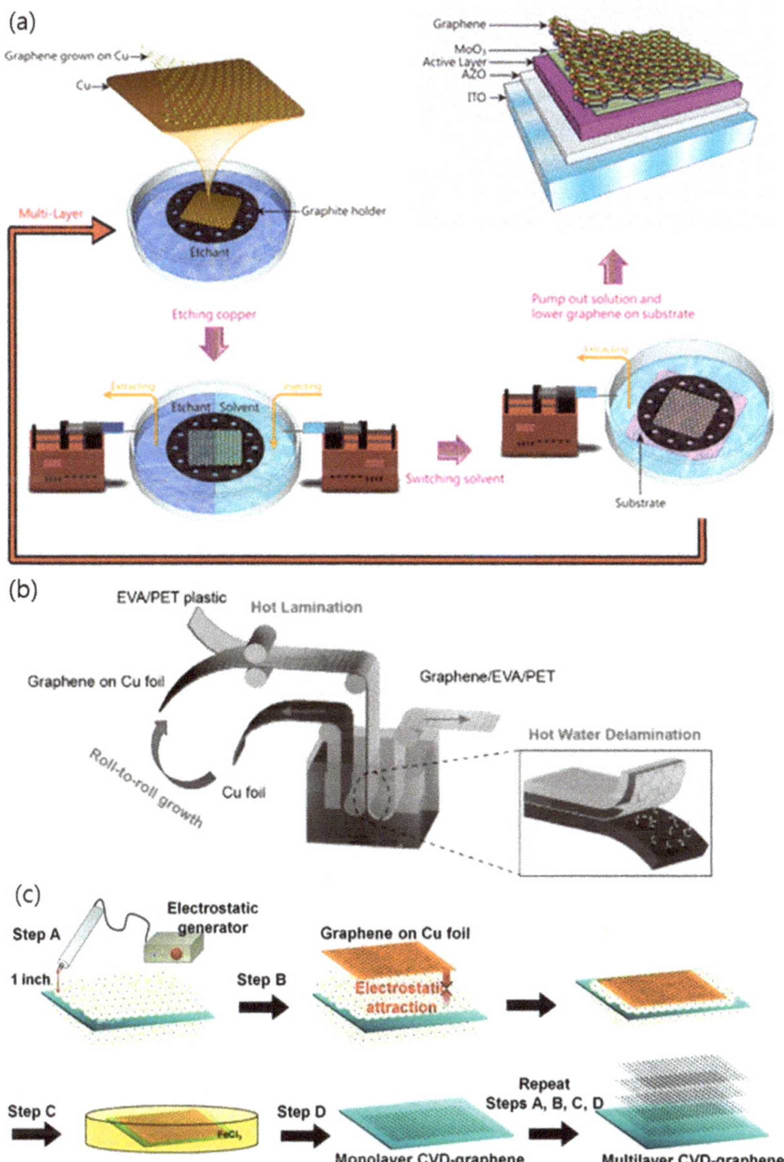

Figure 3. Schematic illustration of (**a**) the polymer-free transfer process, (**b**) the roll-to-roll delamination of copper and graphene onto ethylene-vinyl acetate (EVA)/polyethylene terephthalate (PET) substrate, and (**c**) the clean-lifting transfer process of as-grown graphene on copper foil onto a substrate. Reproduced with the permission of Reference [33,35,81]. Copyright 2014, American Chemical Society. Copyright 2015, 2013, Wiley-VCH.

In general, GO is synthesized by oxidizing graphite with strong acids, followed by intercalation and exfoliation in water, which is represented by the Brodie, Staudenmaier, or Hummers method

(Figure 4a) [88,89]. All three methods involve the oxidation of graphite to various levels [90,91]. The polar oxygen functional groups of GO that are produced by the oxidation process, that is, epoxy and hydroxyl group on the base plane and carboxyl group at the edge, cause it to be exfoliated and disperse well in water and other polar solvents with the assistance of ultrasonic agitation. The prepared GO suspension is then uniformly deposited on an arbitrary substrate while controlling the thickness of GO films for the subsequent TCE production [42,44]. Methods of coating the GO suspension with a film include spin coating, spray coating, dip coating, electrophoretic deposition, Langmuir–Blodgett (LB) assembly, etc. [42,45,92–96]. Spin coating and spray coating are the most convenient ways to deposit a thin film on a substrate from liquid suspension. In particular, spray coating has been proposed as a suitable method for production scale, owing to its fast, scalable, and easy operation [92,93]. Dip coating is also a popular way of coating a GO film on a rigid or flexible substrate, including the immersion of a substrate into GO suspension, the draining of remaining suspension, and the drying of a substrate [94]. However, it was found to be very hard to avoid the partial aggregation and wrinkling or folding of the GO sheet during the spray, spin, or dip coating due to the high flexibility of the sheet. Electrophoretic deposition is performed through the migration of GO sheets in a suspension toward the positive electrode when a direct current (DC) voltage is applied [95]. Despite the many advantages of this technique in film preparation, such as high deposition rate, thickness controllability, and convenience in scaling-up, electrophoretic deposition is limited by the fact that only conductive substrates, such as ITO-coated glass, Al, Ni, and stainless steel are applicable for TCE fabrication. LB assembly is a sophisticated method which allows continuous and uniform film to be deposited on an arbitrary substrate, in which GO sheets floating on water are compressed by LB trough until the desired surface pressure is reached to realize the uniform deposition of GO sheets [96,97].

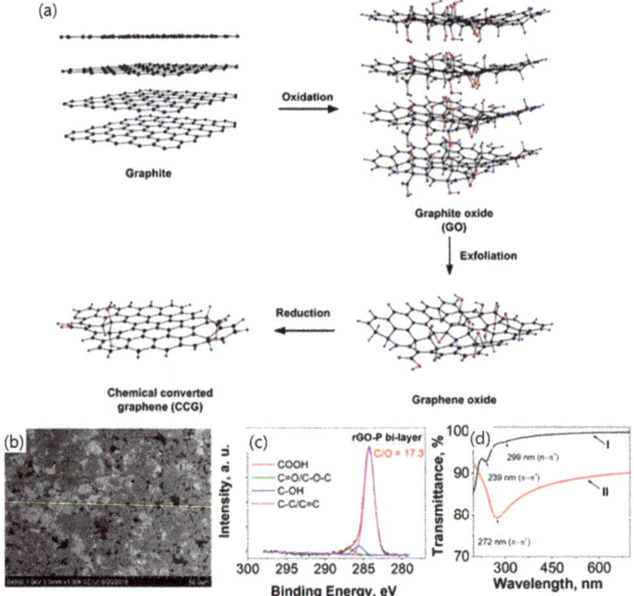

Figure 4. (**a**) Preparation of graphene by chemical reduction of graphene oxide synthesized by Hummers' method. Reproduced with the permission of Reference [98]. Copyright 2012, Royal Society of Chemistry. (**b**) SEM images of reduced graphene oxide-P (rGO-P) bilayer film deposited on undoped Si wafer. (**c**) High-resolution C 1s spectra of rGO-P bilayer film. (**d**) Ultraviolet–visible (UV−vis) transmittance spectra of GP bilayer film (I) and rGO-P bilayer film (II). Reproduced with the permission of Reference [43]. Copyright 2018, American Chemical Society.

Since GO thin film is electrically insulating, it is necessary to reduce it using the thermal and chemical method to recover its conductivity for TCE application. Chemical and thermal reduction is the most common way of reducing GO thin film [97]. In the chemical reduction process, various inorganic and organic reducing agents are applied, such as phenyl hydrazine, hydrazine hydrate, sodium borohydride, ascorbic acid, glucose, hydroxylamine, hydroquinone, pyrrole, amino acids, strongly alkaline solution, and urea [99,100]. However, the chemical reduction is not sufficient to completely recover graphene from GO, and it leaves a substantial amount of residual functionality of epoxy and hydroxyl and carbonyl groups. Thermal reduction by annealing is generally regarded as a more efficient way to reduce GO than chemical reduction [101,102]. Many researchers have reported that the conductivity of GO thin film increases with annealing temperature in a vacuum or reducing atmosphere. In a recent report, rGO thin film exhibited a conductivity of up to 10^4 S·cm^{-1} and a transparency on the level of 90% after annealing at 1100 °C under Ar or N gas flow (Figure 3b–d) [43]. However, since most thermal reduction is carried out at high temperature above 1000 °C for a relatively long time, this approach is not applicable for plastic substrates for flexible TCEs. Other methods have been attempted to reduce GO, such as microwave-assisted heating and the removal of the functional group using a photocatalyst, such as TiO_2 [43,103].

2.3. Chemical Doping of Graphene

Although the graphene films that were prepared using the methods described in Sections 2.1 and 2.2 possess excellent electrical properties and high transparency, their sheet resistance is still too high for the sheets to be used as TCEs. One approach to reduce the sheet resistance of graphene film is post- chemical doping of graphene after transferring it to the desired substrate. The chemical doping of graphene, achieved using chemical species, is classified with surface transfer doping and substitutional doping [104,105]. Surface doping is achieved by charge transfer between graphene and a dopant that is absorbed on the surface of graphene. Graphene can be p-type or n-type doped via chemical doping, depending on the relative position of density of states (DOS) of the highest occupied molecular orbital (HOMO) and lowest unoccupied molecular orbital (LUMO) of the dopant and the Fermi level of graphene. If the HOMO of a dopant is above the Fermi level of graphene, hole transfers take place from the dopant to the graphene, inducing the p-type doping of graphene. If the LUMO of a dopant is below the Fermi level of graphene, then electron transfer takes place from the dopant to the graphene, inducing the n-type doping of graphene.

The resistance of graphene is significantly decreased by charge transfer leading to the p-type doping of graphene when the graphene is exposed to HNO_3, NO_2, $SOBr_2$, Br_2, and I_2 [105–108]. Karsy et al. demonstrated that an eight-stacked layer of graphene which was interlayer-doped with HNO_3 exhibited a sheet resistance of 90 Ω·sq^{-1} at a transmittance of 80% [106]. As described in Figure 5a, through the layer-by-layer doping of four layers of graphene with HNO_3, a sheet resistance of graphene of 80 Ω·sq^{-1} was achieved at a transmittance of 90% [108]. Redox dopants, such as $AuCl_3$ and $AgNO_3$, were also found to lower the sheet resistance of graphene TCEs when graphene was immersed in $AuCl_3$ or $AgNO_3$ solution and Au^{3+} was reduced to form Au nanoparticles on graphene by charge transfer from graphene [109,110]. Surface-adsorbed molecules with electron withdrawing groups can induce a p-type doping effect in graphene. Tetrafluoro-tetra-cyanoquiondimethane (F4-TCNQ) is a strong electron acceptor that is widely used to improve the performance of graphene TCEs for various devices [111–113]. For instance, by using the local density functional theory, Pinto et al. demonstrated that a charge transfer of 0.3 holes/molecule occurs between graphene and F4-TCNQ, which is in agreement with the experimental findings on F4-TCNQ [112]. Graphene can be doped with electrons, that is, n-type doping, via donors such as potassium, ethanol, and ammonia [41,114,115]. Additionally, polymers with amine groups can be used to produce electron-doped graphene. For example, Jo et al. demonstrated a stable and strong n-type doping method with pentaethylene hexamine (PEHA), which reduced the sheet resistance of graphene by up to ~400% compared to pristine graphene [116]. The dual-side n-doping of graphene with diethylenetriamine (DETA) on the top and amine-functionalized self-assembled monolayers (SAMs) at the bottom has been developed to enhance the conductivity of graphene. This method resulted in a sheet resistance as low as ~86 $\Omega \cdot sq^{-1}$ with a transmittance of ~96% (Figure 5b–e) [41].

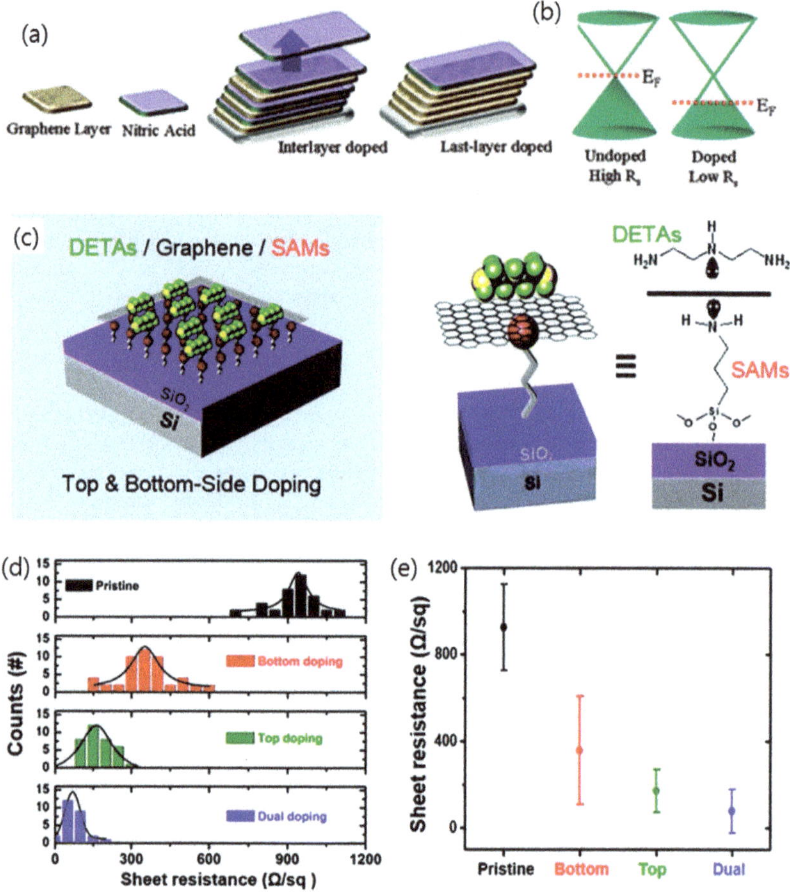

Figure 5. (**a**) Schematic illustrating the interlayer doping methods. The sample is exposed to nitric acid after each layer is stacked. (**b**) Illustration of the graphene band structure, showing the change in the Fermi level due to chemical p-type doping. Reproduced with the permission of Reference [105]. Copyright 2010, American Chemical Society. (**c**) Dual-side doped graphene (**left**) and graphical representation of the molecular structure of the dopants on the both sides of graphene (**right**). (**d**) Histogram of the sheet resistance of graphene doped by NH$_2$-SAMs, diethylenetriamine (DETA), and DETA/NH$_2$-SAMs (dual-side doped). (**e**) Averages and distributions of the sheet resistance plot of four different types of graphene field-effect transistors (FETs). Reproduced with the permission of Reference [41]. Copyright 2014, Royal Society of Chemistry.

3. TCEs of Graphene-Related Materials Hybridized with Other Materials

Combining graphene and other conductive materials can overcome the drawbacks of each individual material. Based on this concept, the hybridization of graphene with various materials, including metal nanowires, carbon nanotubes, and conductive organic and inorganic materials, has been attempted to enhance the conductivity and the optical characteristics of TCEs.

3.1. Hybridization of Graphene with Carbon Nanotubes

CNTs have delivered high axial carrier motilities, making them an obvious choice for use as TCEs. However, TCE films consisting of CNTs have a spiderweb-like network structure with many voids between nanotube bundles; the presence of these voids contributes to the high transparency of CNT films, but restrain the conductance of the films. Meanwhile, CVD graphene presents a fundamental limitation in transmittance and sheet resistance due to its polycrystalline structure, making it difficult to compete with ITO.

To overcome these shortcomings of each substance, a composite that is based on polycrystalline CVD graphene and a subpercolating network of nanowires has been demonstrated, in which nanowires provide the number of electronic pathways to bridge the percolating bottleneck, such as high resistance grain boundary, resulting in reduced resistance while maintaining high transmittance [46,47,51,117]. Kim et al. synthesized a graphene hybrid film by growing graphene using thermal chemical vapor deposition on Cu foil that was coated with single-walled CNTs (SWNTs). The SWNT/graphene hybrid film exhibited superior TCE properties, with a sheet resistance of 300 $\Omega \cdot sq^{-1}$ and transmittance of 96.4%, as compared to graphene spin-coated with SWCNTs, which had a sheet resistance of 1100 $\Omega \cdot sq^{-1}$ and a transmittance of 96.2%. This is presumably due to the low contact resistance between graphene and SWNTs in the hybrid film [47]. Another type of CNT/graphene hybrid film, which has been named as rebar graphene sheets synthesized by annealing the functionalized multi-walled CNT (MWNT) on Cu foil without an exogenous carbon source, has been reported to have ~95.8% transmittance with a sheet resistance of ~600 $\Omega \cdot sq^{-1}$ [118]. Kholmavnov et al. also reported that when CVD graphene is on the top of the MWNT sheet layer (the "G/MWNT" configuration), it significantly modified the MWNT sheet, giving rise to better electrical and optical properties than the reversed structure of the "MWNT/G" configuration [119].

There have also been several attempts to form a nanocomposite comprised of carbon nanotubes and rGO or chemically converted GO (CCG) as shown in Figure 6. The rGO/MWNT double layer was prepared by the sequential electrostatic adsorption of negatively charged GO and positively charged MWNTs on the substrate, followed by the reduction of GO in hydrazine solutions and annealing under an argon atmosphere. The sheet resistance of the rGO/MWNT thin films had the lowest value of 151 $k\Omega \cdot sq^{-1}$ for a 60 µg/mL concentration of aminated MWNT, with a transparency of 93% at a wavelength of 550 nm [48]. Another approach to combine CCG and CNT is to produce hybrid suspension of CCG and CNTs (called G-CNT), as reported by Tung et al. [49]. The stable G-CNT dispersion in hydrazine was readily deposited on a variety of substrates by spin-coating and subsequently heated to 150 °C to remove excess solvents. The G-CNT film had an optical transmittance of 92% and a sheet resistance of 636 $\Omega \cdot sq^{-1}$, which is two orders of magnitude lower than that of the analogous vapor-reduced GO film. This vast improvement in sheet resistance for the G-CNT film is presumably due to the formation of an extended conjugated network with individual CNTs bridging the gap between graphene sheets.

The method of LB assembly was also employed to deposit GO and SWNT in a layer-by-layer manner (Figure 6a). Zheng et al. prepared ultra-large GO sheet/SWNT hybrid films using the LB assembly technique, in which COOH-functionalized SWNTs were crucial for the successful deposition of SWNT layers [50]. The GO/SWNT hybrid film on the substrate was subsequently thermally reduced by heating at 1100 °C to achieve graphitization. A remarkable transmittance, exceeding 90% at a wavelength of 550 nm, was delivered by the 0.5–1.5 bilayer hybrid films and decreased gradually with increasing numbers of bilayers. However, the sheet resistance of the 1.5-bilayer hybrid film was ~600 $\Omega \cdot sq^{-1}$, requiring further improvement by additional acid treatment or doping.

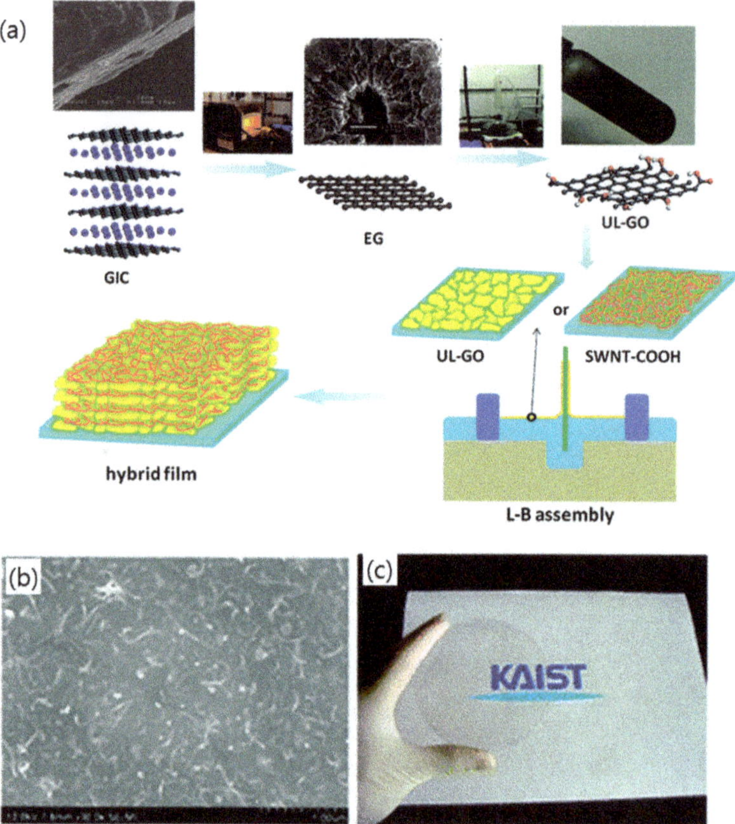

Figure 6. (**a**) Flow chart for the synthesis of ultra-large GO (UL-GO)/SWNT hybrid films using Langmuir–Blodgett (LB) assembly. Reproduced with the permission of Reference [50]. Copyright 2012, Royal Society of Chemistry. (**b**) SEM images of the graphene oxide multi-walled carbon nanotubes (GO/MWNT) double on 500 nm SiO_2/Si substrates that were pretreated with 10 Mm 3-aminopropyltriethoxysilane. (**c**) A photograph of a large, transparent rGO/MWNT electrode fabricated on a 4 in quartz wafer. Reproduced with the permission of Reference [48]. Copyright 2009, American Chemical Society.

3.2. Hybridization of Graphene with Metal Nanostructure

TCEs based on metallic nanostructures, such as metallic nanowires and patterned metal grids, have attracted much attention due to their promising properties of low sheet resistance, high optical transparency, and excellent mechanical durability [17,29]. However, when subjected to conditions of high temperature and current, metallic nanowire networks can experience early failure rates that are caused by the electromigration process and can be destroyed by chemical surface reaction [56,120]. TCEs composed of metallic nanowire networks have the limitation of increasing the electrical conductivity due to the junction resistance between the individual nanowires [121]. Recently, the hybridization of metallic nanostructures and graphene has been devised with the hope that the graphene will complement these drawbacks of TCE of metallic nanostructures, schematically described in Figure 7 [3,16,57,122]. Zhu et al. developed a graphene/metal grid hybrid electrode that was placed onto PET film whose sheet resistance was ~20 $\Omega \cdot sq^{-1}$ at 90% transmittance [123]. The graphene/metal grid hybrid electrode was fabricated by the following sequence. First, metal grids were formed on a transparent substrate, such as PET film. Subsequently, a graphene film grown on Cu foil was transferred to the top of the grid via the wet transfer method and the sacrificial PMMA layer was removed to form the final graphene/metal grid hybrid transparent electrodes. Additionally, high-performance dye-sensitized solar cells fabricated using a hybrid TCE of Pt or Ni grids covered by graphene have been demonstrated, whose efficiencies were comparable to that of the oxide-based transparent electrode [124]. The superior optical and mechanical properties of these graphene/metal grid hybrid electrodes compared to conventional TCEs has motivated the study of hybrid electrodes with various types of graphene and metal nanostructures [60,125].

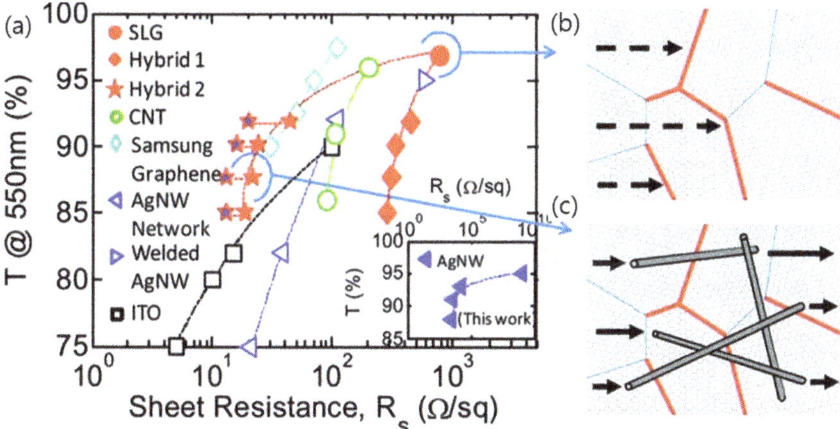

Figure 7. (a) Optical transmittance at wavelength of 550 nm (T@550nm) vs. sheet resistance (R_S) for previous experimental reports, including networks of carbon nanotubes, networks of silver nanowires (AgNW Network and Welded AgNWs), indium tin oxide (ITO), CVD-grown single-layer polycrystalline graphene (SLG) and graphene/NW hybrid structures with various nanowire densities. (**b,c**) illustrate the transport across grain boundaries (GBs) in CVD SLG and hybrid SLG AgNWs networks, respectively. Low-resistance grain boundaries (LGBs, blue lines) and high-resistance grain boundaries (HGBs, red lines) are shown. The HGBs dominate the resistance in SLG. In hybrid structures with appropriate densities of AgNWs, the NWs bridge the HGBs, providing a percolating transport path for the electrons and therefore lowering the sheet resist. Reproduced with the permission of Reference [51]. Copyright 2013, Wiley-VCH.

TCEs composed of a random network of metal nanowires, such as silver or copper nanowires, have the advantage that they can be manufactured in an inexpensive roll-to-roll process while maintaining the high conductivity of the metal [32,51,53,54]. As shown in Figure 8, a hybrid structure employing CVD graphene and a network of silver nanowires has shown a very low sheet resistance of 22 $\Omega\cdot$sq^{-1} at 88% transmittance with excellent stability, and its sheet resistance was stabilized to 13 $\Omega\cdot$sq^{-1}, even after 4 months [54]. The co-percolating conduction model demonstrates that these superior TCE properties of hybrid structure of CVD graphene and a network of metal nanowires are due to the fact that the high-resistance grain boundaries in graphene are bridged by the silver nanowires and the junction between the nanowires are bridged by graphene, and consequently a low sheet resistance was possibly accomplished, even at moderate nanowire densities. Furthermore, this hybrid TCE of graphene and metal nanowires exhibited multiple functionalities, such as robust stability against electrical breakdown and oxidation and superb flexibility [55,77,126,127]. The hybrid silver nanowire and graphene electrode showed the ultimate flexibility without experiencing a significant change of sheet resistance for bending radii of curvature as small as 3.7 μm with a strain of ~27%. Lee et al. presented an inorganic light-emitting diode (ILED) on a soft contact lens that was fabricated with this hybrid metal nanowire and graphene TCE, with the results suggesting a substantial promise for future flexible and wearable electronics and implantable biosensor devices [77]. Moreover, Metha et al. fabricated graphene-encapsulated copper nanowires, whose electrical and thermal conductivity outperformed those of uncoated copper nanowires using a low-temperature plasma-enhanced CVD, with the results suggesting that graphene-encapsulated copper nanowires can be adopted for TCEs in air-stable flexible device applications [55]. On the other hand, Deng et al. demonstrated continuous R2R production of TCEs based on a metal nanowire network that was fully encapsulated between a graphene monolayer and plastic. This low cost and scalable manufacturing method of graphene/metal nanowire hybrid TCEs is expected to accelerate its application to various fields in industry [54].

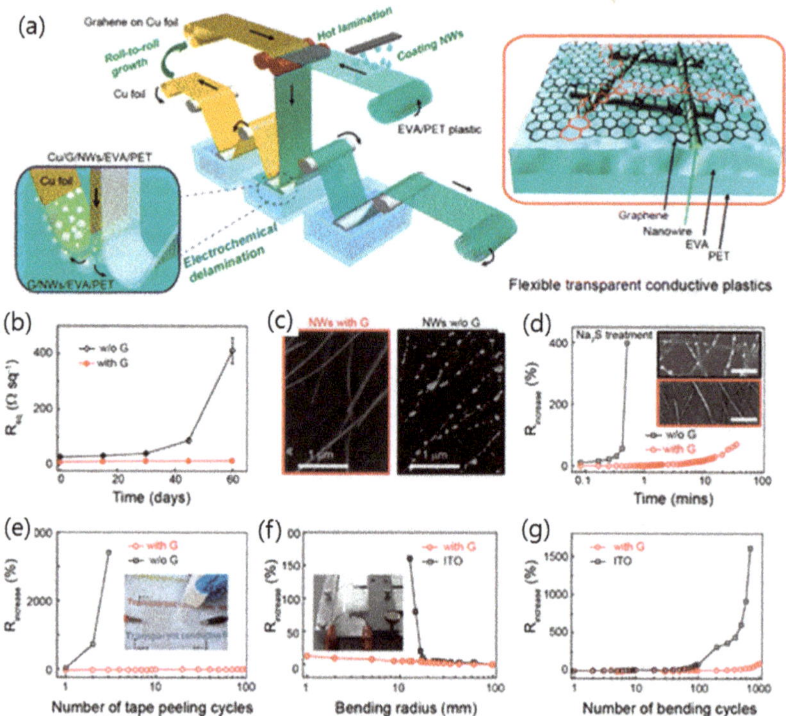

Figure 8. (**a**) Schematic and structure of graphene and metal nanowire hybrid films produced by a continuous roll-to-roll process. (**b**) The durability of graphene and metal nanowire hybrid transparent electrodes. Changes in sheet resistance of pure AgNW films and the graphene/AgNW hybrid films exposed in air at room temperature for 2 months. (**c**) SEM image of the graphene/AgNW hybrid film and pure AgNW films exposed in air for two months, revealing that AgNWs without the protection of graphene were oxidized to break. (**d**) Changes in sheet resistance of pure AgNW films and the graphene/AgNW hybrid films under the attack of aqueous Na$_2$S (4 wt.%). (Inset) Morphologies of AgNWs with or without the graphene coverage attacked for 30 s, respectively. Scale bar: 1 μm. (**e**) Variations in sheet resistance of pure AgNW films and graphene/AgNW hybrid films as a function of the number of cycles of repeated peeling by 3M Scotch tape. (**f**) Variations in sheet resistance versus bending radius for the hybrid transparent plastic electrodes and ITO films on 150 μm thick PET. (**g**) Variations in sheet resistance of the hybrid transparent plastic electrodes and ITO films on PET as a function of the number of cycles of repeated bending to a radius of 20 mm. Reproduced with the permission of Reference [54]. Copyright 2015, American Chemical Society.

3.3. Hybridization of Graphene Oxide with Metal Nanostructure

Hybrid structures of metal nanowires and GO (or rGO) for the production of high-performance transparent electrodes have developed since the low-cost and large-scalable solution process has made it possible to deposit GO (or rGO) film on a plastic substrate. Typically, hybrid TCEs of silver or copper nanowires and GO (or rGO) have been prepared by coating GO (or rGO) onto the silver nanowire film using the dip, spin, and spray coating method [86].

As mentioned above, the issue of the long-term stability of metal nanowire film makes it difficult to use in practical TCEs. However, GO (or rGO)-coated silver nanowire films exhibited highly enhanced long-term stability due to the excellent gas barrier properties of the GO (or rGO) passivation layer on metal nanowire film [78,127–129]. Ahn et al. reported that the sheet resistance of silver nanowire/rGO film was slightly increased, by less than 50%, even at 70 °C and 70% relative humidity (RH) for eight days, while the silver nanowire film showed increased sheet resistance, more than 300% [126]. Additionally, it has been shown that hydrophilic GO nanosheet can be used as a novel adhesive overcoating layer on hydrophilic silver nanowire/PET film that tightly holds the silver nanowires and reduces the sheet resistance in Figure 9 [128,130]. This GO/silver nanowire hybrid TCE also exhibited excellent bendability, showing an almost constant sheet resistance through over 10,000 bending cycles with a ~2 mm curvature radius.

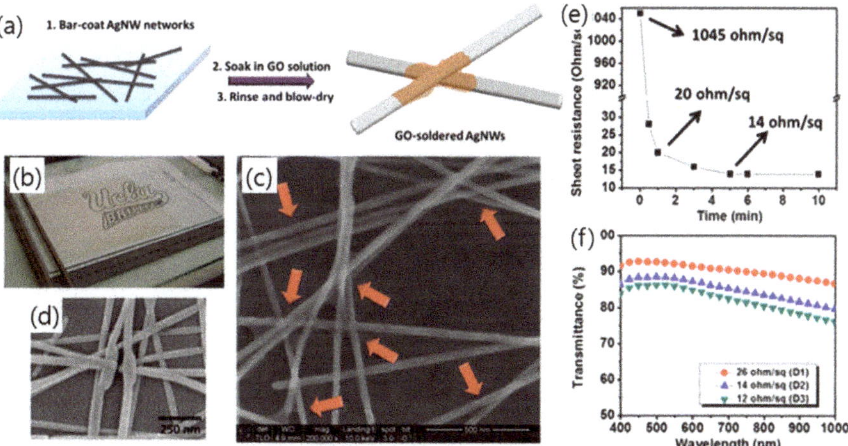

Figure 9. (**a**) Schematic illustration of the fabrication of a GO-AgNW network on a glass substrate at room temperature. (**b**) Optical photograph of a AgNW network bar-coated on a drawdown machine. SEM images of (**c**) GO-soldered AgNW junctions (indicated by red arrows) and (**d**) typical high-temperature fused AgNW junctions. (**e**) Measured sheet resistance of a GO-AgNW network as a function of the soaking time in a GO solution. (**f**) Transmittance spectra of GO-AgNW networks with three different AgNW densities (D1, D2, and D3). Reproduced with the permission of Reference [130]. Copyright 2014, American Chemical Society.

It has also been reported that a high-quality copper/rGO core/shell nanowire could be obtained by wrapping the GO over the surface of the copper nanowire and subsequent mild thermal annealing. These ultrathin core-shell nanowires produced high-performance TCEs with excellent optical and electrical properties, that is, with a sheet resistance of ~28 $\Omega \cdot sq^{-1}$ and a haze of ~2% at a transmittance of ~90%, as reported by Duo et al. [131]. It has also been demonstrated that a film composed of rGO assembled onto copper nanowire film has improved electrical conductivity, oxidation resistance, substrate adhesion, and stability in harsh environments [132]. In particular, an electrochromic device employing the rGO/copper nanowire hybrid TCEs showed reversible coloration/bleaching properties, which cannot be obtained using pure copper nanowire TCEs, since these nanowires form copper hexacyanoferrate compounds during the electrochemical bleaching process, while the rGO protects the copper nanowire from reacting with the harsh solution used for the deposition of electrochromic material [133].

3.4. Hybridization of Graphene with Conducting Polymer

PEDOT:PSS/graphene composite has been fabricated using various methods to give improved performance in chemical and electrical properties [31,58–60,134].

Jo et al. produced a stable aqueous suspension of rGO nanosheet through the chemical reduction of GO in the presence of PEDOT:PSS [59]. The resultant rGO/PEDOT:PSS suspension yielded a hybrid TCE film with a high conductivity of 2.3 $\Omega \cdot sq^{-1}$ with a transmission of 80%. An easy, low cost mass production of PEDOT:PSS/graphene composite has also been developed through the in situ polymerization of PEDOT in the presence of rGO and high-molecular PSS [31]. Since the rGO was used, no additional reduction processes of graphene were required. Liu et al. employed electrochemically exfoliated graphene to prepare the hybrid ink of PEDOT:PSS and graphene, since the electrochemical exfoliation of graphite produces high-quality graphene at a bulk scale (Figure 10) [58]. In order to disperse exfoliated graphene at higher concentrations, PH1000 (Heraeus Clevious, USA) was selected as a surfactant due to its conjugated aromatic chains that can strongly anchor onto the graphene surface via π–π interactions. Subsequently, the TCE films were prepared by the spray-coating method, and their sheet resistance was measured to be 500 $\Omega \cdot sq^{-1}$ at a transmission of 80% after applying 100 cycles of spray-coating.

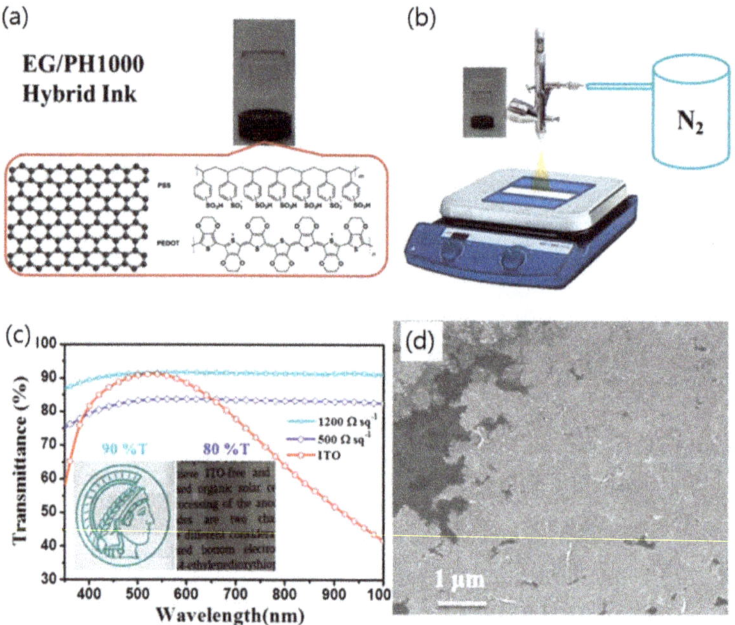

Figure 10. (**a**) Digital image of EG/PH1000 hybrid ink; molecular structures of EG and PH1000. (**b**) Schematic illustration of spray-coating an EG/PH1000 hybrid ink onto desired substrates. (**c**) Transmittance spectrum of both EG/PH1000 hybrid films and ITO on PET substrates. Inset shows the optical images of the EG/PH1000 hybrid films on PET substrates with 90% and 80% transmittance, respectively. (**d**) SEM image of spray-coated EG/PH1000 hybrid film. Reproduced with the permission of Reference [58]. Copyright 2015, Wiley-VCH.

On the other hand, graphene/PEDOT:PSS bilayers have been utilized for the purpose of using PEDOT:PSS as a new supporting layer for the transfer of CVD graphene film without a removal process [124]. An important advantage of this approach is that PEDOT:PSS acts as an effective dopant for CVD graphene film, which exhibits a sheet resistance of $80 \pm 4 \ \Omega \cdot sq^{-1}$ with excellent stability in air. This graphene/PEDOT:PSS bilayer exhibited a high transparency, with a transparency decrease of only 1% being caused by adding a PEDOT:PSS layer.

3.5. Hybridization of Graphene with Oxide

There are not many reports on the fabrication of hybrid metal oxide/graphene TCEs when compared to hybrids using organic materials or nanomaterials with graphene.

Transparent conductive oxides, such as ITO and aluminum-doped zinc oxide (AZO), are widely used in electrodes, however the rigidity of these materials has limited their use as flexible electrodes. There have been attempts to improve the mechanical properties of ITO by fabricating a hybrid electrode using graphene and ITO [61,62]. Liu et al. demonstrated that the graphene/ITO hybrid electrodes showed a resistance change $(\Delta R/R_0)$ of 17.78 after 20% tensile strain; meanwhile, the (non-hybrid) ITO electrode showed a resistance change of 125.91. When the bending radius was 0.1 cm, the resistance change of the ITO electrode fell to ~115.51, while that of the graphene/ITO hybrid electrode fell to ~11.65. These results showed the benefits of the graphene/ITO hybrid electrode over the ITO electrode in terms of mechanical flexibility [61]. Another approach for the hybridization of graphene with ITO was the uniform dispersal of ITO nanoparticles with a size of 25–35 nm on CVD graphene, which is synthesized by the immersion of graphene into aqueous ITO sol-gel. The ITO nanoparticle-decorated graphene exhibited a decrease in sheet resistance of about 28.2% relative to that of CVD graphene, owing to the electron doping of graphene that is induced by the ITO nanoparticles [62].

As shown in Figure 11, a multilayered electrode in which graphene is sandwiched between metal oxide has been demonstrated to have high electrical stability and optical transparency [69]. Since the coating of graphene with metal oxide prevents the desorption of chemical dopants, graphene that is coated with a 60-nm-thick layer of WO_3 showed a lower resistance and slower degradation rate than pristine graphene. Furthermore, the optical transmittance of this WO_3-coated graphene could be enhanced with the addition of a metal oxide layer between the graphene and the glass, which satisfies the zero reflection condition. As a result, the optimal multilayered structure of TiO_2 (62 nm)/graphene (3 layers)/WO_3 (60 nm) on glass showed a transmittance of ~90%, which is same as the highest transmittance of glass/ITO.

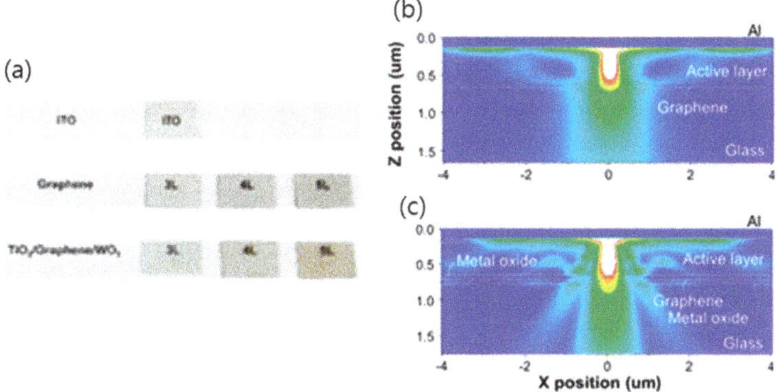

Figure 11. (**a**) Photographs of ITO, graphene, TiO_2/graphene/WO_3 of 3–5 graphene layers. The Finite-Difference Time-Domain (FDTD) simulation. Calculated Poynting vector magnitude of OLEDs (λ = 520 nm) on (**b**) 3 layers graphene (G) and (**c**) TiO_2/3 layers graphene/WO_3 electrodes by FDTD method. Reproduced with the permission of Reference [69]. Copyright 2016, American Chemical Society.

4. Application of Graphene-Based TCEs

The excellent performance of various graphene-based TCEs give graphene a realistic chance of becoming competitive in the production of transparent and bendable device technologies. In particular, the combination of high chemical and thermal stability, high stretchability, and low contact resistance with organic materials offers tremendous advantages for using graphene for TCEs in organic electronic devices, such as solar cells, OLEDs, touch screens, field effect transistors, sensors, and electrochromic devices [70,135].

4.1. Solar Cells

A low-cost, exfoliated graphene oxide followed by thermal reduction was first studied for application as window electrodes in solid-state dye-sensitized solar cell [45]. Eda et al. utilized reduced and doped GO thin films as the cathode in optovoltaics (OPVs), which were fabricated by spin coating a layer of PEDOT:PSS on top of a rGO film and subsequently depositing a poly(3-hexylithiophene) (P3HT) and phenyl-C61-bytyric acid methyl ester (PCBM) nanocomposite layer [136]. Additionally, a rGO film with 1100°C thermal annealing has been utilized as an anode in a dye-sensitized solid solar cell based on spiro-OMeTAD and porous TiO_2, however the short-circuit photocurrent density (I_{sc}) and efficiency of the graphene-based cell was somewhat lower when compared to an fluorine-doped tin oxide (FTO)-based cell, possibly due to the series resistance of the device, the relatively lower transmittance of the electrode, and the electronic interfacial change.

Subsequently, an organic solar cell fabricated with TCEs based on CVD graphene with a low sheet resistance has demonstrated excellent performance with an enhanced power conversion efficiency (PCE) [33,137]. Since CVD graphene offers high conductivity when compared to rGO, it offers a great advantage for the fabrication of OPV devices. A multilayer graphene (MLG) film with a relatively low sheet resistance of 374 $\Omega \cdot sq^{-1}$ at 84.2% transparency was obtained, and the MLG) film-based solar cell with a P3HT:PCBM blend as the active layer exhibited a PCE of 1.17% [137]. An inverted structural solar cell with AZO at the bottom as cathodes, molybdenum-oxide/graphene on top as anode, and P3HT:PCBM as an active layer exhibited a PCE of 2.2%.

Various hybrid TCEs using graphene and a highly conductive material have been employed as a window electrode in the fabrication of OPVs [49,55,60,127,134]. Lee et al. reported that a CVD graphene/PEDOT:PSS bilayer could provide the highest PCE of 5.5% and a fill factor (FF) of 0.67, which is even higher than what is obtainable with the best ITO device (Figure 12) [60]. Additionally, rGO-coated silver nanowire film with excellent thermal and chemical stabilities has been used for the anode layers in bulk heterojunction polymer solar cell [127]. Under illumination, this showed an open-circuit voltage (V_{oc}) of 0.49 V, a short-circuit current density (J_{sc}) of 6.38 mA·cm^{-2}, and an FF of 32.91, resulting in a PEC of 1.03%. Furthermore, a solution-based nanocomposite that is comprised of chemically converted graphene and carbon nanotubes has been used as a platform for the fabrication of P3HT:PCBM photovoltaic (PV) devices. After spin-coating a mixture of graphene and CNT, a thin buffer layer of PEDOT:PSS and P3HT:PCBM with a 1:1 weight ratio was coated on glass [49]. Finally, the Al and Ca were evaporated as the reflective cathode. For these PV devices, a PCE of 0.85% was measured under an illumination of AM 1.5 G and the values of J_{sc}, V_{oc}, and FF were 3.47 mA·cm^{-2}, 0.583 V, and 42.1%, respectively. These relatively low values of J_{sc} and FF, which are detrimental to PCE, were likely due to poor contact at the interface between the graphene/CNT composite and polymer.

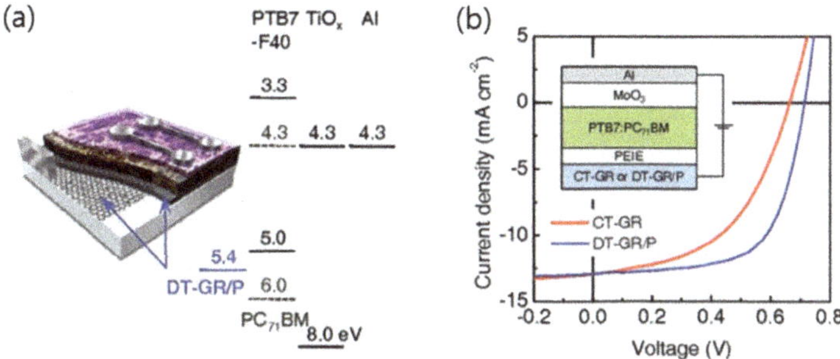

Figure 12. (**a**) Schematic illustrating the structure and corresponding energy-level diagram of the fabricated conventional structure perovskite solar cells (PSC). For some of the devices, an additional PEDOT:PSS layer (AI 4083, 30 nm) was used with the doping transfer-graphene (DT-GR)/P electrode. (**b**) The current density-voltage (J–V) characteristics of the best inverted structure PSCs with the conventional transfer (CT)-GR (red line) and the DT-GR/P (blue line) electrodes. The inset represents the device structure of the inverted PSCs. Reproduced with the permission of Reference [60]. Copyright 2014, Wiley-VCH.

4.2. Organic Light-Emitting Diodes (OLEDs)

At first, a basic OLED structure of anode/PEDOT:PSS/N,N'-di-1-naphthyl-N,N'-diphenyl-1,1'-biphenyl-4,4'diamine (NPD)/tris(8-hydroxyquinoline) aluminum (Alq_3)/LiF/Al was adopted to investigate the performance when a graphene film was used as the transparent electrode [26]. The graphene electrodes were deposited on quartz slides by spin-coating water-based dispersion of functionalized graphene and reduced by high-temperature vacuum annealing; the sheet resistance of the electrodes was ~800 $\Omega \cdot sq^{-1}$ at a transmittance of 82% at 550 nm. An OLED with graphene anode exhibited a turn-on voltage of 4.5 V and it reached a luminance of 300 $cd \cdot m^{-2}$ at 11.7 V, which is slightly higher than the values for ITO. A sky-blue phosphorescent OLED with multilayered CVD graphene anode, consisting of graphene (2–3 nm)/1,1-bis[(di-4-tolylamino)phenyl]cyclohexane (TAPC) (30 nm)/ 1,4,5,8,9,11-hexaazatriphenylene hexacarbonitrile (HAT-CN)(10 nm)/TAPC(30 nm)/HAT-CN (10 nm)/TAPC(30nm)/4,40,400-tri(N-carbazolyl)triphenylamine:iridium(III)bis[(4,6-difluorophenyl)–pyridinato-N,C20] picolinate (TCTA:FIrpic)(5 nm)/2,6-bis[30-(N-carbazole)phenyl] pyridine:iridium (III) bis[(4,6-difluorophenyl)-pyridinato-N,C20] picolinate (DCzPPy:FIrpic) (5 nm)/1,3-bis(3,5-dipyrid-3-yl-phenyl)benzene(BmPyPB) (40 nm)/lithium fluoride (LiF) (1 nm)/aluminum (Al) (100 nm), showed an electron quantum efficiency (EQE) of 15.6% and a power efficiency (PE) of 24.1 lm/W, comparable to the EQE of 18.5% and PE of 28.5 lm/W obtained with an ITO-based OLED [34].

Flexible OLEDs that were fabricated with graphene-based anode shown in Figure 13 have also been demonstrated in several recent reports [70,130,137,138]. A hybrid graphene/silver nanowire/polymer electrode prepared on PET provided not only stable resistance against air exposure, but also superior flexibility with the bending cycles [138]. The optimal OLED device based on a hybrid graphene/silver nanowire/polymer electrode presented the best performance due to its higher conductivity and light transmittance [130]. A GO-soldered silver nanowire network was also provided as an anode of the fully stretchable polymer light-emitting diode (PLED) with the help of GO solder, which improves the stretchability of the silver nanowire network.

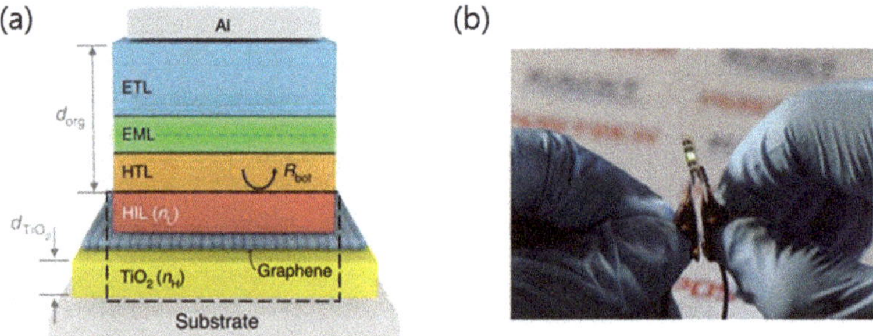

Figure 13. (**a**) Schematic device structure of the proposed flexible TiO_2/graphene Organic light-emitting diodes (OLEDs). (**b**) Photograph of the proposed OLEDs in operation. Reproduced with the permission of Reference [139]. Copyright 2016, Nature Publishing Group.

4.3. Electrochromic Devices

Electrochromic devices are widely used in display devices, including smart windows and mirrors, for improving indoor energy efficiency or personal visual comfort.

Recently, rigid electrochromic devices have advanced, and they now offer flexibility, stretchability, and foldability for deformable devices (Figure 14) [119,133,140,141]. Polat et al. demonstrated a flexible electrochromic device using multilayer graphene, which offers key requirements for practical application, that is, high-contrast optical modulation over a broad spectrum and good electrical conductivity and mechanical flexibility [141]. They showed that the optical transmittance of MLG can be controlled by electrostatic doping via the reversible intercalation of charges into the graphene layers, making it possible to fabricate reflective/transmissive multipixel electrochromic devices.

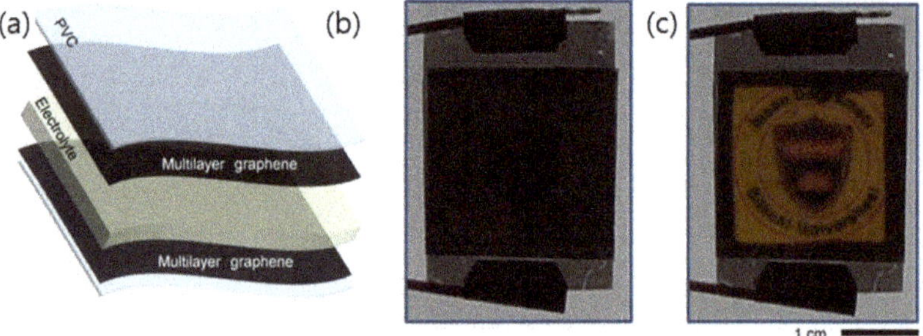

Figure 14. (**a**) Exploded-view illustration of the graphene electrochromic device. The device is formed by attaching two graphene coated polyvinyl chloride (PVC) substrates face to face and imposing ionic liquid in the gap separating the graphene electrodes. (**b,c**), Photographs of the devices under applied bias voltages of 0 V and 5 V, respectively. Reproduced with the permission of Reference [140]. Copyright 2014, Nature Publishing Group.

Smart windows that are based on electrochromic devices were demonstrated by Kim et al., who investigated the electrochemical and electrical characteristics of the PEDOT:PSS polymer-based electrochromic devices s of the different number of graphene layers used as electrodes [142]. The four-layered graphene electrode showed the best electrochemical behaviors, with a fast optical change response of less than 1 s from the dark to the transparent state and 500 ms from the transparent to the dark state and a low bias of $\pm$ 2.5 V for the maximum contrast ratio.

4.4. Transparent Heaters

High performance, transparent, and flexible heaters have been fabricated using graphene-based electrothermal films that are suitable for automobile defogging or deicing systems and heatable smart windows, as shown in Figure 15 [128,143,144]. A multiple-stacked CVD graphene film on PET, being interlayer-doped with $AuCl_3$-CH_3NO_2 and HNO_3 to obtain a low sheet resistance, was provided for flexible heaters, and it was mechanically stable after bending 1000 times with 1.1% strain [142]. Additionally, GO film spin-coated on quartz or polyimide (PI) has also exhibited high transparency and good heating effects.

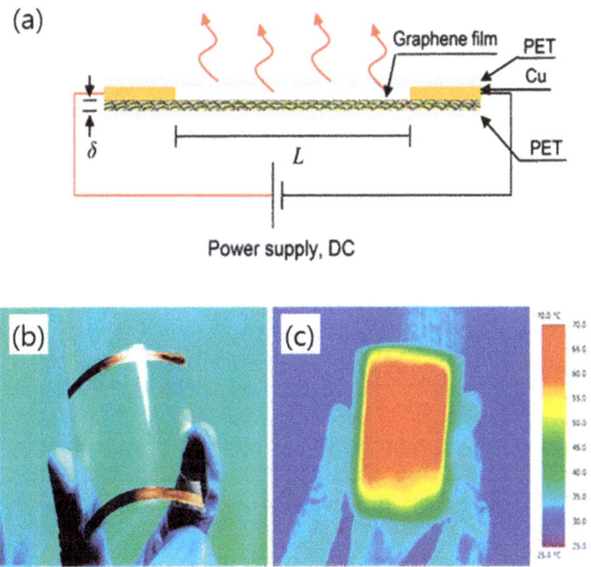

Figure 15. (**a**) A schematic structure of a transparent, flexible graphene heater combined with a plastic substrate and Cu electrodes. (**b**) An optical image of the assembled graphene-based heater showing its outstanding flexibility. (**c**) An infrared picture of the assembled graphene-based heater while applying an input voltage under bending condition. Reproduced with the permission of Reference [143]. Copyright 2011, American Chemical Society.

A new strategy of reduced large-size graphene oxide (rLGO)/silver nanowirehybrid film has been reported to design high-performance transparent film heaters. The thin rLGO provided a protective effect for the silver nanowire network against oxidation as well as the low sheet resistance of rLGO [128].

5. Summary and Conclusions

New TCE materials, including carbon nanotubes, metal grids and nanowires, conductive polymers, and recently graphene, have appeared over the past decades in many fields of application.

In particular, these new materials offer prospective advantages of flexibility, bendability, and even stretchability, enabling them to be applied as TCEs in wearable optoelectronic devices and displays. Among them, graphene has attracted special attention due to its superior electrical conductivity and optical transmission when compared to other TCE materials. Additionally, the excellent moisture barrier and mechanical flexibility of graphene allows it to be hybridized with other TCE materials to further improve the properties of TCE films.

In the past decades, a number of methods for synthesizing high-quality graphene at low cost have been extensively studied, and the development of a large-area synthesis and transfer method using a roll-to-roll process has been a viable step in the practical application of graphene TCEs. Despite the tremendous advances in the graphene synthesis process and quality improvement, TCE films that are based on graphene still have issues that need to be improved, such as poor adhesion to substrates, low abrasion, and poor electrical conductivity as compared to conventional ITO.

Funding: This research was supported by the Basic Science Research Program through the National Research Foundation of Korea (NRF) funded by the Ministry of Education (grant number NRF-2018R1D1A1B07049592).

Conflicts of Interest: The author declares no conflict of interest.

References

1. López-Naranjo, E.J.; González-Ortiz, L.J.; Apátiga, L.M.; Rivera-Muñoz, E.M.; Manzano-Ramírez, A. Transparent electrodes: A review of the use of carbon-based nanomaterials. *J. Nanomater.* **2016**, *2016*, 4928365. [CrossRef]

2. Hofmann, A.I.; Cloutet, E.; Hadziioannou, G. Materials for Transparent Electrodes: From Metal Oxides to Organic Alternatives. *Adv. Electron. Mater.* **2018**, 1700412. [CrossRef]

3. Hecht, D.S.; Hu, L.; Irvin, G. Emerging transparent electrodes based on thin films of carbon nanotubes, graphene, and metallic nanostructures. *Adv. Mater.* **2011**, *23*, 1482–1513. [CrossRef] [PubMed]

4. Luo, M.; Liu, Y.; Huang, W.; Qiao, W.; Zhou, Y.; Ye, Y.; Chen, L.-S. Towards flexible transparent electrodes based on carbon and metallic materials. *Micromachines* **2017**, *8*, 12. [CrossRef]

5. Rowell, M.W.; Topinka, M.A.; McGehee, M.D.; Prall, H.-J.; Dennler, G.; Sariciftci, N.S.; Hu, L.; Gruner, G. Organic solar cells with carbon nanotube network electrodes. *Appl. Phys. Lett.* **2006**, *88*, 233506. [CrossRef]

6. Wu, Z.; Chen, Z.; Du, X.; Logan, J.M.; Sippel, J.; Nikolou, M.; Kamaras, K.; Reynolds, J.R.; Tanner, D.B.; Hebard, A.F. Transparent, conductive carbon nanotube films. *Science* **2004**, *305*, 1273–1276. [CrossRef]

7. Lee, J.-Y.; Connor, S.T.; Cui, Y.; Peumans, P. Solution-processed metal nanowire mesh transparent electrodes. *Nano Lett.* **2008**, *8*, 689–692. [CrossRef]

8. Wu, H.; Hu, L.; Rowell, M.W.; Kong, D.; Cha, J.J.; McDonough, J.R.; Zhu, J.; Yang, Y.; McGehee, M.D.; Cui, Y. Electrospun metal nanofiber webs as high-performance transparent electrode. *Nano Lett.* **2010**, *10*, 4242–4248. [CrossRef]

9. Hu, L.; Kim, H.S.; Lee, J.-Y.; Peumans, P.; Cui, Y. Scalable coating and properties of transparent, flexible, silver nanowire electrodes. *ACS Nano* **2010**, *4*, 2955–2963. [CrossRef]

10. Hu, L.; Wu, H.; Cui, Y. Metal nanogrids, nanowires, and nanofibers for transparent electrodes. *MRS Bull.* **2011**, *36*, 760–765. [CrossRef]

11. Wu, H.; Kong, D.; Ruan, Z.; Hsu, P.-C.; Wang, S.; Yu, Z.; Carney, T.J.; Hu, L.; Fan, S.; Cui, Y. A transparent electrode based on a metal nanotrough network. *Nat. Nanotechnol.* **2013**, *8*, 421. [CrossRef] [PubMed]

12. Bae, S.; Kim, H.; Lee, Y.; Xu, X.; Park, J.-S.; Zheng, Y.; Balakrishnan, J.; Lei, T.; Kim, H.R.; Song, Y.I. Roll-to-roll production of 30-inch graphene films for transparent electrodes. *Nat. Nanotechnol.* **2010**, *5*, 574. [CrossRef] [PubMed]

13. Ellmer, K. Past achievements and future challenges in the development of optically transparent electrodes. *Nat. Photonics* **2012**, *6*, 809. [CrossRef]

14. Ye, S.; Rathmell, A.R.; Chen, Z.; Stewart, I.E.; Wiley, B.J. Metal nanowire networks: The next generation of transparent conductors. *Adv. Mater.* **2014**, *26*, 6670–6687. [CrossRef] [PubMed]

15. Xia, Y.; Sun, K.; Ouyang, J. Solution-processed metallic conducting polymer films as transparent electrode of optoelectronic devices. *Adv. Mater.* **2012**, *24*, 2436–2440. [CrossRef] [PubMed]

16. Lee, D.; Lee, H.; Ahn, Y.; Jeong, Y.; Lee, D.-Y.; Lee, Y. Highly stable and flexible silver nanowire–graphene hybrid transparent conducting electrodes for emerging optoelectronic devices. *Nanoscale* **2013**, *5*, 7750–7755. [CrossRef] [PubMed]

17. De, S.; Higgins, T.M.; Lyons, P.E.; Doherty, E.M.; Nirmalraj, P.N.; Blau, W.J.; Boland, J.J.; Coleman, J.N. Silver nanowire networks as flexible, transparent, conducting films: Extremely high DC to optical conductivity ratios. *ACS Nano* **2009**, *3*, 1767–1774. [CrossRef]

18. Yu, L.; Shearer, C.; Shapter, J. Recent development of carbon nanotube transparent conductive films. *Chem. Rev.* **2016**, *116*, 13413–13453. [CrossRef]

19. Choo, D.C.; Kim, T.W. Degradation mechanisms of silver nanowire electrodes under ultraviolet irradiation and heat treatment. *Sci. Rep.* **2017**, *7*, 1696. [CrossRef]

20. Khaligh, H.; Xu, L.; Khosropour, A.; Madeira, A.; Romano, M.; Pradére, C.; Tréguer-Delapierre, M.; Servant, L.; Pope, M.A.; Goldthorpe, I.A. The Joule heating problem in silver nanowire transparent electrodes. *Nanotechnology* **2017**, *28*, 425703. [CrossRef]

21. Elechiguerra, J.L.; Larios-Lopez, L.; Liu, C.; Garcia-Gutierrez, D.; Camacho-Bragado, A.; Yacaman, M.J. Corrosion at the nanoscale: The case of silver nanowires and nanoparticles. *Chem. Mater.* **2005**, *17*, 6042–6052. [CrossRef]

22. Guo, H.; Lin, N.; Chen, Y.; Wang, Z.; Xie, Q.; Zheng, T.; Gao, N.; Li, S.; Kang, J.; Cai, D. Copper nanowires as fully transparent conductive electrodes. *Sci. Rep.* **2013**, *3*, 2323. [CrossRef] [PubMed]

23. Celle, C.; Cabos, A.; Fontecave, T.; Laguitton, B.; Benayad, A.; Guettaz, L.; Pélissier, N.; Nguyen, V.H.; Bellet, D.; Muñoz-Rojas, D. Oxidation of copper nanowire based transparent electrodes in ambient conditions and their stabilization by encapsulation: Application to transparent film heaters. *Nanotechnology* **2018**, *29*, 085701. [CrossRef] [PubMed]

24. Wassei, J.K.; Kaner, R.B. Graphene, a promising transparent conductor. *Mater. Today* **2010**, *13*, 52–59. [CrossRef]

25. Wu, J.; Becerril, H.A.; Bao, Z.; Liu, Z.; Chen, Y.; Peumans, P. Organic solar cells with solution-processed graphene transparent electrodes. *Appl. Phys. Lett.* **2008**, *92*, 237. [CrossRef]

26. Wu, J.; Agrawal, M.; Becerril, H.A.; Bao, Z.; Liu, Z.; Chen, Y.; Peumans, P. Organic light-emitting diodes on solution-processed graphene transparent electrodes. *ACS Nano* **2009**, *4*, 43–48. [CrossRef] [PubMed]

27. La Notte, L.; Villari, E.; Palma, A.L.; Sacchetti, A.; Giangregorio, M.M.; Bruno, G.; Di Carlo, A.; Bianco, G.V.; Reale, A. Laser-patterned functionalized CVD-graphene as highly transparent conductive electrodes for polymer solar cells. *Nanoscale* **2017**, *9*, 62–69. [CrossRef] [PubMed]

28. Gomez De Arco, L.; Zhang, Y.; Schlenker, C.W.; Ryu, K.; Thompson, M.E.; Zhou, C. Continuous, highly flexible, and transparent graphene films by chemical vapor deposition for organic photovoltaics. *ACS Nano* **2010**, *4*, 2865–2873. [CrossRef] [PubMed]

29. Sannicolo, T.; Lagrange, M.; Cabos, A.; Celle, C.; Simonato, J.P.; Bellet, D. Metallic nanowire-based transparent electrodes for next generation flexible devices: A Review. *Small* **2016**, *12*, 6052–6075. [CrossRef] [PubMed]

30. Kim, C.-L.; Jung, C.-W.; Oh, Y.-J.; Kim, D.-E. A highly flexible transparent conductive electrode based on nanomaterials. *NPG Asia Mater.* **2017**, *9*, e438. [CrossRef]

31. Yoo, D.; Kim, J.; Kim, J.H. Direct synthesis of highly conductive poly (3, 4-ethylenedioxythiophene): Poly (4-styrenesulfonate)(PEDOT: PSS)/graphene composites and their applications in energy harvesting systems. *Nano Res.* **2014**, *7*, 717–730. [CrossRef]

32. Seo, T.H.; Lee, S.; Min, K.H.; Chandramohan, S.; Park, A.H.; Lee, G.H.; Park, M.; Suh, E.-K.; Kim, M.J. The role of graphene formed on silver nanowire transparent conductive electrode in ultra-violet light emitting diodes. *Sci. Rep.* **2016**, *6*, 29464. [CrossRef] [PubMed]

33. Lin, W.-H.; Chen, T.-H.; Chang, J.-K.; Taur, J.-I.; Lo, Y.-Y.; Lee, W.-L.; Chang, C.-S.; Su, W.-B.; Wu, C.-I. A direct and polymer-free method for transferring graphene grown by chemical vapor deposition to any substrate. *ACS Nano* **2014**, *8*, 1784–1791. [CrossRef] [PubMed]

34. Hwang, J.; Kyw Choi, H.; Moon, J.; Yong Kim, T.; Shin, J.-W.; Woong Joo, C.; Han, J.-H.; Cho, D.-H.; Woo Huh, J.; Choi, S.-Y. Multilayered graphene anode for blue phosphorescent organic light emitting diodes. *App. Phys. Lett.* **2012**, *100*, 82. [CrossRef]

35. Chandrashekar, B.N.; Deng, B.; Smitha, A.S.; Chen, Y.; Tan, C.; Zhang, H.; Peng, H.; Liu, Z. Roll-to-roll green transfer of CVD graphene onto plastic for a transparent and flexible triboelectric nanogenerator. *Adv. Mater.* **2015**, *27*, 5210–5216. [CrossRef] [PubMed]

36. Sun, J.; Chen, Z.; Yuan, L.; Chen, Y.; Ning, J.; Liu, S.; Ma, D.; Song, X.; Priydarshi, M.K.; Bachmatiuk, A. Direct chemical-vapor-deposition-fabricated, large-scale graphene glass with high carrier mobility and uniformity for touch panel applications. *ACS Nano* **2016**, *10*, 11136–11144. [CrossRef] [PubMed]

37. Sun, J.; Chen, Y.; Cai, X.; Ma, B.; Chen, Z.; Priydarshi, M.K.; Chen, K.; Gao, T.; Song, X.; Ji, Q. Direct low-temperature synthesis of graphene on various glasses by plasma-enhanced chemical vapor deposition for versatile, cost-effective electrodes. *Nano Res.* **2015**, *8*, 3496–3504. [CrossRef]

38. Xiong, W.; Zhou, Y.S.; Jiang, L.J.; Sarkar, A.; Mahjouri-Samani, M.; Xie, Z.Q.; Gao, Y.; Ianno, N.J.; Jiang, L.; Lu, Y.F. Single-Step Formation of Graphene on Dielectric Surfaces. *Adv. Mater.* **2013**, *25*, 630–634. [CrossRef]

39. Chen, S.; Cai, W.; Piner, R.D.; Suk, J.W.; Wu, Y.; Ren, Y.; Kang, J.; Ruoff, R.S. Synthesis and characterization of large-area graphene and graphite films on commercial Cu–Ni alloy foils. *Nano Lett.* **2011**, *11*, 3519–3525. [CrossRef] [PubMed]

40. Wang, Y.; Tong, S.W.; Xu, X.F.; Özyilmaz, B.; Loh, K.P. Interface engineering of layer-by-layer stacked graphene anodes for high-performance organic solar cells. *Adv. Mater.* **2011**, *23*, 1514–1518. [CrossRef] [PubMed]

41. Kim, Y.; Park, J.; Kang, J.; Yoo, J.M.; Choi, K.; Kim, E.S.; Choi, J.-B.; Hwang, C.; Novoselov, K.; Hong, B.H. A highly conducting graphene film with dual-side molecular n-doping. *Nanoscale* **2014**, *6*, 9545–9549. [CrossRef] [PubMed]

42. Becerril, H.A.; Mao, J.; Liu, Z.; Stoltenberg, R.M.; Bao, Z.; Chen, Y. Evaluation of solution-processed reduced graphene oxide films as transparent conductors. *ACS Nano* **2008**, *2*, 463–470. [CrossRef] [PubMed]

43. Savchak, M.; Borodinov, N.; Burtovyy, R.; Anayee, M.; Hu, K.; Ma, R.; Grant, A.; Li, H.; Cutshall, D.B.; Wen, Y. Highly conductive and transparent reduced graphene oxide nanoscale films via thermal conversion of polymer-encapsulated graphene oxide sheets. *ACS Appl. Mater. Interfaces* **2018**, *10*, 3975–3985. [CrossRef] [PubMed]

44. Eda, G.; Fanchini, G.; Chhowalla, M. Large-area ultrathin films of reduced graphene oxide as a transparent and flexible electronic material. *Nat. Nanotechnol.* **2008**, *3*, 270. [CrossRef] [PubMed]

45. Wang, X.; Zhi, L.; Müllen, K. Transparent, conductive graphene electrodes for dye-sensitized solar cells. *Nano Lett.* **2008**, *8*, 323–327. [CrossRef] [PubMed]

46. Kim, S.H.; Song, W.; Jung, M.W.; Kang, M.A.; Kim, K.; Chang, S.J.; Lee, S.S.; Lim, J.; Hwang, J.; Myung, S. Carbon nanotube and graphene hybrid thin film for transparent electrodes and field effect transistors. *Adv. Mater.* **2014**, *26*, 4247–4252. [CrossRef] [PubMed]

47. King, P.J.; Khan, U.; Lotya, M.; De, S.; Coleman, J.N. Improvement of transparent conducting nanotube films by addition of small quantities of graphene. *ACS Nano* **2010**, *4*, 4238–4246. [CrossRef] [PubMed]

48. Kim, Y.-K.; Min, D.-H. Durable large-area thin films of graphene/carbon nanotube double layers as a transparent electrode. *Langmuir* **2009**, *25*, 11302–11306. [CrossRef]

49. Tung, V.C.; Chen, L.-M.; Allen, M.J.; Wassei, J.K.; Nelson, K.; Kaner, R.B.; Yang, Y. Low-temperature solution processing of graphene−carbon nanotube hybrid materials for high-performance transparent conductors. *Nano Lett.* **2009**, *9*, 1949–1955. [CrossRef]

50. Zheng, Q.; Zhang, B.; Lin, X.; Shen, X.; Yousefi, N.; Huang, Z.-D.; Li, Z.; Kim, J.-K. Highly transparent and conducting ultralarge graphene oxide/single-walled carbon nanotube hybrid films produced by Langmuir–Blodgett assembly. *J. Mater. Chem.* **2012**, *22*, 25072–25082. [CrossRef]

51. Chen, R.; Das, S.R.; Jeong, C.; Khan, M.R.; Janes, D.B.; Alam, M.A. Co-Percolating Graphene-Wrapped Silver Nanowire Network for High Performance, Highly Stable, Transparent Conducting Electrodes. *Adv. Funct. Mater.* **2013**, *23*, 5150–5158. [CrossRef]

52. Lee, M.-S.; Lee, K.; Kim, S.-Y.; Lee, H.; Park, J.; Choi, K.-H.; Kim, H.-K.; Kim, D.-G.; Lee, D.-Y.; Nam, S. High-performance, transparent, and stretchable electrodes using graphene–metal nanowire hybrid structures. *Nano Lett.* **2013**, *13*, 2814–2821. [CrossRef] [PubMed]

53. Kholmanov, I.N.; Magnuson, C.W.; Aliev, A.E.; Li, H.; Zhang, B.; Suk, J.W.; Zhang, L.L.; Peng, E.; Mousavi, S.H.; Khanikaev, A.B. Improved electrical conductivity of graphene films integrated with metal nanowires. *Nano Lett.* **2012**, *12*, 5679–5683. [CrossRef] [PubMed]

54. Deng, B.; Hsu, P.-C.; Chen, G.; Chandrashekar, B.; Liao, L.; Ayitimuda, Z.; Wu, J.; Guo, Y.; Lin, L.; Zhou, Y. Roll-to-roll encapsulation of metal nanowires between graphene and plastic substrate for high-performance flexible transparent electrodes. *Nano Lett.* **2015**, *15*, 4206–4213. [CrossRef] [PubMed]

55. Ahn, Y.; Jeong, Y.; Lee, D.; Lee, Y. Copper nanowire–graphene core–shell nanostructure for highly stable transparent conducting electrodes. *ACS Nano* **2015**, *9*, 3125–3133. [CrossRef] [PubMed]

56. Im, H.-G.; Jung, S.-H.; Jin, J.; Lee, D.; Lee, J.; Lee, D.; Lee, J.-Y.; Kim, I.-D.; Bae, B.-S. Flexible transparent conducting hybrid film using a surface-embedded copper nanowire network: A highly oxidation-resistant copper nanowire electrode for flexible optoelectronics. *ACS Nano* **2014**, *8*, 10973–10979. [CrossRef]

57. Min, J.-H.; Jeong, W.-L.; Kwak, H.-M.; Lee, D.-S. High-performance metal mesh/graphene hybrid films using prime-location and metal-doped graphene. *Sci. Rep.* **2017**, *7*, 10225. [CrossRef]

58. Liu, Z.; Parvez, K.; Li, R.; Dong, R.; Feng, X.; Müllen, K. Transparent conductive electrodes from graphene/ PEDOT: PSS hybrid inks for ultrathin organic photodetectors. *Adv. Mater.* **2015**, *27*, 669–675. [CrossRef]
59. Jo, K.; Lee, T.; Choi, H.J.; Park, J.H.; Lee, D.J.; Lee, D.W.; Kim, B.-S. Stable aqueous dispersion of reduced graphene nanosheets via non-covalent functionalization with conducting polymers and application in transparent electrodes. *Langmuir* **2011**, *27*, 2014–2018. [CrossRef]
60. Lee, B.H.; Lee, J.H.; Kahng, Y.H.; Kim, N.; Kim, Y.J.; Lee, J.; Lee, T.; Lee, K. Graphene-Conducting Polymer Hybrid Transparent Electrodes for Efficient Organic Optoelectronic Devices. *Adv. Funct. Mater.* **2014**, *24*, 1847–1856. [CrossRef]
61. Liu, J.; Yi, Y.; Zhou, Y.; Cai, H. Highly stretchable and flexible graphene/ITO hybrid transparent electrode. *Nanoscale Res. Lett.* **2016**, *11*, 108. [CrossRef] [PubMed]
62. Hemasiri, B.W.N.H.; Kim, J.-K.; Lee, J.-M. Synthesis and Characterization of Graphene/ITO Nanoparticle Hybrid Transparent Conducting Electrode. *Nano-Micro Lett.* **2018**, *10*, 18. [CrossRef] [PubMed]
63. Zhang, Y.; Zhang, L.; Zhou, C. Review of chemical vapor deposition of graphene and related applications. *Acc. Chem. Res.* **2013**, *46*, 2329–2339. [CrossRef] [PubMed]
64. Li, X.; Cai, W.; An, J.; Kim, S.; Nah, J.; Yang, D.; Piner, R.; Velamakanni, A.; Jung, I.; Tutuc, E. Large-area synthesis of high-quality and uniform graphene films on copper foils. *Science* **2009**, *324*, 1312–1314. [CrossRef] [PubMed]
65. Li, X.; Magnuson, C.W.; Venugopal, A.; Tromp, R.M.; Hannon, J.B.; Vogel, E.M.; Colombo, L.; Ruoff, R.S. Large-area graphene single crystals grown by low-pressure chemical vapor deposition of methane on copper. *J. Am. Chem. Soc.* **2011**, *133*, 2816–2819. [CrossRef] [PubMed]
66. Hofmann, S.; Braeuninger-Weimer, P.; Weatherup, R.S. CVD-enabled graphene manufacture and technology. *J. Phys. Chem. Lett.* **2015**, *6*, 2714–2721. [CrossRef]
67. Chen, Y.; Gong, X.L.; Gai, J.G. Progress and Challenges in Transfer of Large-Area Graphene Films. *Adv. Sci.* **2016**, *3*, 1500343. [CrossRef] [PubMed]
68. Reina, A.; Jia, X.; Ho, J.; Nezich, D.; Son, H.; Bulovic, V.; Dresselhaus, M.S.; Kong, J. Large area, few-layer graphene films on arbitrary substrates by chemical vapor deposition. *Nano Lett.* **2008**, *9*, 30–35. [CrossRef] [PubMed]
69. Kim, S.; Kwon, K.C.; Park, J.Y.; Cho, H.W.; Lee, I.; Kim, S.Y.; Lee, J.-L. Challenge beyond Graphene: Metal Oxide/Graphene/Metal Oxide Electrodes for Optoelectronic Devices. *ACS Appl. Mater. Interfaces* **2016**, *8*, 12932–12939. [CrossRef] [PubMed]
70. Pang, S.; Hernandez, Y.; Feng, X.; Müllen, K. Graphene as transparent electrode material for organic electronics. *Adv. Mater.* **2011**, *23*, 2779–2795. [CrossRef] [PubMed]
71. Bhaviripudi, S.; Jia, X.; Dresselhaus, M.S.; Kong, J. Role of kinetic factors in chemical vapor deposition synthesis of uniform large area graphene using copper catalyst. *Nano Lett.* **2010**, *10*, 4128–4133. [CrossRef] [PubMed]
72. Wood, J.D.; Schmucker, S.W.; Lyons, A.S.; Pop, E.; Lyding, J.W. Effects of polycrystalline Cu substrate on graphene growth by chemical vapor deposition. *Nano Lett.* **2011**, *11*, 4547–4554. [CrossRef] [PubMed]
73. Losurdo, M.; Giangregorio, M.M.; Capezzuto, P.; Bruno, G. Graphene CVD growth on copper and nickel: Role of hydrogen in kinetics and structure. *Phys. Chem. Chem. Phys.* **2011**, *13*, 20836–20843. [CrossRef] [PubMed]
74. Guermoune, A.; Chari, T.; Popescu, F.; Sabri, S.S.; Guillemette, J.; Skulason, H.S.; Szkopek, T.; Siaj, M. Chemical vapor deposition synthesis of graphene on copper with methanol, ethanol, and propanol precursors. *Carbon* **2011**, *49*, 4204–4210. [CrossRef]
75. Wu, T.; Zhang, X.; Yuan, Q.; Xue, J.; Lu, G.; Liu, Z.; Wang, H.; Wang, H.; Ding, F.; Yu, Q. Fast growth of inch-sized single-crystalline graphene from a controlled single nucleus on Cu–Ni alloys. *Nat. Mater.* **2016**, *15*, 43. [CrossRef]
76. Liu, X.; Fu, L.; Liu, N.; Gao, T.; Zhang, Y.; Liao, L.; Liu, Z. Segregation growth of graphene on Cu–Ni alloy for precise layer control. *J. Phys. Chem. C* **2011**, *115*, 11976–11982. [CrossRef]
77. Wu, Y.; Chou, H.; Ji, H.; Wu, Q.; Chen, S.; Jiang, W.; Hao, Y.; Kang, J.; Ren, Y.; Piner, R.D. Growth mechanism and controlled synthesis of AB-stacked bilayer graphene on Cu–Ni alloy foils. *ACS Nano* **2012**, *6*, 7731–7738. [CrossRef]
78. Moon, I.K.; Kim, J.I.; Lee, H.; Hur, K.; Kim, W.C.; Lee, H. 2D graphene oxide nanosheets as an adhesive over-coating layer for flexible transparent conductive electrodes. *Sci. Rep.* **2013**, *3*, 1112. [CrossRef]

79. Cho, J.H.; Gorman, J.J.; Na, S.R.; Cullinan, M. Growth of monolayer graphene on nanoscale copper-nickel alloy thin films. *Carbon* **2017**, *115*, 441–448. [CrossRef]

80. Sun, J.; Chen, Y.; Priydarshi, M.K.; Chen, Z.; Bachmatiuk, A.; Zou, Z.; Chen, Z.; Song, X.; Gao, Y.; Rümmeli, M.H. Direct chemical vapor deposition-derived graphene glasses targeting wide ranged applications. *Nano Lett.* **2015**, *15*, 5846–5854. [CrossRef]

81. Wang, D.Y.; Huang, I.S.; Ho, P.H.; Li, S.S.; Yeh, Y.C.; Wang, D.W.; Chen, W.L.; Lee, Y.Y.; Chang, Y.M.; Chen, C.C. Clean-Lifting Transfer of Large-area Residual-Free Graphene Films. *Adv. Mater.* **2013**, *25*, 4521–4526. [CrossRef]

82. Suk, J.W.; Kitt, A.; Magnuson, C.W.; Hao, Y.; Ahmed, S.; An, J.; Swan, A.K.; Goldberg, B.B.; Ruoff, R.S. Transfer of CVD-grown monolayer graphene onto arbitrary substrates. *ACS Nano* **2011**, *5*, 6916–6924. [CrossRef] [PubMed]

83. Yan, C.; Cho, J.H.; Ahn, J.-H. Graphene-based flexible and stretchable thin film transistors. *Nanoscale* **2012**, *4*, 4870–4882. [CrossRef] [PubMed]

84. Zaretski, A.V.; Lipomi, D.J. Processes for non-destructive transfer of graphene: Widening the bottleneck for industrial scale production. *Nanoscale* **2015**, *7*, 9963–9969. [CrossRef] [PubMed]

85. Zaretski, A.V.; Moetazedi, H.; Kong, C.; Sawyer, E.J.; Savagatrup, S.; Valle, E.; O'Connor, T.F.; Printz, A.D.; Lipomi, D.J. Metal-assisted exfoliation (MAE): Green, roll-to-roll compatible method for transferring graphene to flexible substrates. *Nanotechnology* **2015**, *26*, 045301. [CrossRef] [PubMed]

86. Zheng, Q.; Li, Z.; Yang, J.; Kim, J.-K. Graphene oxide-based transparent conductive films. *Prog. Mater. Sci.* **2014**, *64*, 200–247. [CrossRef]

87. Loh, K.P.; Bao, Q.; Eda, G.; Chhowalla, M. Graphene oxide as a chemically tunable platform for optical applications. *Nat. Chem.* **2010**, *2*, 1015. [CrossRef] [PubMed]

88. Singh, R.K.; Kumar, R.; Singh, D.P. Graphene oxide: Strategies for synthesis, reduction and frontier applications. *RSC Adv.* **2016**, *6*, 64993–65011. [CrossRef]

89. Zhu, Y.; Murali, S.; Cai, W.; Li, X.; Suk, J.W.; Potts, J.R.; Ruoff, R.S. Graphene and graphene oxide: Synthesis, properties, and applications. *Adv. Mater.* **2010**, *22*, 3906–3924. [CrossRef]

90. Compton, O.C.; Nguyen, S.T. Graphene oxide, highly reduced graphene oxide, and graphene: Versatile building blocks for carbon-based materials. *Small* **2010**, *6*, 711–723. [CrossRef]

91. Marcano, D.C.; Kosynkin, D.V.; Berlin, J.M.; Sinitskii, A.; Sun, Z.; Slesarev, A.; Alemany, L.B.; Lu, W.; Tour, J.M. Improved synthesis of graphene oxide. *ACS Nano* **2010**, *4*, 4806–4814. [CrossRef] [PubMed]

92. Pham, V.H.; Cuong, T.V.; Hur, S.H.; Shin, E.W.; Kim, J.S.; Chung, J.S.; Kim, E.J. Fast and simple fabrication of a large transparent chemically-converted graphene film by spray-coating. *Carbon* **2010**, *48*, 1945–1951. [CrossRef]

93. Li, D.; Müller, M.B.; Gilje, S.; Kaner, R.B.; Wallace, G.G. Processable aqueous dispersions of graphene nanosheets. *Nat. Nanotechnol.* **2008**, *3*, 101. [CrossRef] [PubMed]

94. Kim, J.; Cote, L.J.; Kim, F.; Yuan, W.; Shull, K.R.; Huang, J. Graphene oxide sheets at interfaces. *J. Am. Chem. Soc.* **2010**, *132*, 8180–8186. [CrossRef] [PubMed]

95. An, S.J.; Zhu, Y.; Lee, S.H.; Stoller, M.D.; Emilsson, T.; Park, S.; Velamakanni, A.; An, J.; Ruoff, R.S. Thin film fabrication and simultaneous anodic reduction of deposited graphene oxide platelets by electrophoretic deposition. *J. Phys. Chem. Lett.* **2010**, *1*, 1259–1263. [CrossRef]

96. Cote, L.J.; Kim, F.; Huang, J. Langmuir−Blodgett assembly of graphite oxide single layers. *J. Am. Chem. Soc.* **2008**, *131*, 1043–1049. [CrossRef] [PubMed]

97. Acik, M.; Chabal, Y.J. A review on thermal exfoliation of graphene oxide. *J. Mater. Sci. Res.* **2013**, *2*, 101–112.

98. Xiang, Q.; Yu, J.; Jaroniec, M. Graphene-based semiconductor photocatalysts. *Chem. Soc. Rev.* **2012**, *41*, 782–796. [CrossRef]

99. Moon, I.K.; Lee, J.; Ruoff, R.S.; Lee, H. Reduced graphene oxide by chemical graphitization. *Nat. Commun.* **2010**, *1*, 73. [CrossRef]

100. Chua, C.K.; Pumera, M. Chemical reduction of graphene oxide: A synthetic chemistry viewpoint. *Chem. Soc. Rev.* **2014**, *43*, 291–312. [CrossRef]

101. Gao, X.; Jang, J.; Nagase, S. Hydrazine and thermal reduction of graphene oxide: Reaction mechanisms, product structures, and reaction design. *J. Phys. Chem. C* **2009**, *114*, 832–842. [CrossRef]

102. Chen, W.; Yan, L.; Bangal, P.R. Preparation of graphene by the rapid and mild thermal reduction of graphene oxide induced by microwaves. *Carbon* **2010**, *48*, 1146–1152. [CrossRef]

103. Williams, G.; Seger, B.; Kamat, P.V. TiO$_2$-graphene nanocomposites. UV-assisted photocatalytic reduction of graphene oxide. *ACS Nano* **2008**, *2*, 1487–1491. [CrossRef] [PubMed]
104. Liu, H.; Liu, Y.; Zhu, D. Chemical doping of graphene. *J. Mater. Chem.* **2011**, *21*, 3335–3345. [CrossRef]
105. Kasry, A.; Kuroda, M.A.; Martyna, G.J.; Tulevski, G.S.; Bol, A.A. Chemical doping of large-area stacked graphene films for use as transparent, conducting electrodes. *ACS Nano* **2010**, *4*, 3839–3844. [CrossRef] [PubMed]
106. Wehling, T.; Novoselov, K.; Morozov, S.; Vdovin, E.; Katsnelson, M.; Geim, A.; Lichtenstein, A. Molecular doping of graphene. *Nano Lett.* **2008**, *8*, 173–177. [CrossRef] [PubMed]
107. Crowther, A.C.; Ghassaei, A.; Jung, N.; Brus, L.E. Strong charge-transfer doping of 1 to 10 layer graphene by NO$_2$. *ACS Nano* **2012**, *6*, 1865–1875. [CrossRef]
108. Jung, N.; Kim, N.; Jockusch, S.; Turro, N.J.; Kim, P.; Brus, L. Charge transfer chemical doping of few layer graphenes: Charge distribution and band gap formation. *Nano Lett.* **2009**, *9*, 4133–4137. [CrossRef]
109. Shi, Y.; Kim, K.K.; Reina, A.; Hofmann, M.; Li, L.-J.; Kong, J. Work function engineering of graphene electrode via chemical doping. *ACS Nano* **2010**, *4*, 2689–2694. [CrossRef]
110. Shin, D.H.; Lee, K.W.; Lee, J.S.; Kim, J.H.; Kim, S.; Choi, S.-H. Enhancement of the effectiveness of graphene as a transparent conductive electrode by AgNO$_3$ doping. *Nanotechnology* **2014**, *25*, 125701. [CrossRef]
111. Pinto, H.; Jones, R.; Goss, J.; Briddon, P. p-type doping of graphene with F4-TCNQ. *J. Phys. Condens. Matter* **2009**, *21*, 402001. [CrossRef] [PubMed]
112. Park, J.; Jo, S.B.; Yu, Y.J.; Kim, Y.; Yang, J.W.; Lee, W.H.; Kim, H.H.; Hong, B.H.; Kim, P.; Cho, K. Single-Gate Bandgap Opening of Bilayer Graphene by Dual Molecular Doping. *Adv. Mater.* **2012**, *24*, 407–411. [CrossRef] [PubMed]
113. Chen, W.; Chen, S.; Qi, D.C.; Gao, X.Y.; Wee, A.T.S. Surface transfer p-type doping of epitaxial graphene. *J. Am. Chem. Soc.* **2007**, *129*, 10418–10422. [CrossRef] [PubMed]
114. Dong, X.; Fu, D.; Fang, W.; Shi, Y.; Chen, P.; Li, L.J. Doping single-layer graphene with aromatic molecules. *Small* **2009**, *5*, 1422–1426. [CrossRef] [PubMed]
115. Wang, X.; Li, X.; Zhang, L.; Yoon, Y.; Weber, P.K.; Wang, H.; Guo, J.; Dai, H. N-doping of graphene through electrothermal reactions with ammonia. *Science* **2009**, *324*, 768–771. [CrossRef] [PubMed]
116. Jo, I.; Kim, Y.; Moon, J.; Park, S.; San Moon, J.; Park, W.B.; Lee, J.S.; Hong, B.H. Stable n-type doping of graphene via high-molecular-weight ethylene amines. *Phys. Chem. Chem. Phys.* **2015**, *17*, 29492–29495. [CrossRef] [PubMed]
117. Li, C.; Li, Z.; Zhu, H.; Wang, K.; Wei, J.; Li, X.; Sun, P.; Zhang, H.; Wu, D. Graphene nano-"patches" on a carbon nanotube network for highly transparent/conductive thin film applications. *J. Phys. Chem. C* **2010**, *114*, 14008–14012. [CrossRef]
118. Yan, Z.; Peng, Z.; Casillas, G.; Lin, J.; Xiang, C.; Zhou, H.; Yang, Y.; Ruan, G.; Raji, A.-R.O.; Samuel, E.L. Rebar graphene. *ACS Nano* **2014**, *8*, 5061–5068. [CrossRef] [PubMed]
119. Kholmanov, I.N.; Magnuson, C.W.; Piner, R.; Kim, J.Y.; Aliev, A.E.; Tan, C.; Kim, T.Y.; Zakhidov, A.A.; Sberveglieri, G.; Baughman, R.H. Optical, electrical, and electromechanical properties of hybrid graphene/carbon nanotube films. *Adv. Mater.* **2015**, *27*, 3053–3059. [CrossRef] [PubMed]
120. Khaligh, H.H.; Goldthorpe, I.A. Failure of silver nanowire transparent electrodes under current flow. *Nanoscale Res. Lett.* **2013**, *8*, 235. [CrossRef] [PubMed]
121. Garnett, E.C.; Cai, W.; Cha, J.J.; Mahmood, F.; Connor, S.T.; Christoforo, M.G.; Cui, Y.; McGehee, M.D.; Brongersma, M.L. Self-limited plasmonic welding of silver nanowire junctions. *Nat. Mater.* **2012**, *11*, 241. [CrossRef] [PubMed]
122. Jeong, C.; Nair, P.; Khan, M.; Lundstrom, M.; Alam, M.A. Prospects for nanowire-doped polycrystalline graphene films for ultratransparent, highly conductive electrodes. *Nano Lett.* **2011**, *11*, 5020–5025. [CrossRef] [PubMed]
123. Zhu, Y.; Sun, Z.; Yan, Z.; Jin, Z.; Tour, J.M. Rational design of hybrid graphene films for high-performance transparent electrodes. *ACS Nano* **2011**, *5*, 6472–6479. [CrossRef]
124. Dong, P.; Zhu, Y.; Zhang, J.; Peng, C.; Yan, Z.; Li, L.; Peng, Z.; Ruan, G.; Xiao, W.; Lin, H. Graphene on metal grids as the transparent conductive material for dye sensitized solar cell. *J. Phys. Chem. C* **2014**, *118*, 25863–25868. [CrossRef]
125. Cho, E.; Kim, M.; Sohn, H.; Shin, W.; Won, J.; Kim, Y.; Kwak, C.; Lee, C.; Woo, Y. A graphene mesh as a hybrid electrode for foldable devices. *Nanoscale* **2018**, *10*, 628–638. [CrossRef] [PubMed]

126. Mehta, R.; Chugh, S.; Chen, Z. Enhanced electrical and thermal conduction in graphene-encapsulated copper nanowires. *Nano Lett.* **2015**, *15*, 2024–2030. [CrossRef] [PubMed]

127. Ahn, Y.; Jeong, Y.; Lee, Y. Improved thermal oxidation stability of solution-processable silver nanowire transparent electrode by reduced graphene oxide. *ACS Appl. Mater. Interfaces* **2012**, *4*, 6410–6414. [CrossRef] [PubMed]

128. Zhang, X.; Yan, X.; Chen, J.; Zhao, J. Large-size graphene microsheets as a protective layer for transparent conductive silver nanowire film heaters. *Carbon* **2014**, *69*, 437–443. [CrossRef]

129. Zhu, Z.; Mankowski, T.; Balakrishnan, K.; Shikoh, A.S.; Touati, F.; Benammar, M.A.; Mansuripur, M.; Falco, C.M. Ultrahigh aspect ratio copper-nanowire-based hybrid transparent conductive electrodes with PEDOT: PSS and reduced graphene oxide exhibiting reduced surface roughness and improved stability. *ACS Appl. Mater. Interfaces* **2015**, *7*, 16223–16230. [CrossRef] [PubMed]

130. Liang, J.; Li, L.; Tong, K.; Ren, Z.; Hu, W.; Niu, X.; Chen, Y.; Pei, Q. Silver nanowire percolation network soldered with graphene oxide at room temperature and its application for fully stretchable polymer light-emitting diodes. *ACS Nano* **2014**, *8*, 1590–1600. [CrossRef] [PubMed]

131. Dou, L.; Cui, F.; Yu, Y.; Khanarian, G.; Eaton, S.W.; Yang, Q.; Resasco, J.; Schildknecht, C.; Schierle-Arndt, K.; Yang, P. Solution-processed copper/reduced-graphene-oxide core/shell nanowire transparent conductors. *ACS Nano* **2016**, *10*, 2600–2606. [CrossRef] [PubMed]

132. Kholmanov, I.N.; Domingues, S.H.; Chou, H.; Wang, X.; Tan, C.; Kim, J.-Y.; Li, H.; Piner, R.; Zarbin, A.J.; Ruoff, R.S. Reduced graphene oxide/copper nanowire hybrid films as high-performance transparent electrodes. *ACS Nano* **2013**, *7*, 1811–1816. [CrossRef] [PubMed]

133. Qiu, T.; Luo, B.; Liang, M.; Ning, J.; Wang, B.; Li, X.; Zhi, L. Hydrogen reduced graphene oxide/metal grid hybrid film: Towards high performance transparent conductive electrode for flexible electrochromic devices. *Carbon* **2015**, *81*, 232–238. [CrossRef]

134. Hong, W.; Xu, Y.; Lu, G.; Li, C.; Shi, G. Transparent graphene/PEDOT–PSS composite films as counter electrodes of dye-sensitized solar cells. *Electrochem. Commun.* **2008**, *10*, 1555–1558. [CrossRef]

135. Parvez, K.; Li, R.; Müllen, K. Graphene as Transparent Electrodes for Solar Cells. In *Nanocarbons for Advanced Energy Conversion*; John Wiley & Sons: Hoboken, NJ, USA, 2015; Volume 2.

136. Eda, G.; Lin, Y.-Y.; Miller, S.; Chen, C.-W.; Su, W.-F.; Chhowalla, M. Transparent and conducting electrodes for organic electronics from reduced graphene oxide. *Appl. Phys. Lett.* **2008**, *92*, 209. [CrossRef]

137. Choi, Y.-Y.; Kang, S.J.; Kim, H.-K.; Choi, W.M.; Na, S.-I. Multilayer graphene films as transparent electrodes for organic photovoltaic devices. *Sol. Energy Mater. Sol. Cells* **2012**, *96*, 281–285. [CrossRef]

138. Dong, H.; Wu, Z.; Jiang, Y.; Liu, W.; Li, X.; Jiao, B.; Abbas, W.; Hou, X. A flexible and thin graphene/silver nanowires/polymer hybrid transparent electrode for optoelectronic devices. *ACS Appl. Mater. Interfaces* **2016**, *8*, 31212–31221. [CrossRef] [PubMed]

139. Lee, J.; Han, T.-H.; Park, M.-H.; Jung, D.Y.; Seo, J.; Seo, H.-K.; Cho, H.; Kim, E.; Chung, J.; Choi, S.-Y. Synergetic electrode architecture for efficient graphene-based flexible organic light-emitting diodes. *Nat. Commun.* **2016**, *7*, 11791. [CrossRef] [PubMed]

140. Lin, F.; Bult, J.B.; Nanayakkara, S.; Dillon, A.C.; Richards, R.M.; Blackburn, J.L.; Engtrakul, C. Graphene as an efficient interfacial layer for electrochromic devices. *ACS Appl. Mater. Interfaces* **2015**, *7*, 11330–11336. [CrossRef] [PubMed]

141. Polat, E.O.; Balcı, O.; Kocabas, C. Graphene based flexible electrochromic devices. *Sci. Rep.* **2014**, *4*, 6484. [CrossRef] [PubMed]

142. Kim, J.Y.; Cho, N.S.; Cho, S.; Kim, K.; Cheon, S.; Kim, K.; Kang, S.-Y.; Cho, S.M.; Lee, J.-I.; Oh, J.-Y. Graphene Electrode Enabling Electrochromic Approaches for Daylight-Dimming Applications. *Sci. Rep.* **2018**, *8*, 3944. [CrossRef] [PubMed]

143. Kang, J.; Kim, H.; Kim, K.S.; Lee, S.-K.; Bae, S.; Ahn, J.-H.; Kim, Y.-J.; Choi, J.-B.; Hong, B.H. High-performance graphene-based transparent flexible heaters. *Nano Lett.* **2011**, *11*, 5154–5158. [CrossRef] [PubMed]

144. Sui, D.; Huang, Y.; Huang, L.; Liang, J.; Ma, Y.; Chen, Y. Flexible and transparent electrothermal film heaters based on graphene materials. *Small* **2011**, *7*, 3186–3192. [CrossRef] [PubMed]

 *micromachines*

Article

Graphene Oxide Decorated Nanometal-Poly(Anilino-Dodecylbenzene Sulfonic Acid) for Application in High Performance Supercapacitors

Nomxolisi R. Dywili [1,2,*], Afroditi Ntziouni [2], Chinwe Ikpo [1], Miranda Ndipingwi [1], Ntuthuko W. Hlongwa [1], Anne L. D. Yonkeu [1], Milua Masikini [1], Konstantinos Kordatos [2,*] and Emmanuel I. Iwuoha [1,*]

[1] SensorLab, Department of Chemistry, University of the Western Cape, Private Bag X17, Bellville, Cape Town 7535, South Africa; cikpo@uwc.ac.za (C.I.); 3318577@myuwc.ac.za (M.N.); 2962477@myuwc.ac.za (N.W.H.); 3116018@myuwc.ac.za (A.L.D.Y.); mmasikini@uwc.ac.za (M.M.)
[2] School of Chemical Engineering, Section I: Chemical Sciences, Lab of Inorganic and Analytical Chemistry, National Technical University of Athens, 9 Heroon Polytechniou Str., 15773 Athens, Greece; ntziouni@mail.ntua.gr
* Correspondence: 2762713@myuwc.ac.za (N.R.D.); kordatos@central.ntua.gr (K.K.); eiwuoha@uwc.ac.za (E.I.I.)

Received: 1 December 2018; Accepted: 6 February 2019; Published: 11 February 2019

Abstract: Graphene oxide (GO) decorated with silver (Ag), copper (Cu) or platinum (Pt) nanoparticles that are anchored on dodecylbenzene sulfonic acid (DBSA)-doped polyaniline (PANI) were prepared by a simple one-step method and applied as novel materials for high performance supercapacitors. High-resolution transmission electron microscopy (HRTEM) and high-resolution scanning electron microscopy (HRSEM) analyses revealed that a metal-decorated polymer matrix is embedded within the GO sheet. This caused the M/DBSA–PANI (M = Ag, Cu or Pt) particles to adsorb on the surface of the GO sheets, appearing as aggregated dark regions in the HRSEM images. The Fourier transform infrared (FTIR) spectroscopy studies revealed that GO was successfully produced and decorated with Ag, Cu or Pt nanoparticles anchored on DBSA–PANI. This was confirmed by the appearance of the GO signature epoxy C–O vibration band at 1040 cm^{-1} (which decreased upon the introduction of metal nanoparticle) and the PANI characteristic N–H stretching vibration band at 3144 cm^{-1} present only in the GO/M/DBSA–PANI systems. The composites were tested for their suitability as supercapacitor materials; and specific capacitance values of 206.4, 192.8 and 227.2 F·g^{-1} were determined for GO/Ag/DBSA–PANI, GO/Cu/DBSA–PANI and GO/Pt/DBSA–PANI, respectively. The GO/Pt/DBSA–PANI electrode exhibited the best specific capacitance value of the three electrodes and also had twice the specific capacitance value reported for Graphene/MnO$_2$//ACN (113.5 F·g^{-1}). This makes GO/Pt/DBSA–PANI a very promising organic supercapacitor material.

Keywords: supercapacitors; graphene oxide; metal nanoparticles; dodecylbenzene sulfonic acid (DBSA) doped polyaniline; capacitance

---

1. Introduction

The world has been facing global warming and energy problems with Earth's natural resources depleting at a very rapid rate. In 2007, the International Energy Agency published the World Energy Outlook which estimated that by 2030 there will be 55% more energy demand as compared to today [1]. Global economic development and prosperity have been built on cheap and abundant fossil fuels with petroleum standing at 39%, natural gas at 24% and coal at 23%, but there is a limited amount of fossil fuels and they are non-renewable [2]. Estimates state that there is a 2% annual growth in global oil demand along with a natural decline in production from existing reserves [3]. Therefore, there is need to invest in alternative sources of energy both in terms of energy conversion, as well as storage devices such as electrostatic capacitors, electrochemical capacitors (supercapacitors), batteries, and fuel cells. Metal oxides and transition metal oxides have been utilized for the development and utilization of smart materials in applications such as gas sensors [4], energy storage smart materials [5,6] and advanced energy conversion devices. Perovskite solar cells (PSCs) have attracted a great deal of attention in the photovoltaic cell field of study, due to their high photo-to-electric power conversion efficiency (PCE) and low cost. The high PCE is due to the high and excellent physical properties of organic–inorganic hybrid perovskite materials. These include a long charge diffusion length and high absorption coefficient in the visible range [7]. Titanium oxide (TiO_2) nanostructures are excellent anode materials for sodium ion batteries due to their inherent safety, low cost and structural stability. When tested as a binder and conducting additive-free electrodes in sodium cells, TiO_2 nanotubular arrays, obtained from simple anodic oxidation, exhibited different electrochemical responses which then render TiO_2 as a good anodic material [8]. Perreault et al., developed a spray-dried mesoporous mixed Cu–Ni oxide and graphene nanocomposite microspheres for high power and durable Li-ion battery anodes. They exhibited unprecedented electrochemical behavior such as high reversible specific capacity, excellent coulombic efficiency and long-term stability at high current density that are very remarkable when compared to most traditional metal oxides and nanocomposites prepared by conventional techniques [9].

Supercapacitors are electrical energy storage devices that store and discharge energy at the electrochemical interface and utilize the three-electrode system, i.e. working electrode, counter electrode and reference electrode [10,11]. Supercapacitors have very high capacities and low resistance and are able to store energy at relatively higher rates, when compared to other energy storage devices. This is due to the mechanism of energy storage which involves a simple charge separation at the interface between the electrode and electrolyte [12]. Supercapacitors have attracted attention due to their unique and wide potential in a variety of applications such as electric vehicles, power back-up in mobile phones, digital cameras, radio tuners, laptops, etc. and they find application as power back-ups for uninterruptible power system (UPS) applications and other high-power apparatus [13,14]. Supercapacitors are classified into three main general categories: (i) electric double-layer capacitors (EDLC), (ii) Faradaic pseudocapacitors and (iii) hybrid capacitors [2]. Electric double-layer capacitors (EDLC) are known to utilize carbon-based materials such as graphene oxide, carbon black, activated carbon, etc. Faradaic pseudocapacitors are known to utilize metal oxide and/or conducting polymers. Hybrid capacitors are known to utilize carbon materials and metal oxides and/or conducting polymers [15,16]. This work uses graphene oxide (GO), metal nanoparticles (Ag, Cu and Pt) and dodecylbenzene sulfonic acid (DBSA)-doped polyaniline (PANI) as the materials to use in the development of high capacitance supercapacitors. GO has considerable exceptional properties such as mechanical, optical, electronic, electrical, etc. which then render GO as a good candidate for use in energy storage devices [17,18]. Metal nanoparticles have attracted attention due to the performance in electronic, optical, magnetic, and catalytic applications and, recently, these metal nanoparticles have been supported on the surface of GO [19–21]. It is expected that small sized and well-dispersed nanoparticles will enhance activity and selectivity for catalytic applications [22,23]. Conducting polymers have been used to maintain the need for high electrical conductivity of materials for a myriad of applications including energy-related devices [24,25]. There has been much interest in PANI-based materials because of their low cost and easy synthesis [26]. Since PANI is an excellent organic conductor with good environmental stability, good electronic and optical properties, is highly stable in air, and is soluble in several solvents [27], it has been used often to produce materials or composites with carbon materials for supercapacitor electrodes or sensor applications [28]. The combination of the properties of these materials is expected to increase the capacitance of the developed supercapacitors [29]. Furthermore, some organic polymers have been found to be suitable materials for supercapacitor applications, for which the incorporation of cations, such as metal molybdates, leads to increased specific capacitance [30–32].

Table 1 is a comparison of specific capacitances, energies and powers of different supercapacitor electrode materials reported by researchers over the past years. From the table, it can be deduced that the material used in this work exhibited higher capacitance, energy and power values when compared to graphene-containing supercapacitor material [33]. Metal oxides combined with carbon-based materials and conducting polymers give the best capacitance ranging from 200 to 400 F·g^{-1} and low resistances. They also give highly specific energy and power. From the table, we can observe that manganese oxide combined with e-CMG and tantalum (IV) oxide combined with PANI–PSSA gives higher specific capacitances of 389 and 318.4 F·g^{-1} respectively. Lithium manganese oxide combined with aluminum oxide gives the best specific energy of 864.3 Wh kg^{-1} and titanium oxide combined with carbon nanotubes gives the best specific power of 6428 W kg^{-1}.

Table 1. Comparison table of specific capacitances, energy and power of different supercapacitor electrode materials over the years.

Material	Specific Capacitance (F·g^{-1})	Specific Energy (Wh·kg^{-1})	Specific Power (W·kg^{-1})	References
GO/Pt/DBSA–PANI	227.2	126.2	178.4	This work
Graphene/MnO$_2$//CAN	113.5	51.1	102.2	[33]
TaO$_2$–PANI–PSSA	318.4	157.0	435.0	[34]
MnO$_2$/e–CMG	389.0	44.0	250.0	[35]
Li$_2$MnSiO$_4$/Al$_2$O$_3$	117.5	864.3	104.0	[36]
TiO$_2$/CNT	176.5	40.0	6428.0	[37]

To the best of our knowledge, no work has been published on GO decorated with Ag, Cu and Pt nanoparticles that are anchored on DBSA–PANI, for application in high performance supercapacitors.

2. Materials and Methods

2.1. Materials

All chemicals used in the experiments were of analytical grade and were used as purchased without further purification. Graphite (1–2 μm); sodium nitrate, NaNO$_3$ ($\geq$99%); sulfuric acid, H$_2$SO$_4$ ($\geq$98%); potassium permanganate, KMnO$_4$ ($\geq$99%); hydrogen peroxide, H$_2$O$_2$ ($\geq$30%); hydrochloric acid, HCl ($\geq$37%); hexachloroplatinic acid, H$_2$PtCl$_6$6H$_2$O, ACS reagent grade ($\geq$37.50%); Pt wire; silver nitrate AgNO$_3$, ACS reagent grade ($\geq$99.0%); sodium hydroxide, NaOH, BioXtra ($\geq$98%) acidimetric, pellets (anhydrous); cetyl trimethylammonium bromide, CTAB (98%); poly(sodium 4-styrenesulfonate), PSS ($\sim$70000) powder; sodium acetate, NaAc; poly(ethylene glycol), PEG (400 powder); ethylene glycol anhydrous, EG (99.8%); acetone ($\geq$99.9%); dodecylbenzenesulfonic acid, DBSA ($\geq$98%); aniline, ACS reagent grade ($\geq$99.5%); ammonium persulfate, ACS reagent grade ($\geq$98.0%); methanol ($\geq$99.9%); and isopropanol ($\geq$99.7%), were all purchased from Sigma-Aldrich (Modderfontein, South Africa).

2.2. Instrumentation

The cell system was fabricated using 1 M H$_2$SO$_4$ solution as the electrolyte and tested for the supercapacitor parameters using the BST8-3 eight-channel battery testing machine. Fourier transform infrared (FTIR) spectra were recorded on a 100 spectrophotometer, Perkin Elmer Fourier Transform Infrared model (USA), operating between 400 and 4000 cm^{-1} in order to characterize the presence of specific features of the materials. The high-resolution scanning electron microscopy (HRSEM) of GO measurements were made with a Ziess Auriga, Hitachi S3000N, Quorum Technology (Lewes, England), operating at 50 kV and high-resolution transmission electron microscopy (HRTEM) measurements were made with Tecnai G2 F20X-Twin MAT Field Emission Transmission Microscopy. FEI (Eindhoven, Netherlands) equipped with an energy-dispersive spectroscopy (EDS) detector was used to study the size and morphology of the samples. Copper grid (Cu) was used as a sample holder for the immobilization of (2 µL) solution of GO, GO/Ag/DBSA–PANI and GO/Pt/DBSA–PANI and a nickel grid for GO/Cu/DBSA–PANI, and the micrographs were recorded at room temperature. Cyclic voltammetry (CV) and electrochemical impedance spectroscopy (EIS) measurements were made with VMP 300 Bio-Logic SAS, Biologic Science Instruments (Seyssinet-Pariset, France), where all cyclic voltammograms and EIS graphs were recorded with a computer interfaced with VMP 300 Bio-Logic SAS using a 10 mL electrochemical cell that has a three-electrode system. The electrodes used were: (1) 0.071 cm^2 glassy carbon electrode (GCE) as the working electrode, (2) 500 mm × 0.635 mm platinum wire electrode (Pt) from Sigma-Aldrich as the counter electrode, and (3) Ag/AgCl electrode (with a 3 M NaCl salt-bridge) as the reference electrode. Alumina micropolishing pads were obtained from Buehler, LL, USA and were used for polishing the glassy carbon electrode before modification. Galvanostatic charge-discharge measurements were taken with 8-Channels Battery analyzer BST8-3, MTI Corporation (Richmond, VA, USA) using an electrochemical cell and the three-electrode system with a potential sweep rate of 0.9 mV s^{-1}.

2.3. Experimental

2.3.1. Synthesis of GO

GO was prepared using a modified Hummers method. Specifically, 0.5 g of graphite powder was added to a cold solution of 40 mL of concentrated sulfuric acid (H_2SO_4) and 0.375 g of sodium nitrate ($NaNO_3$) under vigorous stirring for 1 h in an ice bath. A mass of 2.25 g potassium permanganate ($KMnO_4$) was added portion wise to the solution while it was stirring and the mixture remained in the ice bath for a further 2 h to cool the mixture below 10 °C. The mixture took on a green brown color and stirring continued for 5 days in order to ensure complete oxidation of the graphite. After completion of the reaction, 70 mL of dilute aqueous solution of 5% H_2SO_4 was added to the mixture to cleave the formed precipitate salts due to oxidation. The mixture was heated and stirred at 98 °C for 1 h. The heating was removed and 2 mL 30% H_2O_2 peroxide was added to the mixture (after it was let to cool down to 60 °C). The mixture was stirred for a further 2 h. Subsequently, in order to remove the residues of $KMnO_4$ and derivatives such as Mn_2O_7, the following procedure was followed. The mixture was centrifuged for 10 min at 4000 rpm, washed with 600 mL aqueous solution of 3% H_2SO_4 and 0.5% H_2O_2 and then placed in an ultrasonic bath for 10 min. The process was repeated a number of times to get rid of the salts. Then, the mixture was washed and purified with 150 mL of aqueous 3% HCl two to three times by mixing and centrifugation, to eliminate any metal ions. Then, the mixture was washed with distilled water until (average of four washings) the pH increased to the value of 7, and thereby ensuring the removal of any remaining acidic. Finally, the solution was washed with acetone and dried at 60 °C in a vacuum oven for 12 h. After drying, the GO was obtained in the form of a shell, followed by grinding, weighing and collecting the product [38].

2.3.2. Synthesis of Graphene Oxide Loaded with Pt, Ag and Cu NPs

The loading of platinum, silver and copper NPs onto GO nanosheets was carried out by electrostatic self-assembly (scheme 1). Initially, GO was functionalized by cetyl trimethylammonium bromide (CTAB), a cationic polyelectrolyte which acts as a surfactant and poly(sodium 4-styrenesulfonate) (PSS) an anionic polyelectrolyte. Thirty milligrams of GO was homogeneously dispersed in 40 mL of an aqueous solution of 1% wt. CTAB using ultra-sonication for 30 min followed by centrifugation to remove the remaining excess CTAB. The functionalized GO was then dispersed in 40 mL of an aqueous solution of 1% wt. PSS by stirring and ultra-sonication for 30 min and the mixture was stored overnight. After 12 h, the excess PSS was removed by centrifugation and the prepared material was subjected to ultrasonic agitation in 40 mL of ethylene glycol (EG) for 30 min. Consequently, 0.2 g of hexachloroplatinic acid ($H_2PtCl_6 6H_2O$) for platinum nanoparticles (scheme 1), silver nitrate ($AgNO_3$) for silver nanoparticles and copper acetate $Cu(CH_3COO)_2$ for copper nanoparticles, respectively, were dissolved in the 40 mL dispersion of EG/functionalized GO and the mixture was sonicated for 30 min to form a stable suspension. At this point, 3.6 g of sodium acetate (NaAc) and 1.0 g of poly-ethylene glycol (PEG) were added under continuous stirring for a further 30 min. The suspension was then sealed in a Teflon autoclave of stainless steel (capacity 100 mL) and heated at 200 °C for 12 h followed by natural cooling to room temperature. A black precipitate was obtained by filtration, washed by de-ionized water and acetone and dried in a vacuum oven at 60 °C for 12 h [39].

2.3.3. Synthesis of GO/Pt NPs, GO/Ag NPs and GO/Cu NPs Anchored DBSA-Doped PANI

A mass of 0.5 g GO/Ag NPs, GO/Cu NPs and GO/Pt NPs was sonicated in 400 mL of 1 M DBSA solution prepared in 1 M HCl for 5 h. DBSA, an anionic surfactant, was used as the dopant as well as to disperse GO/Ag NPs, GO/Cu NPs and GO/Pt NPs in the solution. DBSA is an anionic surfactant used to blend polyaniline with the graphene oxide-loaded nanoparticles. Thereafter, 5 mL of double distilled aniline and solution of ammonium persulfate (12.53 g of $(NH_4)_2S_2O_8$ in 100 mL of 1 M HCl) were added drop-wise to the previous solution of DBSA for in situ oxidative polymerization of aniline with GO/Ag NPs, GO/Cu NPs and GO/Pt NPs under stirring conditions in an ice bath for 6.5 h (scheme 1). A greenish black precipitate was obtained, which was washed thoroughly with double distilled water and methanol to remove any traces of reactants and PANI oligomers until the filtrate became transparent. Thus, the prepared nanocomposite was dried at 60 °C and stored in a desiccator for further experiments [40]. The materials were named GO/Ag/DBSA/PANI, GO/Cu/DBSA–PANI and GO/Pt/DBSA–PANI, respectively.

2.4. Characterization of Electrode Materials

Characterization of the electrode materials was carried out using the following techniques: HRTEM, HRSEM, EDS and FTIR. In summary, elemental compositions of the GO, GO/Ag/DBSA–PANI, GO/Cu/DBSA–PANI, and GO/Pt/DBSA–PANI were quantitatively studied by the EDS. The morphological properties and degree of agglomerations of the nanostructures were studied qualitatively by HRSEM. The structural properties and composition of materials were studied using FTIR. Particle shape, particle size and morphological distribution of GO, GO/Ag/DBSA–PANI, GO/Cu/DBSA–PANI, and GO/Pt/DBSA–PANI were qualitatively studied by HRTEM and HRSEM. The stability and efficiency of the electrode material synthesized and used for supercapacitor electrodes was tested using a potentiostatic-galvanostatic charge-discharge test.

2.5. Fabrication of the Electrode Material

2.5.1. Preparation of the Electrode Materials

The materials used in the experiment for fabrication of the electrode consisted of 40 mg of active material, 5 mg of carbon black, 5 mg (3 drops) of isopropanol, and 8 mg of polytetrafluoroethylene (PTFE) binder. The active material consisted of GO/Ag/DBSA–PANI. The carbon black and GO/Ag/DBSA–PANI were mixed together and crushed to ensure that they were correctly mixed. For a given electrode, relevant materials were mixed together in a 10 mL small beaker to form dough. The dough was transferred onto a flat glass plate. A stainless steel/Teflon rod was used to roll the dough into 1 mm-thick flexible thin films. When making the thin film, the dough was rolled many times with the constant addition of three drops of isopropanol to ensure that the material was correctly mixed in the thin film. The thin film was then placed in an oven and was allowed to bake at 80 °C under vacuum. Once the thin film was dry it was then cut into small wafers for the construction of the electrode. The same principle was used for GO/Cu/DBSA–PANI and GO/Pt/DBSA–PANI composites [17].

2.5.2. Construction of Supercapacitor Cell

A single electrode was assembled with three parts electrode material, stainless steel mesh current collector and stainless-steel wire. The electrode was assembled by cutting the stainless-steel mesh current collector into a 1 cm × 4 cm rectangular shape. The collector was then cleaned by shaking it in ethanol, drying it and then weighing. The approximately 1 cm^2 wafer was placed on the stainless-steel mesh and pressed at a pressure of 20 MPa for 5 min. The electrode was then weighed and the difference in mass was used as the active mass of the electrode (which was 0.021 g for the anode and 0.0190 g for the cathode). The stainless-steel wire was tightly held onto the current collector for external circuit connection and acted as a cathode. The active material was GO/Ag/DBSA–PANI and acted as an anode. The stainless-steel wire and active material were used to make a two-electrode asymmetric supercapacitor cell. The cell system was fabricated using 1 M H_2SO_4 solution as the electrolyte and tested for the supercapacitor parameters using the BST8-3 eight-channel battery testing machine. The cell was fabricated by holding together the two single electrodes (cathode and anode) with a porous and electronically non-conductive separator sandwiched between them to form the cell configuration [34]. The same principle used GO/Cu/DBSA–PANI and GO/Pt/DBSA–PANI.

3. Results and Discussion

The synthesis route for producing GO/Pt/DBSA–PANI nanocomposite is illustrated in Figure 1. A similar experimental process was followed for obtaining GO/Ag/DBSA–PANI and GO/Cu/DBSA–PANI by using the corresponding salts of Ag and Cu as starting materials.

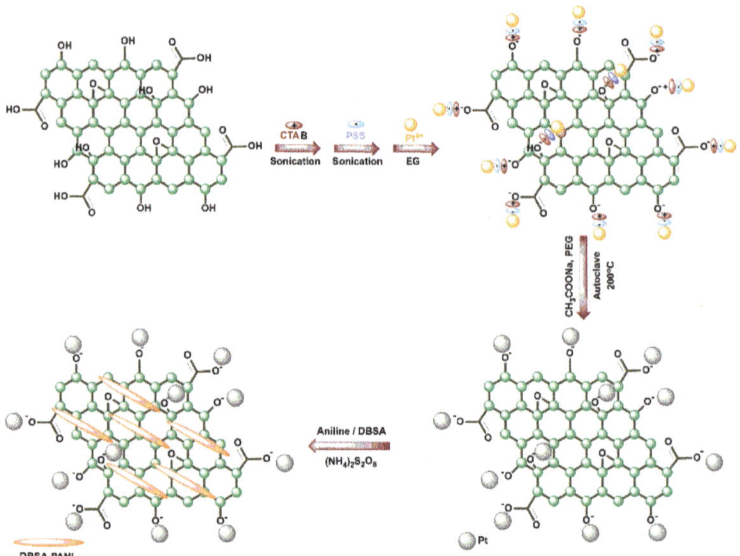

Figure 1. Schematic illustration for the synthesis of a GO/Pt/DBSA–PANI nanocomposite.

Figure 2 shows the schematic diagram of the supercapacitor cell which consists of an electrolyte (KOH), two electrodes and a separator that electrically separate the electrodes. The active material (GO/Pt/DBSA–PANI) of the electrodes is considered one of the most important components of supercapacitors, as the capacitance depends on the type and properties of the electrode material.

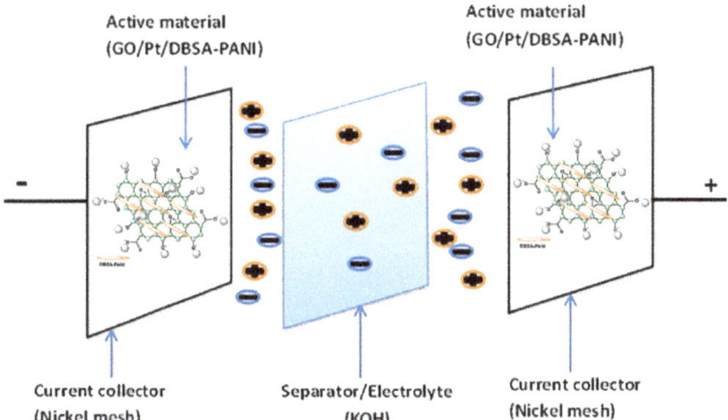

Figure 2. Schematic illustration of the GO/Pt/DBSA–PANI-based supercapacitor cell.

3.1. Morphology Characterization

Figure 3a demonstrates that GO retains a graphene-like lattice substructure which is due to the ultra-sonication of GO [41]. The nanosheets are observed to be flat, light, and transparent and they appear to be larger than 1.5 μm and to be situated on top of the copper grid which is used during HRTEM analysis [42]. The wrinkles and the bends that are observed are due to the abundant defects and functional groups during the oxidation process which takes place over a period of 5 days [43,44]. These nanosheets also appear to be extremely dispersed in water due to the existence of topological features along the overlapping of the nanosheets [45–47]. Figure 3b shows the SEM image of GO, highlighting that GO was efficiently exfoliated to form thin wrinkled sheets with porous structures [48,49]. The images also resemble sponge-like structures due to the well-defined and interlinked three-dimensional graphene sheets [50]. The EDS spectrum in Figure 3c shows the elemental composition of GO and confirms that GO was oxidized due to the presence of the GO functional groups i.e. carbon and oxygen [50,51].

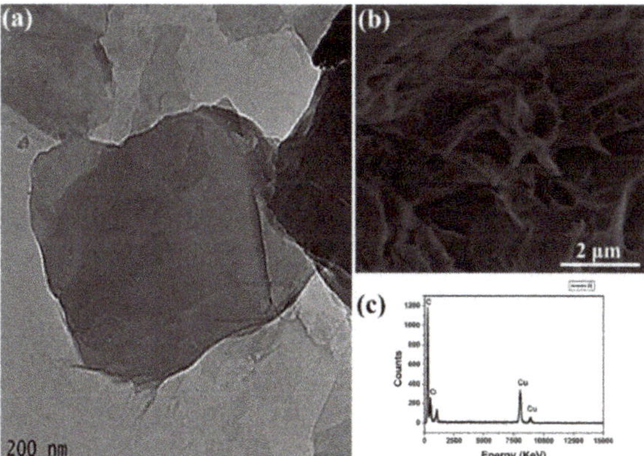

Figure 3. High-resolution transmission electron microscopy (HRTEM) image (**a**), scanning electron microscope (SEM) image (**b**) and energy-dispersive spectroscopy (EDS) spectrum of GO (**c**), respectively.

Figure 4a–c shows the HRSEM images of GO–Ag NPs, GO–Cu NPs and GO–Pt NPs, respectively. As observed from the images, when the GO surface was loaded with the nanoparticles, the surface changed from smooth to rough with small particles observed to be situated on the surface of GO. Upon magnification, the nanoparticles appeared to be spread out on the surface of GO and, therefore, this confirms that GO was loaded with the nanoparticles [52–54]. Figure 4d–f shows the HRSEM images of GO/Ag/DBSA–PANI, GO/Cu/DBSA–PANI and GO/Pt/DBSA–PANI, respectively. The M/DBSA–PANI (M = Ag, Cu or Pt) nanocomposites are observed as aggregated particles that are adsorbed on the GO surface. The surfaces and the edges are toothed, rough and very much agglomerated [55–58].

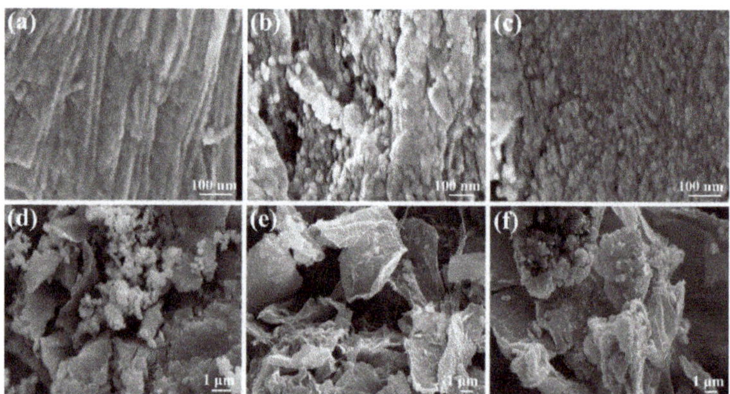

Figure 4. HRSEM images of (**a**) GO–Ag NPs, (**b**) GO–Cu NPs, (**c**) GO–Pt NPs, (**d**) GO/Ag/DBSA–PANI, (**e**) GO/Cu/DBSA–PANI, and (**f**) GO/Pt/DBSA–PANI.

Figure 5a–c shows the EDS spectra of GO/Ag NPs, GO/Cu NPs and GO/Pt NPs, respectively. This confirms the presence of the metal nanoparticles loaded on the surface of GO [59–61]. However, the presence of copper is attributed to the copper grid used upon sample preparation for the GO–Ag NPs and GO-Pt NPs, but in the case of GO–Cu NPs, a nickel grid was used so as to observe the presence of Cu NPs. Figure 5d–f shows the EDS spectra of GO/Ag/DBSA–PANI, GO/Cu/DBSA–PANI and GO/Pt/DBSA–PANI and the quantitative analysis result indicates the presence of carbon, sulfur, platinum, silver, and copper in the polymer composites. Also, the result confirms the formation of platinum, silver and copper nanoparticles [62]. The presence of the sulfur is due to DBSA and this confirms that GO was anchored with DBSA–PANI [63].

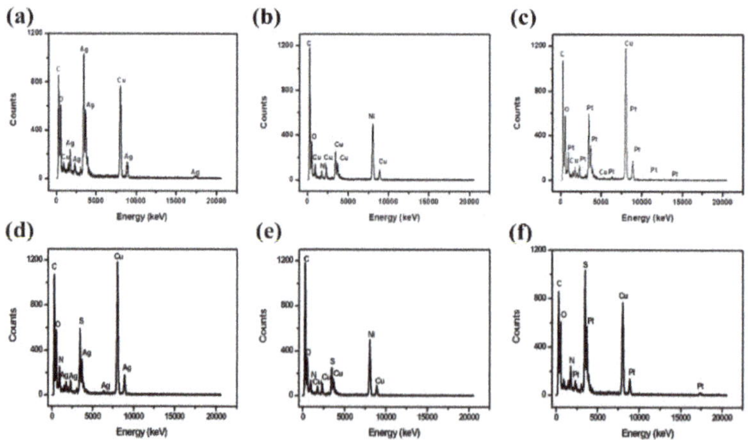

Figure 5. EDS images of (**a**) GO–Ag NPs, (**b**) GO–Cu NPs, (**c**) GO–Pt NPs, (**d**) GO/Ag/DBSA–PANI, (**e**) GO/Cu/DBSA–PANI and (**f**) GO/Pt/DBSA–PANI.

Figure 6a–c shows HRTEM images of GO–Ag NPs, GO–Cu NPs and GO–Pt NPs, respectively. The nanoparticles are small-sized and well-dispersed on the surface of GO with mean particle sizes of 2.6 ± 0.3 nm, 3.5 ± 0.5 nm and 2.3 ± 0.2 nm for Ag, Cu and Pt NPs, respectively. Upon heat treatment, there was no aggregation of the nanoparticles, hence the nanoparticles are small-sized. There is also a strong interaction between the nanoparticle atoms and GO [64]. Figure 6d–f shows the surface structures of GO/Ag/DBSA–PANI, GO/Cu/DBSA–PANI and GO/Pt/DBSA–PANI, respectively. They are observed to be very dark due to the presence of GO in the polymer matrix. DBSA acts as a surfactant and binding agent and assists in binding the polymer and the GO [65]. The single GO sheets may also be embedded into the polymer matrix, which causes the DBSA–PANI particles to become adsorbed on their surfaces and this process then appears as dark surfaces in the HRTEM images of the materials [66]. Agglomerates appear as dark regions in the HRTEM images, thus supporting that single GO sheets are embedded into the polymer matrix and DBSA–PANI particles are adsorbed onto the surface of GO [67–70].

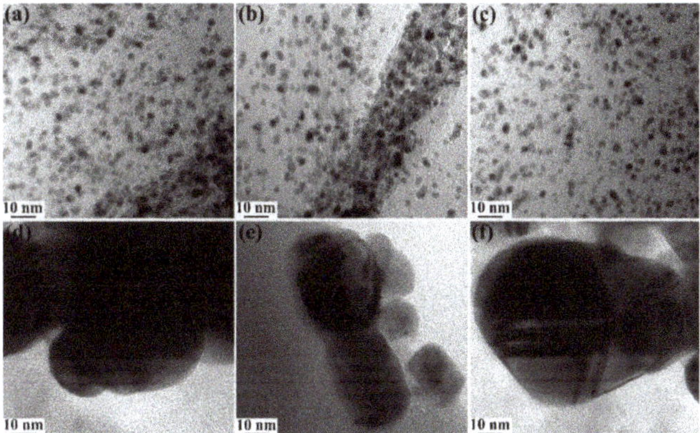

Figure 6. HRTEM images of (**a**) GO-Ag NPs, (**b**) GO-Cu NPs, (**c**) GO-Pt NPs, (**d**) GO/Ag/DBSA–PANI, (**e**) GO/Cu/DBSA–PANI, and (**f**) GO/Pt/DBSA–PANI.

3.2. Molecular Structure Characterization—Fourier Transform Infrared (FTIR)

Figure 7a shows the FTIR spectrum of GO/Pt NPs (ii) compared with the GO spectrum (i). GO contains carbon and oxygen functional groups, mainly O–H at 3436 cm^{-1} attributed to hydroxyl and carboxylic acid functionalities, C=O at 1740 cm^{-1}, C=C at 1636 cm^{-1}, C=C-O at 1390 cm^{-1}, CO-H at 1220 cm^{-1} attributed to the functionality of graphene sheets and C–O at 1040 cm^{-1} was related to the vibration of epoxide functionality [71]. The appearance of all these vibrational bands indicates the presence of rich oxygen-containing functionalities in graphene oxide [72]. When the ethylene glycol was introduced in the synthesis process, it reduced the platinum precursor and GO and the band intensities revealed a decrease in the functional groups of the epoxide and carbonyl which indicates the incomplete reduction of GO [73,74]. GO–Ag NPs and GO–Cu NPs behaved the same way. FTIR studies of GO/Pt/DBSA–PANI observed in Figure 7b revealed a broad absorption band at around ~3144 cm^{-1}, which corresponds to polyaniline N–H stretching vibrations [27,75]. The vibrational bands centred at 1590 and 1408 cm^{-1} can be attributed to the stretching frequencies of quinoid and benzenoid rings of polyaniline, respectively [76]. The band that is produced at 1223 cm^{-1} belongs to the C–N stretching of the secondary amide group [52]. The band at 1180 cm^{-1} is attributed to in-plane bending of the C–H bond, and bands at 1084 and 1004 cm^{-1} are due to the SO$_3^-$ group (O=S=O and S-O stretching) of DBSA [77–79]. FTIR studies of GO/Ag/DBSA–PANI and GO/Cu/DBSA–PANI revealed the same pattern. Table 2 shows and confirms the frequencies and their respective bonds found in (a) (ii) GO/Pt NPs and (b) GO/Pt/DBSA–PANI.

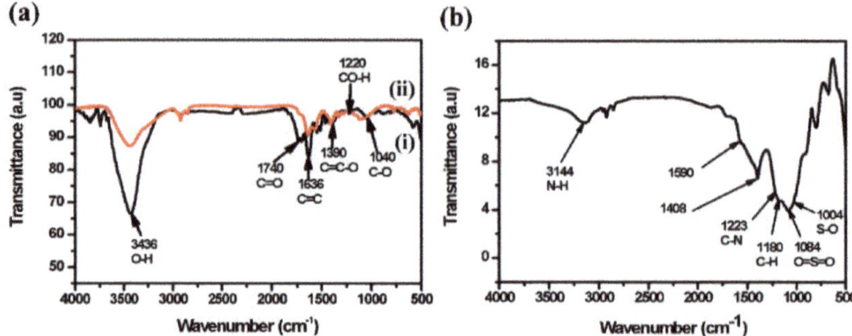

Figure 7. Fourier transform infrared (FTIR) spectra of (**a**) GO (i) and GO/Pt NPs (ii) and (**b**) GO/Pt/DBSA–PANI.

Table 2. Table showing the wavenumbers and their respective bonds found in GO/Pt NPs and GO/Pt/DBSA–PANI.

GO–Pt NPs		GO/Pt/DBSA–PANI	
Wavenumber (cm^{-1})	Bond	Wavenumber (cm^{-1})	Bond
3436	O-H	3144	N-H
1740	C=O	1590	Quinoid
1636	C=C	1408	Benzenoid
1390	C=C-O	1223	C-N
1220	CO-H	1180	C-H
1040	C-O	1084	O=S=O
-		1004	S-O

3.3. Electrochemistry

3.3.1. Cyclic Voltammetry

The electrochemical performances of GO were measured in a symmetrical two-electrode cell in the potential range of −1.0 V to +1.0 V in 1.0 M KOH. Figure 8a exhibits the cyclic voltammogram of GO in which one anodic peak (i) and one cathodic peak (ii) are observed during the process. The anodic peak (i) and cathodic peak (ii) are credited to the electrochemically active oxygen functional groups of reduced planes of GO [34,80,81]. The persistent increase of the peak currents shown with successive potential scans as observed in Figure 8a, indicates that the deposition of GO on glassy carbon electrode (GCE) has been achieved. Figure 8b shows that the deposited glassy carbon electrode (GCE) modified with the material GO displays an anodic peak (i) and cathodic peak (ii), which are attributed to the large number of electrochemically active oxygen-containing groups of GO planes that are very stable [82,83]. The capacitive current covering from −30 to 0 μA in the CV of Figure 7b, indicates that the materials exhibit pseudocapacitance behavior. The current under the curve is slowly increased with the scan rate of CV, which then reveals that voltammetric current is directly proportional to the scan rates of CV [75]. The capacitance, C, can be calculated by the following equations where Q is the positive voltammetric charge in (V); I is the average current in (A) and $\frac{dV}{dt}$ is the voltage scanning rate in (mV·s^{-1}); m is the mass of the active materials in (g); and t is the time in (s).

$$C = \frac{Q}{V} \tag{1}$$

$$C = \frac{I}{\frac{dV}{dt} \times m} \tag{2}$$

The calculated capacitance for graphene oxide using Equation (2) was determined as 180 $F \cdot g^{-1}$, falling within the ranges of 100–200 $F \cdot g^{-1}$ determined by other researchers. Cyclic voltammetry was carried out a number of times in different electrolytes such as Na_2SO_4, Li_2SO_4 and H_2SO_4. The reason for this was to compare and calculate the experimental error. The experimental error was calculated to be very small and this could be due to the experimental conditions, i.e. current density, scan rates and concentration remaining constant.

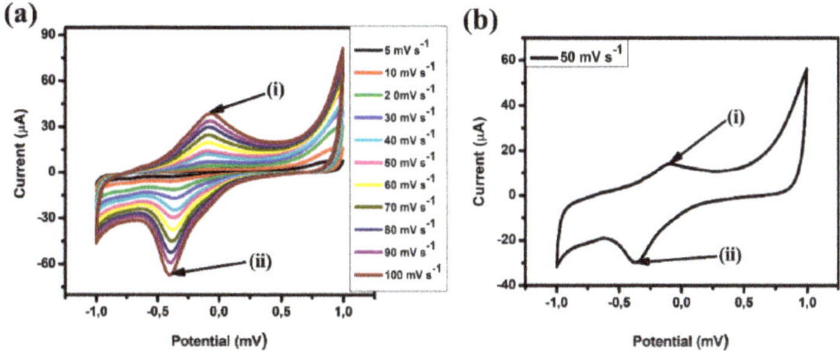

Figure 8. Cyclic voltammetry (CV) behaviour of (**a**) GO at 5–100 $mV \cdot s^{-1}$ and (**b**) GO at 50 $mV \cdot s^{-1}$. The electrochemical behaviour of GO was studied in the potential range of -1.0 V to $+1.0$ V in 0.1 M KOH.

3.3.2. Electrochemical Impedance Spectroscopy (EIS)

Electrochemical impedance spectroscopy (EIS) is an effective method to investigate the interfacial electron transfer characteristics of modified electrodes. The diameter of the semicircles of the Nyquist plots is usually equal to the electron transfer resistance (R_{et}) value of 95.45 Ω cm^2. The relationship between R_{et} and exchange current density (i_0) is consistent with the equation $R_{et} = (RT)/Fi_0$. Here, R is the ideal gas constant (8.314 J mol^{-1} K^{-1}) and T is the room temperature (298.15 K). Figure 9 shows the impedance of GO, GO–Ag NPs, GO–Cu NPs, GO–Pt NPs, and GO–Pt–DBSA/PANI, respectively, obtained in aqueous solution of 0.1 M Na_2SO_4. The GO showed low resistance compared to the other materials. Moreover, when the GO was modified with the other materials, the resistance increased. These results demonstrate that the nanocomposites effectively enhance the electron transfer efficiency.

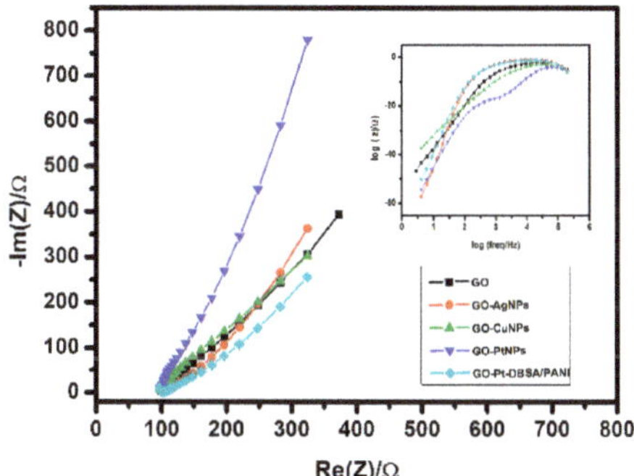

Figure 9. Electrochemical impedance spectroscopy (EIS) Nyquist plot from impedance testing of GO, GO–Ag NPs, GO–Cu NPs, GO–Pt NPs and GO–Pt–DBSA/PANI obtained in 0.1 M Na_2SO_4.

3.4. Galvanostatic-Charge Discharge Test

The capacitive behaviour of GO, GO/Ag/DBSA–PANI, GO/Cu/DBSA–PANI and GO/Pt/DBSA–PANI is given by the galvanostatic/charge discharge curve as shown in Figure 10a–d respectively, done at a potential range of 0 to -0.9 V. The choice of the potential range is dictated by the choice of the electrolyte. The choice for the electrolyte was aqueous potassium hydroxide (KOH). Its decomposition voltage limit is theoretically 1.23 V, or practically, in kinetic terms, between 1.3 V and 1.4 V. KOH is very soluble in water and because of the OH^- anion, it has very good conductivities and advantageously high equivalent conductivities in aqueous medium owing to the special mechanism of proton transport (proton hop-ping) that determines their conductance. The materials showed a large current density in CV curves and a longer charge–discharge time in V-t curves which then implies a larger capacitance. The mass of the thin film of GO was 5.2 g and the first cycle charge and discharge capacities obtained were 44.2 Ah·g^{-1}. The voltage range was 0 -0.9 V vs. Ag/AgCl. Cycling was done at a current density of 50 A·g^{-1} in an aqueous solution of 1 M KOH. The specific charge/discharge current density was found to be 45.7 A·g^{-1}. The specific capacitance Csp can be calculated, where I is the current density measured in (A), t is the time measured in (s), and V is the voltage scanning rate in (mV·s^{-1}):

$$C = \frac{It}{V} \tag{3}$$

Therefore, specific capacitance, C_{sp}, can be calculated: C_{sp} = (45.7 Ag^{-1} × 34.6 s)/0.9 V = 182.8 F·g^{-1}. The Csp for GO was calculated as 182.8 F·g^{-1}, 206.4 F·g^{-1} for GO/Ag/DBSA–PANI, 192.8 F·g^{-1} for GO/Cu/DBSA–PANI and 227.2 F·g^{-1} for GO/Pt/DBSA–PANI. Comparing these values with the specific capacitance of GO, we can conclude that the materials are good materials and therefore can be used for supercapacitor applications. Platinum is more ductile, is stable at high temperatures and has stable electrical properties when compared to both silver and copper. Hence, platinum has a higher response when compared to the other metals. The experiment was carried out for a number of times under the same conditions which gave reproducible results with non-significant error. Also, coulombic efficiency of the supercapacitor as a function of the number of cycles was performed for 1500 cycles and a retention of up to 96% efficiency was observed.

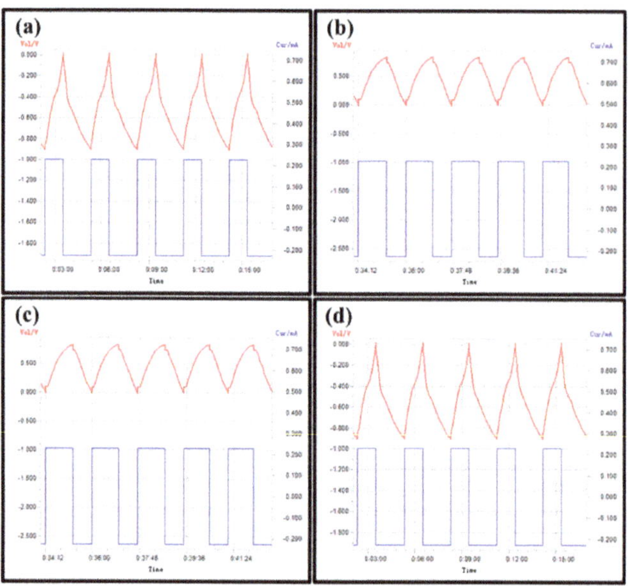

Figure 10. Voltage/time cycling plot of (**a**) GO, (**b**) GO/AgDBSA–PANI (**c**) GO/Cu/DBSA–PANI, (**d**) GO/Pt/DBSA–PANI; voltage range (0 to -0.9 V vs. Ag/AgCl) at 50 A·g^{-1} in 1 M KOH (*aq*). Specific charge/discharge current density = 56.8 A·g^{-1}.

4. Conclusions

The main aim of this work was to combine the properties of graphene oxide (GO) with the properties of metal nanoparticles (Ag, Cu, Pt) and dodecylbenzene sulfonic acid (DBSA)-doped polyaniline (PANI) to enhance the capacitance, energy and power of developed supercapacitor electrodes. GO has good electrical, mechanical and thermal properties and a high surface area, which makes it a good candidate in applications such as polymer composites and energy-related materials. Metal nanoparticles are exceptionally important due to their unique performance in electronic, magnetic, optical and catalytic applications. Recently, carbon material/conducting polymer/metal oxide nanoparticle composites and carbon material/conducting polymer/metal nanoparticle composites have been used as a new class of composite supercapacitor material with improved properties, when compared to conducting polymer or metal oxide/metal nanoparticle alone. The addition of metal oxides or metal nanoparticles improves the size, morphology and conducting properties of the polymer. GO was successfully synthesized by the modified Hummers method and loaded with Ag, Cu and Pt nanoparticles by the electrostatic self-assembly and finally treated with aniline and DBSA to form conducting polymer composites. Characterization by microscopy techniques, HRTEM and HRSEM, revealed thin, flat, bended, and wrinkled nanosheets, with a size of 1.5 μm, and darker surfaces were shown for the conducting polymer composites. The presence of the functional groups was revealed by EDS. The quantitative analysis of the polymer composites indicates the presence of carbon and sulfur along with silver, copper and platinum for the composites that contain Ag, Cu and Pt. Structural analysis by FTIR confirms the successful loading of the nanoparticles on the surface of graphene oxide and the formation of conducting polymer composites. Prominent bands that confirm the loading of nanoparticles and formation of conducting polymer composites were present. The materials were tested for supercapacitor applications by galvanostatic charge discharge and the specific capacitance of GO was determined as 182.8 $F \cdot g^{-1}$, while for the polymer composites, the specific capacitance was determined as 206.4 for GO/Ag/DBSA–PANI, 192.8 $F \cdot g^{-1}$ for GO/Cu/DBSA–PANI and 227.2 for GO/Pt/DBSA–PANI. There are no reports on graphene oxide loaded with metal nanoparticles combined with DBSA-doped polyaniline for application in supercapacitors, thereby making the use of GO/M/DBSA–PANI a novel concept for the development of high-performance organic supercapacitors.

Author Contributions: K.K. and E.I.I. supervised the research; N.R.D. conceived the idea, designed and performed the experiments, analyzed the data and wrote the original draft; A.N. performed material's characterization and analysis; C.I. co-supervised and edited manuscript; M.N. performed electrochemical data analysis; N.W.H., performed analysis data for supercapacitor cell; A.L.D.Y. performed electroanalysis validation studies; M.M. performed data analysis; K.K. reviewed and edited the final manuscript; E.I.I. reviewed the final manuscript.

Funding: This research was partly funded by [National Research Foundation (NRF) of South Africa Research Initiative (SARChI) Chair Grant for NanoElectrochemistry and Sensor Technology] grant number [UID 85102].

Acknowledgments: N.R.D. would like to thank National Research Foundation (NRF) of South Africa and European Union's Erasmus Mundus for European and South African Partnership on Heritage and Past (AESOP) for funding.

References

1.	Colgan, J. *The International Energy Agency*; Global Public Policy Institute: Berlin Germany, 2009.
2.	Simon, P.; Gogotsi, Y. Materials for electrochemical capacitors. *Nat. Mater.* **2008**, *7*, 845–854. [CrossRef] [PubMed]
3.	Shafiee, S.; Topal, E. When will fossil fuel reserves be diminished? *Energy Policy* **2009**, *37*, 181–189. [CrossRef]
4.	Kim, Y.K.; Hwang, S.-H.; Jeong, S.M.; Son, K.Y.; Lim, S.K. Colorimetric hydrogen gas sensor based on PdO /metal oxides hybrid nanoparticles. *Talanta* **2018**, *188*, 356–364. [CrossRef] [PubMed]

5. Li, H.; McRae, L.; Firby, C.J.; Al-Hussein, M.; Elezzabi, A.Y. Nanohybridization of molybdenum oxide with tungsten molybdenum oxide nanowires for solution-processed fully reversible switching of energy storing smart windows. *Nanoenergy* **2018**, *47*, 130–139. [CrossRef]

6. Xie, S.; Bi, Z.; Chen, Y.; He, X.; Guo, X.; Gao, X.; Li, X. Electrodeposited Mo-doped WO_3 film with large optical modulation and high areal capacitance toward electrochromic energy-storage applications. *Appl. Surf. Sci.* **2018**, *459*, 774–781. [CrossRef]

7. Chen, Y.; Meng, Q.; Zhang, L.; Han, C.; Gao, H.; Zhang, Y.; Yan, H. SnO_2-based electron transporting layer materials for perovskite solar cells: A review of recent progress. *J. Energy. Chem.* **2019**, *35*, 144–167. [CrossRef]

8. Bella, F.; Muñoz-García, A.B.; Meligrana, G.; Lamberti, A.; Destro, M.; Pavone, M.; Gerbaldi, C. Unveiling the controversial mechanism of reversible Na storage in TiO_2 nanotube arrays: Amorphous versus anatase TiO_2. *Nano Res.* **2017**. [CrossRef]

9. Perreault, L.L.; Colò, F.; Meligrana, G.; Kim, K.; Fiorilli, S.; Bella, F.; Nair, J.R.; Vitale-Brovarone, V.; Florek, J.; Kleitz, F.; et al. Spray-Dried Mesoporous Mixed Cu-Ni Oxide@Graphene Nanocomposite Microspheres for High Power and Durable Li-Ion Battery Anodes. *Adv. Energy Mater.* **2018**, *8*, 1802438. [CrossRef]

10. Iro, Z.S.; Subramani, C.; Dash, S.S. A Brief Review on Electrode Materials for Supercapacitor. *Int. J. Electrochem. Sci.* **2016**, 10628–10643. [CrossRef]

11. Chen, Y.T.; Ma, C.W.; Chang, C.M.; Yang, Y.J. Micromachined Planar Supercapacitor with Interdigital Buckypaper Electrodes. *Micromachines* **2018**, *9*, 242. [CrossRef]

12. Winter, M.; Brodd, R.J. What Are Batteries, Fuel Cells, and Supercapacitors? *Chem. Rev.* **2004**, *104*, 4245–4270. [CrossRef]

13. Lu, P.; Xue, D.; Yang, H.; Liu, Y. Supercapacitor and nanoscale research towards electrochemical energy storage. *Int. J. Smart Nano Mater.* **2013**, *4*, 2–26. [CrossRef]

14. Jayalakshmi, M.; Balasubramanian, K. Simple Capacitors to Supercapacitors—An Overview. *Int. J. Electrochem. Sci.* **2008**, *3*, 1196–1217.

15. Snook, G.A.; Peng, C.; Fray, D.J.; Chen, G.Z. Chieving high electrode specific capacitance with materials of low mass specific capacitance: Potentiostatically grown thick micro-nanoporous PEDOT films. *Electrochem. Commun.* **2007**, *9*, 83–88. [CrossRef]

16. Burke, A.; Zhao, H. Present and future applications of supercapacitors in electric and hybrid vehicles. In Proceedings of the 2014 IEEE International Electric Vehicle Conference (IEVC), Florence, Italy, 17–19 December 2014; pp. 1–8. [CrossRef]

17. Gao, W. The Chemistry of Graphene Oxide. *Graphene Oxide Reduct. Recipes, Spectrosc. Appl.* **2015**, 61–95. [CrossRef]

18. Hang, L.; Zhao, Y.; Zhang, H.; Liu, G.; Cai, W. Copper nanoparticle@graphene composite arrays and their enhanced catalytic performance. *Acta Mater.* **2016**, *105*, 59–67. [CrossRef]

19. Hu, J.; He, B.; Lu, J.; Hong, L.; Yuan, J.; Song, J.; Niu, L. Facile Preparation of Pt/Polyallylamine/Reduced Graphene Oxide Composites and Their Application in the Electrochemical Catalysis on Methanol Oxidation. *Int. J. Electrochem. Sci.* **2012**, *7*, 10094–10107.

20. Shen, C.; Barrios, E.; McInnis, M.; Zuyus, J.; Zhai, L. Fabrication of graphene aerogols with heavily loaded metallic nanoparticles. *Micromachines* **2017**, *8*, 47. [CrossRef]

21. Luo, M.; Liu, Y.; Huan, W.; Qiao, W.; Zhou, Y.; Ye, Y.; Chen, L.S. Towards flexible transparent electrodes based on carbon and metallic materials. *Micromachines* **2017**, *8*, 12. [CrossRef]

22. Grinou, A.; Yun, Y.S.; Cho, S.Y.; Park, H.H.; Jin, H.J. Dispersion of Pt Nanoparticle-Doped Reduced Graphene Oxide Using Aniline as a Stabilizer. *Materials* **2012**, *5*, 2927–2936. [CrossRef]

23. Bonet, F.; Delmas, V.; Grugeon, S.; Herrera Urbina, R.; Silvert, P.Y.; Tekaia-Elhsissen, K. Synthesis of Monodisperse Au, Pt, Pd, Ru and Ir Nanoparticles in Ethylene Glycol. *Nanostructured Mater.* **1999**, *11*, 1277–1284. [CrossRef]

24. Snook, G.A.; Kao, P.; Best, A.S. Conducting-polymer-based supercapacitor devices and electrodes. *J. Power Sources* **2011**, *196*, 1–12. [CrossRef]

25. Wang, H.; Liu, Y.; Li, M.; Huang, H.; Xu, H.-M.; Hong, R.-J.; Shen, H. Multifunctional TiO_2 nanowires-modified nanoparticles bilayer film for 3D dye-sensitized solar cells. *Optoelectron. Adv. Mater. Rapid Commun.* **2010**, *4*, 1166–1169.

26. Zou, Y.; Wang, Q.; Xiang, C.; Tang, C. Doping composite of polyaniline and reduced graphene oxide with palladium nanoparticles for room-temperature hydrogen-gas sensing. *Int. J. Hydrogen Energy* **2016**, *41*, 5396–5404. [CrossRef]

27. Trchová, M.; Stejskal, J. Polyaniline: The infrared spectroscopy of conducting polymer nanotubes (IUPAC Technical Report). *Pure Appl. Chem.* **2011**, *83*, 1803–1817. [CrossRef]

28. Halper, M.J. Ellenbogen, Supercapacitors: A brief overview. *Rep. No. MP 05W0000272* **2006**, 1–29.

29. Ansari, M.O.; Khan, M.M.; Ansari, S.A.; Amal, I.; Lee, J.; Cho, M.H. pTSA doped conducting graphene/polyaniline nanocomposite fibers: Thermoelectric behavior and electrode analysis. *Chem. Eng. J.* **2014**, *242*, 155–161. [CrossRef]

30. Ramkumar, R.; Sundaram, M.M. A biopolymer gel-decorated cobalt molybdate nanowafer: effective graft polymer cross-linked with an organic acid for better energy storage. *New J. Chem.* **2016**, *40*, 2863–2877. [CrossRef]

31. Ramkumar, R.; Sundaram, M.M. Electrochemical synthesis of polyaniline cross-linked $NiMoO_4$ nanofibre dendrites for energy storage devices. *New J. Chem.* **2016**, *40*, 7456–7464. [CrossRef]

32. Ramkumar, R.; Minakshi, M. Fabrication of ultrathin $CoMoO_4$ nanosheets modified with chitosan and their improved performance in energy storage device. *Dalton Trans.* **2015**, *44*, 6158–6168, doi101039/c5dt00622h. [CrossRef]

33. Yan, J.; Fan, Z.; Wei, T.; Qian, W.; Zhang, M.; Wei, F. Fast and reversible surface redox reaction of graphene–MnO_2 composites as supercapacitor electrodes. *Carbon* **2010**, *48*, 3825–3833. [CrossRef]

34. Njomo, N.; Waryo, T.; Masikini, M.; Ikpo, C.O.; Mailu, S.; Tovide, O.; Ross, N.; Williams, A.; Matinise, N.; Sunday, C.E.; et al. Graphenated tantalum(IV) oxide and poly(4-styrene sulfonic acid)-doped polyaniline nanocomposite as cathode material in an electrochemical capacitor. *Electrochim. Acta* **2014**, *128*, 226–237. [CrossRef]

35. Li, B.; Cheng, J.; Wang, Z.; Li, Y.; Ni, W.; Wang, B. Highly-wrinkled reduced graphene oxide-conductive polymer fibers for flexible fiber-shaped and interdigital-designed supercapacitors. *J. Power Sources* **2018**, *376*, 117–124. [CrossRef]

36. Ndipingwi, M.M.; Ikpo, C.; Hlongwa, N.W.; Ross, N.; Masikini, M.; John, S.V.; Baker, P.; Roos, W.; Iwuoha, E.I. Orthorombic nanostructured Li_2MnSiO_4/Al_2O_3 supercapattery electrode with efficient Lithium-ion migratory pathway. *Batteries & Supercaps* **2018**, *1*, 223–235. [CrossRef]

37. Hsieh, C.-T.; Chen, Y.-C.; Chen, Y.-F.; Huq, M.M.; Chen, P.-Y.; Jang, B.-S. Microwave synthesis of titania-coated carbon nanotube composites for electrochemical capacitors. *J. Power Sources* **2014**, *269*, 526–533. [CrossRef]

38. Karabatsos, D.; Ntziouni, A.; Efthymiou, G.; Fujisawa, K.; Lei, Y.; Terrones, M.; Kordatos, K. Controlled anchoring of Fe_3O_4 nanoparticles on graphene oxide and thermally reduced graphene oxide. *Nanoscale Adv.* (under review).

39. Fang, Y.; Guo, S.; Zhu, C.; Zhai, Y.; Wang, E. Self-assembly of cationic polyelectrolyte-functionalized graphene nanosheets and gold nanoparticles: A two-dimensional heterostructure for hydrogen peroxide sensing. *Langmuir* **2010**, *26*, 11277–11282. [CrossRef]

40. Kumar, R.; Ansari, M.O.; Barakat, M.A. DBSA doped polyaniline/multi-walled carbon nanotubes composite for high efficiency removal of Cr(VI) from aqueous solution. *Chem. Eng. J.* **2013**, *228*, 748–755. [CrossRef]

41. Karthika, P.; Rajalakshmi, N.; Dhathathreyan, K.S. Functionalized exfoliated graphene oxide as supercapacitor electrodes. *Soft Nanosci. Lett.* **2012**, *2*, 59–66. [CrossRef]

42. Sarlak, N.; Meyer, T.J. Fabrication of completely water-soluble graphene oxides graft poly citric acid using different oxidation methods and comparison of them. *J. Mol. Liq.* **2017**, *243*, 654–663. [CrossRef]

43. Mane, A.T.; Navale, S.T.; Mane, R.S.; Naushad, M.; Patil, V.B. Synthesis and structural, morphological, compositional, optical and electrical properties of DBSA-doped PPy–WO_3 nanocomposites. *Prog. Org. Coatings* **2015**, *87*, 88–94. [CrossRef]

44. Zaaba, N.I.; Foo, K.L.; Hashim, U.; Tan, S.J.; Liu, W.W.; Voon, C.H. Synthesis of Graphene Oxide using Modified Hummers Method: Solvent Influence. *Procedia Eng.* **2017**, *184*, 469–477. [CrossRef]

45. Marcano, D.C.; Kosynkin, D.V.; Berlin, J.M.; Sinitskii, A.; Sun, Z.; Slesarev, A.; Alemany, L.B.; Lu, W.; Tour, J.M. Improved synthesis of graphene oxide. *ACS Nano* **2010**, *4*, 4806–4814. [CrossRef] [PubMed]

46. Shalaby, A.; Nihtianova, D.; Markov, P.; Staneva, A.D.; Iordanova, R.S.; Dimitriev, Y.B. Structural analysis of reduced graphene oxide by transmission electron microscopy. *Bulg. Chem. Commun.* **2015**, *47*, 291–295.

47. Wilson, N.R.; Pandey, P.A.; Beanland, R.; Young, R.J.; Kinloch, I.A.; Gong, L.; Liu, Z.; Suenaga, K.; Rourke, J.P.; York, S.J.; et al. Graphene oxide: Structural analysis and application as a highly transparent support for electron microscopy. *ACS Nano* **2009**, *3*, 2547–2556. [CrossRef]

48. Dikin, D.A.; Stankovich, S.; Zimney, E.J.; Piner, R.D.; Dommett, G.H.; Evmenenko, G.; Nguyen, S.T.; Ruoff, R.S. Preparation and characterization of graphene oxide paper. *Nature* **2007**, *448*, 457–460. [CrossRef] [PubMed]

49. Xu, B.; Yue, S.; Sui, Z.; Zhang, X.; Hou, S.; Cao, G.; Yang, Y. What is the choice for supercapacitors: graphene or graphene oxide? *Energy Environ. Sci.* **2011**, *4*, 2826–2830. [CrossRef]

50. Xin, L.; Yang, F.; Rasouli, S.; Qiu, Y. Understanding Pt Nanoparticle Anchoring on Graphene Supports through Surface Functionalization. *ACS Catal.* **2016**, *6*, 2642–2653. [CrossRef]

51. Calheiros, L.F.; Soares, B.G.; Barra, G.M.O. DBSA-CTAB mixture as the surfactant system for the one step inverse emulsion polymerization of aniline: Characterization and blend with epoxy resin. *Synth. Met.* **2017**, *226*, 139–147. [CrossRef]

52. Afzal, A.B.; Javed Akhtar, M. Effects of silver nanoparticles on thermal properties of DBSA-doped polyaniline/PVC blends. *Iran. Polym. J.* **2012**, *21*, 489–496. [CrossRef]

53. Haba, Y.; Segal, E.; Narkis, M.; Titelman, G.I.; Siegmann, A. Polymerization of aniline in the presence of DBSA in an aqueous dispersion. *Synth. Met.* **1999**, *106*, 59–66. [CrossRef]

54. Haba, Y.; Segal, E.; Narkis, M.; Titelman, G.I.; Siegmann, A. Polyaniline–DBSA/polymer blends prepared via aqueous dispersions. *Synth. Met.* **2000**, *110*, 189–193. [CrossRef]

55. Long, N.V.; Chien, N.D.; Hayakawa, T.; Hirata, H.; Lakshminarayana, G.; Nogami, M. The synthesis and characterization of platinum nanoparticles: A method of controlling the size and morphology. *Nanotechnology* **2010**, *21*, 035605. [CrossRef] [PubMed]

56. Del Castillo-Castro, T.; Castillo-Ortega, M.M.; Villarreal, I.; Brown, F.; Grijalva, H.; Perez-Tello, M.; Nuno-Donlucas, S.M.; Puig, J.E. Synthesis and characterization of composites of DBSA-doped polyaniline and polystyrene-based ionomers. *Compos. Part A Appl. Sci. Manuf.* **2007**, *38*, 639–645. [CrossRef]

57. Shahriary, L.; Athawale, A. Graphene Oxide Synthesized by using Modified Hummers Approach. *Int. J. Renew. Energy Environ. Eng.* **2014**, *2*, 58–63.

58. Dong, L.; Gari, R.R.S.; Li, Z.; Craig, M.M.; Hou, S. Graphene-supported platinum and platinum–ruthenium nanoparticles with high electrocatalytic activity for methanol and ethanol oxidation. *Carbon* **2010**, *48*, 781–787. [CrossRef]

59. Yue, Y.; Zhou, B.; Shi, J.; Chen, C.; Li, N.; Xu, Z.; Liu, L.; Kuang, L.; Ma, M.; Fu, H. γ-Irradiation assisted synthesis of graphene oxide sheets supported Ag nanoparticles with single crystalline structure and parabolic distribution from interlamellar limitation. *Appl. Surf. Sci.* **2017**, *403*, 282–293. [CrossRef]

60. Mallakpour, S.; Abdolmaleki, A.; Karshenas, A. Graphene oxide supported copper coordinated amino acids as novelheterogeneous catalysts for epoxidation of norbornene. *Catal. Commun.* **2017**, *92*, 109–113. [CrossRef]

61. Chew, T.; Daik, R.; Hamid, M. Thermal Conductivity and Specific Heat Capacity of Dodecylbenzenesulfonic Acid-Doped Polyaniline Particles—Water Based Nanofluid. *Polymers* **2015**, *7*, 1221–1231. [CrossRef]

62. Jianming, J.; Wei, P.; Shenglin, Y.; Guang, L. Electrically conductive PANI-DBSA/Co-PAN composite fibers prepared by wet spinning. *Synth. Met.* **2005**, *149*, 181–186. [CrossRef]

63. Yin, H.; Yamamoto, T.; Wada, Y.; Yanagida, S. Large-scale and size-controlled synthesis of silver nanoparticles under microwave irradiation. *Mater. Chem. Phys.* **2004**, *83*, 66–70. [CrossRef]

64. Yin, W.; Ruckenstein, E. Soluble polyaniline co-doped with dodecyl benzene sulfonic acid and hydrochloric acid. *Synth. Met.* **2000**, *108*, 39–46. [CrossRef]

65. Fakhri, P.; Jaleh, B.; Nasrollahzadeh, M. Synthesis and characterization of copper nanoparticles supported on reduced graphene oxide as a highly active and recyclable catalyst for the synthesis of formamides and primary amines. *J. Mol. Catal. A Chem.* **2014**, *383*, 17–22. [CrossRef]

66. Khan, S.; Ali, J.; Husain, M.; Zulfequar, M. Synthesis of reduced graphene oxide and enhancement of its electrical and optical properties by attaching Ag nanoparticles. *Phys. E Low-dimensional Syst. Nanostructures* **2016**, *81*, 320–325. [CrossRef]

67. Xiao, Y.; Liu, J.; Xie, K.; Wang, W.; Fang, Y. Aerobic oxidation of cyclohexane catalyzed by graphene oxide: Effects of surface structure and functionalization. *Mol. Catal.* **2017**, *431*, 1–8. [CrossRef]

68. Yang, S.-D.; Shen, C.-M.; Tong, H.; He, W.; Zhang, X.-G.; Gao, H.-J. Highly dispersed Pd nanoparticles on chemically modified graphene with aminophenyl groups for formic acid oxidation. *Chinese Phys. B* **2011**, *20*, 113301. [CrossRef]

69. Han, D.; Chu, Y.; Yang, L.; Liu, Y.; Lv, Z. Reversed micelle polymerization: a new route for the synthesis of DBSA–polyaniline nanoparticles. *Colloids Surfaces A Physicochem. Eng. Asp.* **2005**, *259*, 179–187. [CrossRef]
70. Nasehnia, F.; Mohammadpour Lima, S.; Seifi, M.; Mehran, E. First principles study on optical response of graphene oxides: From reduced graphene oxide to the fully oxidized surface. *Comput. Mater. Sci.* **2016**, *114*, 112–120. [CrossRef]
71. Wang, J.; Caliskan Salihi, E.; Šiller, L. Green reduction of graphene oxide using alanine. *Mater. Sci. Eng.* **2017**, *72*, 1–6. [CrossRef] [PubMed]
72. Singh, K.; Ohlan, A.; Dhawan, S.K.; Kong, B.S. Nanostructured graphene/Fe_3O_4 incorporated polyaniline as a high performance shield against electromagnetic pollution. *Nanoscale* **2013**, *5*, 2411–2420. [CrossRef] [PubMed]
73. Zoladek, S.; Rutkowska, I.A.; Blicharska, M.; Miecznikowski, K.; Ozimek, W.; Orlowska, J.; Negro, E.; Noto, V.D.; Kulesza, P.J. Evaluation of reduced-graphene-oxide-supported gold nanoparticles as catalytic system for electroreduction of oxygen in alkaline electrolyte. *Electrochim. Acta* **2017**, *233*, 113–122. [CrossRef]
74. Bhadra, J.; Madi, N.K.; Al-Thani, N.J.; Al-Maadeed, M.A. Polyaniline/polyvinyl alcohol blends: Effect of sulfonic acid dopants on microstructural, optical, thermal and electrical properties. *Synth. Met.* **2014**, *191*, 126–134. [CrossRef]
75. Xie, H.Q.; Ma, Y.M.; Guo, J.S. Secondary doping phenomena of two conductive polyaniline composites. *Synth. Met.* **2001**, *123*, 47–52. [CrossRef]
76. Guo, Q.; Yi, C.; Zhu, L.; Yang, Q.; Xie, Y. Chemical synthesis of cross-linked polyaniline by a novel solvothermal metathesis reaction of p-dichlorobenzene with sodium amide. *Polymer* **2005**, *46*, 3185–3189. [CrossRef]
77. Chang, K.C.; Yang, G.-W.; Peng, C.-W.; Lin, C.-Y.; Shieh, J.-C.; Yeh, J.M.; Yang, J.-C.; Li, W.T. Comparatively electrochemical studies at different operational temperatures for the effect of nanoclay platelets on the anticorrosion efficiency of DBSA-doped polyaniline/Na+–MMT clay nanocomposite coatings. *Electrochim. Acta* **2007**, *52*, 5191–5200. [CrossRef]
78. Chen, C.H. Thermal Studies of Polyaniline Doped with Dodecyl Benzene Sulfonic Acid Directly Prepared via Aqueous Dispersions. *J. Polym. Res.* **2002**, *9*, 195–200. [CrossRef]
79. Fu, C.; Zhao, G.; Zhang, H.; Li, S. Evaluation and Characterization of Reduced Graphene Oxide Nanosheets as Anode Materials for Lithium-Ion Batteries. *Int. J. Electrochem. Sci.* **2013**, *8*, 6269–6280.
80. Sharma, V.V.; Gualandi, I.; Vlamidis, Y.; Tonelli, D. Electrochemical behavior of reduced graphene oxide and multi-walled carbon nanotubes composites for catechol and dopamine oxidation. *Electrochim. Acta* **2017**, *246*, 415–423. [CrossRef]
81. Chen, L.; Tang, Y.; Wang, K.; Liu, C.; Luo, S. Direct electrodeposition of reduced graphene oxide on glassy carbon electrode and its electrochemical application. *Electrochem. Commun.* **2011**, *13*, 133–137. [CrossRef]
82. Jang, H.; Kim, J.; Kang, H.; Bae, D.; Chang, H.; Choi, H. Reduced graphene oxide as a protection layer for Al. *Appl. Surf. Sci.* **2017**, *407*, 1–7. [CrossRef]
83. Wang, J.; Engelhard, M.H.; Wang, C.; Lin, Y. Facile and controllable electrochemical reduction of grapheme oxide and its applications. *J. Mater. Chem.* **2010**, *20*, 743–748. [CrossRef]

 *micromachines*

Article

Nano-Graphitic based Non-Volatile Memories Fabricated by the Dynamic Spray-Gun Deposition Method

Paolo Bondavalli [1,*]**, Marie Blandine Martin** [1]**, Louiza Hamidouche** [1]**, Alberto Montanaro** [1]**, Aikaterini-Flora Trompeta** [2] **and Costas A. Charitidis** [2]

[1] Thales Research and Technology, 1 Av Augustin Fresnel, 91767 Palaiseau, France; marie-blandine.martin@thalesgroup.com (M.B.M.); hamidouche.louiza@gmail.com (L.H.); alberto.montanaro@cnit.it (A.M.)

[2] Research Unit of Advanced, Composite, Nanomaterials and Nanotechnology, School of Chemical Engineering, National Technical University of Athens, 9 Heroon Polytechneiou st., Zografos, GR-15773 Athens, Greece; ktrompeta@chemeng.ntua.gr (A.-F.T.); charitidis@chemeng.ntua.gr (C.A.C.)

* Correspondence: paolo.bondavalli@thalesgroup.com

Received: 7 December 2018; Accepted: 24 January 2019; Published: 29 January 2019

Abstract: This paper deals with the fabrication of Resistive Random Access Memory (ReRAM) based on oxidized carbon nanofibers (CNFs). Stable suspensions of oxidized CNFs have been prepared in water and sprayed on an appropriate substrate, using the dynamic spray-gun deposition method, developed at Thales Research and Technology. This technique allows extremely uniform mats to be produced while heating the substrate at the boiling point of the solvent used for the suspensions. A thickness of around 150 nm of CNFs sandwiched between two metal layers (the metalized substrate and the top contacts) has been achieved, creating a Metal-Insulator-Metal (MIM) structure typical of ReRAM. After applying a bias, we were able to change the resistance of the oxidized layer between a low (LRS) and a high resistance state (HRS) in a completely reversible way. This is the first time that a scientific group has produced this kind of device using CNFs and these results pave the way for the further implementation of this kind of memory on flexible substrates.

Keywords: ReRAM; carbon nanofibers; spray-gun deposition

---

1. Introduction

The dynamic spray-gun deposition method is a new, revolutionary technique that, thanks to its versatility, can be implemented for a large range of applications. In this context, an extremely promising application is the fabrication of low-cost graphite-based memories that can be integrated with flexible, plastic, or paper-based substrates. In the International Technology Roadmap for Semiconductors (ITRS) of 2011 [1], within the chapter concerning the Emerging Research Devices (ERDs) and, more specifically, memory devices, it was stated that ultrathin graphite layers are "interesting materials for macromolecular memories, thanks to the potential fabrication costs that are considered as the primary driver for this type of memory, while extreme scaling is de-emphasized". Indeed, these memories can be implemented in low-cost applications and integrated, for example, in ID cards, driver licenses, smart-cards, and on paper, as well as potentially being patched for health applications; the implementation of these memories demonstrate the possibility of storing information and changing it in an easy way. Up until now, there has been no commercialized technology available for flexible, low-cost memories. In this piece, we will describe how Metal-Insulator-Metal (MIM) structures were fabricated, where the carbon-based materials were sandwiched between two metal contacts. This kind of structure is a Resistive Random Access Memory (ReRAM), in which the resistance of the sandwiched material is changed by applying a bias between the top and bottom contact.

The first work highlighting the utilization of graphene-based mats for flexible, non-volatile ReRAM was published in 2010 by Jeong et al. [2]. In this paper, by exploiting the hydrophilic character of Graphene Oxide (GO) flakes, they achieved stable suspensions and deposited them on large surfaces through a spin coating process. The work of Jeong et al. demonstrated the non-volatile effect on the resistance of a 70-nm-thick layer of GO flakes consisting of a layered structure of Al/GO/Al in a cross-bar configuration. Another interesting piece of work was produced by He et al. in 2009 [3], when reliable and reproducible resistive switching behaviors were observed in GO thin films prepared by the vacuum filtration method, a common technique used to fabricate Carbon Nanotube (CNT)-based bucky papers. The main hypothesis regarding the physical principle at the origin of the switching effect was surrounding the absorption/desorption of oxygen-related groups on the GO sheets, as well as the diffusion of the top electrodes. Indeed, the alignment of the oxygen vacancies creates conducting paths that reduce the resistance of the sandwiched layers moving from a high-resistance state (HRS) to a low-resistance state (LRS) [4]. Another hypothesis concerned the oxidation of the top contact. The pioneering work of Hong et al. [5] included a deep analysis of the switching mechanism of this kind of device, underlining that these structures' performances depend on the origin of the top contact. For example, in the case of Au-based top electrodes, there was no oxygen migration in opposition to Al electrodes. This can be explained because Au contacts cannot be oxidized. To evaluate the effect of the oxidation, Hong et al. performed measurements by fabricating contacts with different dimensions. As suspected, if the change in the resistance is related to the formation of dendrites, it does not depend on the contact dimensions [6]. Hong and coworkers demonstrated that the current between the two contacts was a function of the dimensions of the top contact and was therefore proportional to the number of potential conductive patches built-up by the oxygen vacancies. All of these experiments underlined the fact that GO was potentially useful for future non-volatile memory applications.

According to the above, our work focused on oxidized carbon nanofibers (CNFs) together with GO, a combination that could overcome the challenges that were faced in the previous studies.

2. Materials and Methods

2.1. Definition of the CNFs' Specifications

Taking into account the aforementioned preliminary results, this study aimed to fabricate memories based on oxidized CNFs [7]. CNFs are less expensive in comparison with CNTs and are easier to be fabricated. In this case, graphene worked only as the "vehicle" of the oxygen atoms, i.e., graphene was the carrier of the necessary oxygen [8]. Considering that the final thickness of the insulator layer needed to be around 50 nm, extremely diluted suspensions using oxidized CNFs in de-ionized water were used, as described in the following.

2.2. Carbon Nanofibers' Growth, Purification, and Functionalization

CNFs were synthesized in the Research Unit of Advanced, Composite, Nanomaterials, and Nanotechnology, of the National Technical University of Athens, through the Thermal Catalytic Chemical Vapor Deposition technique (TC-CVD). The specifications of the synthesized materials were defined as described in 2.1, to fit the application. The TC-CVD method is a versatile technique, through which structures with different geometries can be produced [9]. Thus, a length of ~5 μm and a diameter of 50 $\pm$ 5 nm were aimed for, which was achieved with the use of the appropriate catalyst and was synthesized in-house. The catalyst consisted of Fe^{3+} particles, embedded on Al_2O_3, which had been prepared via the impregnation method [10]. The parameters of the CVD (Chemical Vapor Deposition) reaction also played a crucial role in the obtained structures. Thus, after a series of preliminary experiments, a temperature range of 700–750 °C was chosen for the reaction to take place, together with a total flow rate of 330 mL/min for the reaction gases (acetylene and nitrogen of a volume ratio of 1:2, respectively).

After their synthesis, CNFs were purified to remove any amorphous by-product remaining, which could be observed as spherical agglomerates in a fluffy form. Moreover, the purification enabled the removal of the metallic catalyst residues. In order to purify the sample, a step-by-step chemical procedure took place. Initially, the sample was stirred in NaOH (1 mol/L) for 2 h at 80 °C to remove the Al_2O_3 substrate. Afterward, the dried sample was extracted with 5 mol/L of HCl overnight in a Soxhlet extractor at the boiling point of the acid. Washing took place with distilled water extensively, in order to remove the dissolved catalyst. The success of the purification was confirmed with Energy Dispersive X-Ray Spectroscopy (EDS) analysis, the results of which are presented below (Figure 1). As can be seen, Fe has been totally removed (0.0 wt.%), while Al from the Al_2O_3 substrate is up to 1.8 wt.%. The 5 wt.% Si refers to the Si wafer substrate, where the CNFs' growth took place. In Figure 1, the morphology of the purified CNFs is also presented, which was studied through Scanning Electron Microscopy (SEM). The CNFs produced form entangled bundles with straight structures of less than 5 μm. Their surface is smooth and their diameters are 50 ± 5 μm. Figure 2 presents the CNFs' internal structure, which has been studied through Transmission Electron Microscopy (TEM). The interconnections of the inner sidewalls are clearly visible. Their internal cavity is also able to be observed and these nanofibers are classified as being stacked-cup.

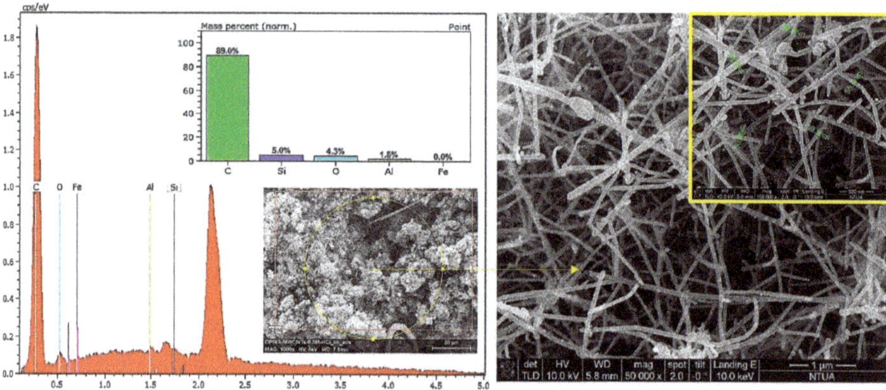

Figure 1. SEM images of the produced carbon nanofibers (CNFs) in (50,000× and 100,000× magnification) and an Energy Dispersive X-Ray Spectroscopy (EDS) analysis of CNFs for the purification assessment.

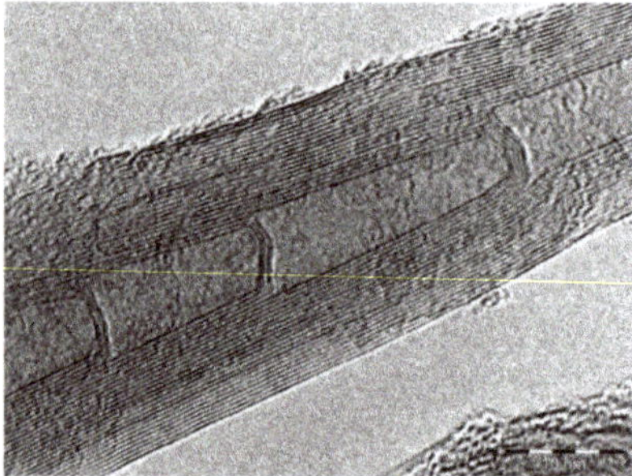

Figure 2. A Transmission Electron Microscopy (TEM) image of the produced carbon nanofibers (CNFs), showing the interactions of the inner sidewalls.

The final step prior to the use of the CNFs in the spray-gun deposition is their oxidation. Chemical functionalization of their surface was carried out with the treatment of strong acids [11]. Specifically, a mixture of 12 mol/L H_2SO_4/HNO_3 in a volume ratio of 3:1 was used. The CNFs were boiled under reflux (~120 °C) in the acid mixture overnight. Then, intense washing with distilled water took place through several repetitions of centrifugation and a final step of vacuum filtration. By this point, oxidation of their surface was achieved, since the oxygen groups were anchored (–COOH, –OH, =O). After the described chemical functionalization, the CNFs were dispersed in de-ionized water and the stabilized suspensions were ready to be sprayed. Considering that the CNFs were oxidized, extremely stable suspensions were obtained. This enabled the development of a green-type process and a reduction in the heating temperature on the heating chuck; in the case of N-Methyl-Pyrrolidone (NMP, the most common solvent for carbon-based nanomaterials), the heating temperature is larger than 202 °C, but in the case of water, it is only 100 °C.

2.3. Carbon Nanofiber-based Suspensions for the Spray-Gun Deposition Method

The suspensions were obtained using 10mg of oxidized CNFs in 500 mL of de-ionized water. Centrifugation was performed for 20 min, in two phases, at 3000 rpm. The top parts of the suspensions that were centrifuged was removed. This step is important in order to avoid the utilization of the bundles of high thickness; a critical parameter for the final fabrication of the memories. Indeed, the increased roughness can create dendrites (aligned metal particles linking the two electrodes) and, therefore, can lead to the short-circuiting of the device. The parts of the suspensions that were removed were further diluted in 500 mL of de-ionized water and sonicated in a bath for 6 h. Afterward, the deposition of a 2″ silicon-based substrate, that was preliminarily metalized using a Ti/Pt layer, was performed. The equipment used to spray the suspensions has been developed at Thales Research and Technology, using a patented process [12], and it constitutes the first prototype of this type of machine. This technique has already been implemented for other applications, such as gas sensing [13–16] and energy storage [17–19]. A picture of the machine set-up is shown in Figure 3 [18,20]. To perform the spray, the nozzle can move in two directions (x and y, see Figure 3), while the z-direction is set-up at the beginning. Due to this, samples with dimensions of 15 cm × 15 cm can be sprayed. The substrate is positioned on a heated chuck that can reach 250 °C. The temperature adjustment is an important parameter; in order to avoid the "coffee ring effect", the drops reaching the substrate are heated at the boiling point of the solvent, which is 100 °C in this case, with water. Due to this, the nanomaterials in the drops are stuck in the impact zone and cannot move to the borders of the drop, achieving a uniform deposition.

In the following pictures, the top view of the sample is shown at different time points of the deposition of the suspensions. The last picture corresponds to one liter of the suspension having been deposited. We can see that, in Figure 4f, the couverture of the sample is optimized. The final thickness is around 150 nm.

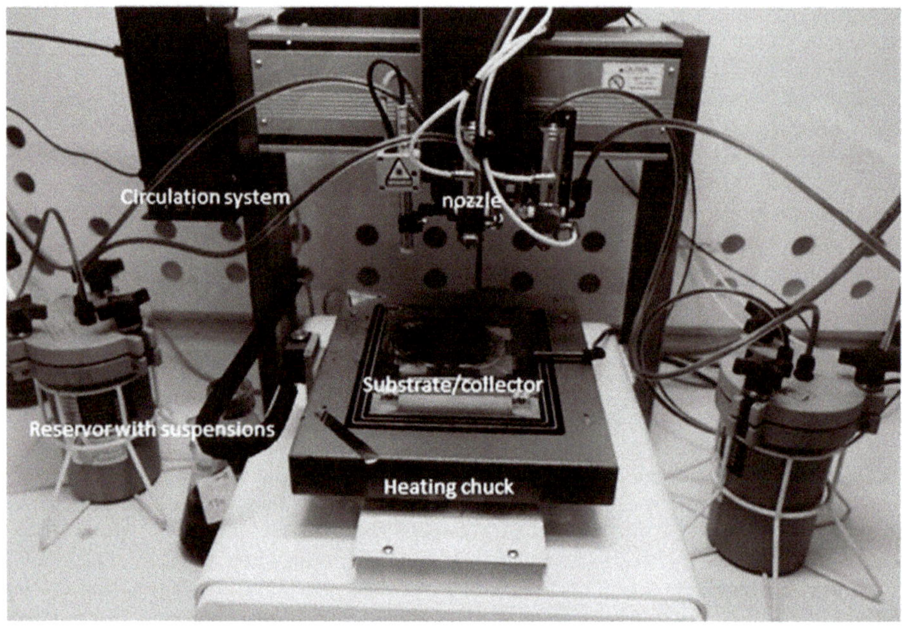

Figure 3. The dynamic spray-gun set-up developed at Thales Research and Technology.

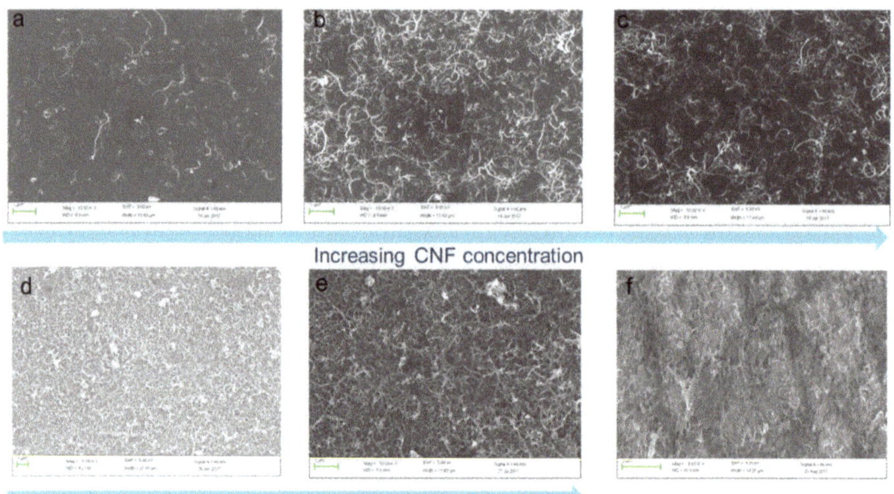

Figure 4. SEM images of carbon nanofibers (CNFs) deposited with the spray technique. The coverage rate increases until full coverage is reached in Figure 5e (single nozzle) and 5f (double nozzle). The concentration of the CNFs is increased due to the deposition time (**a**) 3 μm^{-1} (**b**) 6 μm^{-1} (**c**) 10 μm^{-1} (**d**) 15 μm^{-1} (**e**) 30 μm^{-1} (**f**) 60 μm^{-1} (the number corresponds to the CNFs, but an error of around 20% must be taken into account).

On top of the deposited CNFs layer, we patterned 100 μm × 100 μm squares where we then deposited AlCu/Au contacts. The active memory stack is depicted in Figure 5:

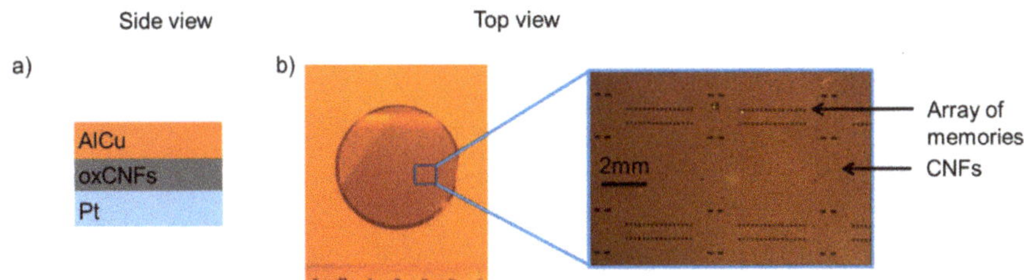

Figure 5. The side and top views of the sample.

3. Results and Discussion

To investigate the change in the resistance of the sandwiched CNF samples, an I-V (current-voltage) measurement was made using a probe station and a Keithly SMU module. The samples were placed on the chuck of the probe station, with two probes being used, one to contact the top electrode and the other to contact the back one. A bias was applied and the current flowing through the CNFs was measured. The sweep of the bias was made firstly from 0 V to −1.5 V, then from −1.5 V to 1.5 V, before moving back from 1.5 V to 0 V again.

One hundred cycles like the one described were performed. The measured current for one cycle is shown in Figure 6. The oxidized CNFs are in contact with the inert bottom electrode of Pt and the easily-oxidized top electrode of AlCu. By applying a positive voltage between the bottom and top electrode, the oxygen atoms migrate from the oxidized CNF layer to the AlCu electrode. By applying the opposite voltage, the oxidized AlCu electrode gives back the oxygen atoms to the CNF-based layer. This back and forth movement of the oxygen atoms leads to a hysteretic change.

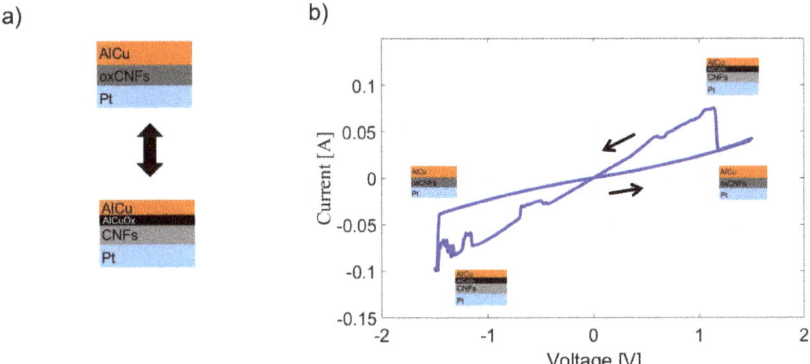

Figure 6. (**a**) The schematic side view of the two states of resistance of the memory stack (**b**) An image of a cycle obtained by applying a voltage between the inert Pt electrode and the AlCu electrode. The hysteretic change of current results from the migration of oxygen atoms between the carbon nanofiber (CNF) layer and the AlCu metal.

One must note that the purity of the CNFs is essential to obtain this kind of behavior. In the case of non-purified CNFs, pinning of the memory has been observed, meaning that after a few cycles, the memory stays stuck in one state. This issue has been overcome by using purified CNFs, in which the memory stands show typical hysteresis cycles without being trapped in one state, as depicted in Figure 7. The tests showed that the memory behavior was retained throughout the 100 cycles.

a) Non-purified CNF

b) Purified CNF

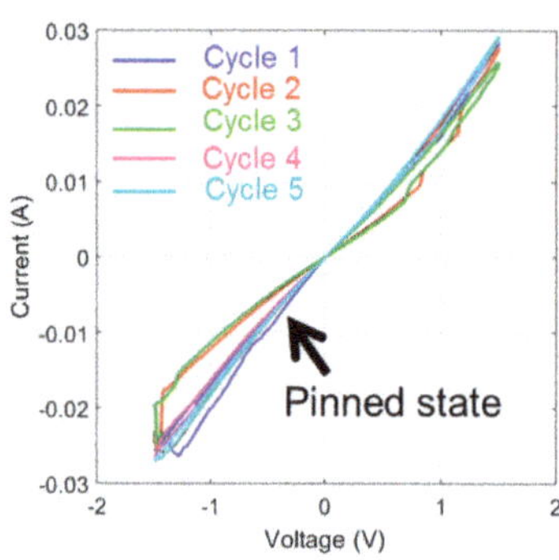

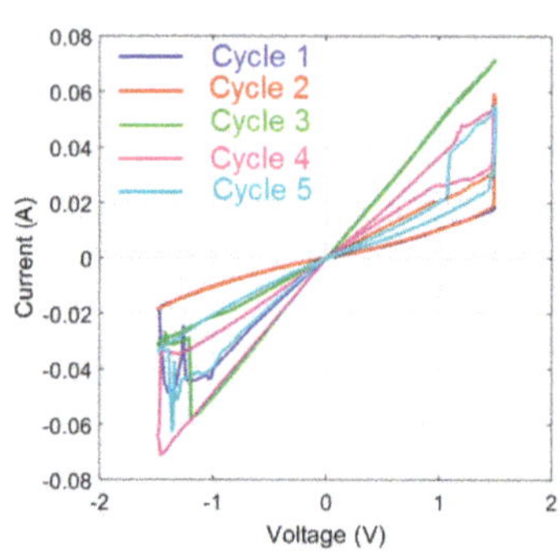

Figure 7. (**a**) Non-purified carbon nanofibers (CNFs) lead to memories being pinned in one state after a few cycles. (**b**) A memory fabricated with purified CNFs withstands many cycles without being pinned in one state.

If the principle of the memory has been understood correctly, many aspects are still to be investigated, such as the number of cycles the memory can withstand and the aging of the memory. By carefully observing Figure 7, it is clear that we have to understand why a sort of instability of the cycles exists, causing them to have the tendency to shift, even if their on/off ratio is maintained. However, their easy fabrication and their high rate of success, in terms of cycles observed for each memory row, make us believe that this new type of memory is very encouraging.

4. Conclusions

This paper describes, for the first time, the fabrication of memories based on oxidized CNFs. This is the first time that a scientific group has demonstrated the possibility of producing memories using these materials. Moreover, these Resistive Random Access Memories (ReRAMs) have been fabricated using a completely new technique based on the dynamic spray-gun deposition method, patented by Thales. The importance of these results has to be taken into account because the CNF memories obtained on hard substrates can potentially pave the way for the revolution of memories on flexible substrates which, at the moment, do not exist as a commercial component in the market. Due to this, we would be able to widen the implementation of the technique to large market applications, such as ID cards, smart cards, memories for health monitoring on specific patches (e.g., diabetes monitoring), and ticketing. CNF-based memories can be fabricated roll-to-roll using the spray-gun deposition method, thereby dramatically reducing the cost of the devices. They can also be fabricated on paper-based substrates, creating disposable components, or integrated into radio-frequency identification (RFID) on plastics to add to the capability of storing information. However, a lot of work has to be done. Indeed, we have to understand why the cycles are unstable; the on/off ratio for each cycle is nearly the same but the current is increased. In addition, the metal contact appears to have a very important influence, considering that if we perform more measurements on the same byte, we do not obtain exactly the same graphs, even if the overall hysteresis behavior is always present. The other main problem is the identification of the exact formatting bias that activates the hysteresis cycle. For this reason, we need to do more systematic measurements on more samples, even if the phenomenon identified is very impressive considering the equipment used to achieve these kinds of structures and their potential implementation on a roll-to-roll production pilot line.

Micromachines **2019**, *10*, 95

Author Contributions: Conceptualization, P.B.; Methodology, P.B.; Validation, P.B.; Investigation, A.-F.T.; Data Curation, L.H., A.M. and M.B.M.; Writing—Original Draft Preparation, P.B.; Writing—Review & Editing, A.-F.T.; Visualization, A.-F.T.; Supervision, C.A.C.; Project Administration, C.A.C.; Funding Acquisition, C.A.C.

Funding: This research was partially funded by European Union's Horizon 2020 Research and Innovation Programme MODCOMP (Modified cost effective fibre-based structures with improved multi-functionality and performance), under grant number 685844.

Conflicts of Interest: The authors declare no conflicts of interest.

References

1. ITRS Homepage. Available online: http://www.itrs2.net/itrs-reports.html (accessed on 27 December 2018).
2. Jeong, H.Y.; Yun Kim, O.; Won Kim, J.; Hwang, J.O.; Kim, J.-E.; Yong Lee, J.; Yoon, T.H.; Cho, B.J.; Kim, S.O.; Ruoff, R.S.; et al. Graphene oxide thin films for flexible nonvolatile memory applications. *Nanoletters* **2010**, *10*, 4381–4386. [CrossRef] [PubMed]
3. He, C.L.; Zhuge, F.; Zhou, X.F.; Li, M.; Zhou, G.C. Nonvolatile resistive switching in graphene oxide thin films. *Appl. Phys. Lett.* **2009**, *95*, 232101. [CrossRef]
4. Waser, R.; Masakazu, A. Nanoionics-based resistive switching memories. *Nat. Mater.* **2007**, *6*, 833–840. [CrossRef] [PubMed]
5. Hong, S.K.; Kim, J.E.; Kim, S.O.; Choi, S.-Y.; Cho, B.J. Flexible resistive switching memory device based on graphene oxide. *J. Appl. Phys.* **2010**, *31*, 1005–1007.
6. Agraït, N.; Levy Yeyati, A.; Van Ruitenbeek, J.M. Quantum properties of atomic-sized conductors. *Phys. Rep.* **2003**, *377*, 81–279. [CrossRef]
7. Serrano-Aroca, A.; LLorens-Gamez, M. Low-Cost Advanced Hydrogels of Calcium Alginate/Carbon Nanofibers with Enhanced Water Diffusion and Compression Properties. *Polymers* **2018**, *10*, 405.
8. Serrano-Aroca, A.; Ruiz-Pividal, J.-F.; LLorens-Gamez, M. Enhancement of water diffusion and compression performance of crosslinked alginate films with a minuscule amount of graphene oxide. *Sci Rep.* **2017**, *7*, 11684. [CrossRef] [PubMed]
9. Trompeta, A.F.; Koklioti, M.A.; Perivoliotis, D.K.; Lynch, I.; Charitidis, C.A. Towards a holistic environmental impact assessment of carbon nanotube growth through chemical vapour deposition. *J. Clean. Prod.* **2016**, *129*, 384–394. [CrossRef]
10. Abdulkareem, A.S.; Suleiman, B.; Abdulazeez, A.T.; Kariim IAbubakre, O.K.; Kariim, I.; Afolabi, A.S. Synthesis and Characterization of Carbon Nanotubes on Fe/Al2O3 Composite Catalyst by Chemical Vapour Deposition Method. In Proceedings of the WCECS 2016, San Francisco, CA, USA, 19–21 October 2016.
11. Allegri, M.; Charitidis, C.A.; Perivoliotis, D.K.; Bianchi, M.G.; Chiu, M.; Pagliaro, A.; Koklioti, M.A.; Trompeta, A.F.; Bergamaschi, E.; Bussolati, O. Toxicity determinants of multi-walled carbon nanotubes: the relationship between functionalization and agglomeration. *Toxicol. Rep.* **2016**, *3*, 230–243. [CrossRef] [PubMed]
12. Bondavalli, P.; Legagneux, P.; Gorintin, L. Method for Depositing Nanoparticles on a Surface and Corresponding Nanoparticle Depositing Appliance. WO2012049428. 2012.
13. Bondavalli, P.; Legagneux, P.; Lebarny, P.; Pribat, D.; Nagle, J. Conductive Nanotube of Nanowire FET Transistor Network and Corresponding Electronic Device, for Detecting Analytes. WO2006128828. 2006.
14. Bondavalli, P.; Gorintin, L.; Feugnet, G.; Lehouc, G.; Pribat, D. Selective gas detection using CNTFET arrays fabricated using air-brush technique, with different metal as electrodes. *Sens. Actuators B Chem.* **2014**, *202*, 1290–1297. [CrossRef]
15. Bondavalli, P.; Gorintin, L.; Legagneux, P. Carbon Nanotube-Based Gas Sensors: A State of the Art. In *Carbon Nanotubes and their Applications*; Zhang, Q., Ed.; Pan Stanford Publishing: Singapore, 2011; ISBN 9789814303187.
16. Bondavalli, P.; Legagneux, P.; Pribat, D. Carbon nanotubes based transistors as gas sensors: State of the art and critical review. *Sens. Actuators B Chem.* **2009**, *140*, 304–318. [CrossRef]
17. Bondavalli, P.; Pognon, G.; Galindo, C. Method of Depositing Oxidized Carbon-Based Micro-Particles and Nanoparticles. WO2016124756. 2016.

18. Bondavalli, P.; Delfaure, C.; Legagneux, P.; Pribat, D. Supercapacitor Electrode Based on Mixtures of Graphite and Carbon Nanotubes Deposited Using a Dynamic Air-Brush Deposition Technique. *JECS* **2013**, *160*, A1–A6. [CrossRef]

19. Ansaldo, A.; Bondavalli, P.; Bellani, S.; Del Rio Castillo, A.E.; Prato, M.; Pellegrini, V.; Pognon, G.; Bonaccorso, F. High-Power Graphene-Carbon Nanotube Hybrid Supercapacitors. *ChemNanoMat* **2017**, *3*, 436–446. [CrossRef]

20. Bondavalli, P.; Pognon, G.; Koumoulos, E.; Charitidis, C. Dynamic Air-Brush Deposition Method for the New Generation of Graphene Based Supercapacitors. *MRS Adv.* **2018**, *3*, 79–84. Available online: https://doi.org/10.1557/adv.2018.65 (accessed on 15 January 2018). [CrossRef]

 *micromachines*

Article

Laser Treatments for Improving Electrical Conductivity and Piezoresistive Behavior of Polymer–Carbon Nanofiller Composites

Andrea Caradonna [1], Claudio Badini [1,*], Elisa Padovano [1], Antonino Veca [2], Enea De Meo [2] and Mario Pietroluongo [1]

[1] Department of Applied Science and Technology, Politecnico di Torino, Corso Duca degli Abruzzi 24, 10129 Torino, Italy; andrea.caradonna@polito.it (A.C.); elisa.padovano@polito.it (E.P.); mario.pietroluongo@polito.it (M.P.)

[2] CRF, Centro Ricerche FIAT, Strada Torino 50, Orbassano, 10043 Torino, Italy; antonino.veca@crf.it (A.V.); enea.de-meo@external.fcagroup.com (E.D.M)

* Correspondence: claudio.badini@polito.it; Tel.: +39-011-0904635

Received: 13 December 2018; Accepted: 15 January 2019; Published: 18 January 2019

Abstract: The effect of carbon nanotubes, graphene-like platelets, and another carbonaceous fillers of natural origin on the electrical conductivity of polymeric materials was studied. With the aim of keeping the filler content and the material cost as low as possible, the effect of laser surface treatments on the conductivity of polymer composites with filler load below the percolation threshold was also investigated. These treatments allowed processing in situ conductive tracks on the surface of insulating polymer-based materials. The importance of the kinds of fillers and matrices, and of the laser process parameters was studied. Carbon nanotubes were also used to obtain piezoresistive composites. The electrical response of these materials to a mechanical load was investigated in view of their exploitation for the production of pressure sensors and switches based on the piezoresistive effect. It was found that the piezoresistive behavior of composites with very low filler concentration can be improved with proper laser treatments.

Keywords: carbon nanofillers; electrical conductivity; piezoresistive behavior

1. Introduction

1.1. Polymer/Carbon Filler Conductive Composites

Polymer composites with carbonaceous micro and nanofillers were widely investigated during recent years because of their mechanical, electrical, and thermal properties. In particular, nanocomposites show noticeable structural and functional properties that can be exploited for a broad range of applications in every field. The potential of these materials as low-weight structural materials and functional materials for optical devices, electromagnetic shields, electric components, and medical devices (body-attachable adhesives for measuring biosignals) attracted increasing interest [1–5].

Actually, carbon nanofillers can be used for greatly enhancing the electrical conductivity of both thermoplastic and thermoset resins and improving their mechanical behavior at the same time. For these reasons, the potential of carbon nanotubes (CNTs) and graphene-like nanoplatelets (GNPs) for processing conductive polymer-based composites was investigated [1–4]. The electrical conductivity of CNTs ranges between 10^5 and 10^7 S/m, and the typical conductivity value for GNPs is around 10^5 S/m. These conductivity values are similar to those of graphite (10^5 S/m on the graphene sheets constituting the graphite structure), which is traditionally used as filler for polymeric composites. On the other hand, the high aspect ratio of CNTs and GNPs can be exploited for achieving more easily the percolation threshold for electrical conductivity. This threshold for electrical conductivity is observed when the concentration of carbon filler is high enough to allow the formation, inside the polymer matrix, of a continuous network of filler particles, which causes a sudden jump in electrical conductivity.

However, the electrical conductivity of polymer–nanocarbon composites is also greatly affected by the kind of matrix and the fabrication process. For instance, a comparison of literature data [1,5] shows that the percolation path forms after the addition of very different amounts of CNTs to different matrices (from less than 1 wt.% to more than 10 wt.%). Likewise, the addition of similar loads of GNPs to different matrices, as well as the fabrication of the same composite with different methods, can result in dramatically different electrical conductivity values, placed in a range even ten orders of magnitude wide [3]. As a consequence, the filler concentration required for achieving the threshold depends not only on the kind of composite, but also on the composite processing method. In fact, very different values of filler concentration at the percolation threshold were reported in the literature for composites with the same matrix and filler [1–3], which makes it very hard to establish a filler concentration of general validity corresponding to the threshold.

1.2. Relevance of Filler Morphology, Synergetic Effects, and Laser Treatments on Electrical Conductivity

Huang et al. [6] modeled the effect of the geometric factor on the conductivity of composites containing CNTs and GNPs. In general, due to the higher aspect ratio, CNTs are believed more effective than GNPs for achieving the percolation threshold [7,8], but the method adopted for the fabrication of the composite in this case also entails great importance. In fact, the alignment of CNTs along a direction favors the formation of a conductive pathway for both electrons and phonons and, therefore, can be exploited for enhancing conductivity. Goh et al. [9] reviewed the methods that can be used for obtaining a preferential orientation of fillers (e.g., the application of electrical or magnetic fields and shear forces during the composite fabrication). Both the orientation and size of GNPs were also found to appreciably affect the electrical, thermal, and mechanical behavior of polypropylene–GNP composites [10]. The combination of fillers with different aspect ratios can be exploited to obtain a conductive network inside a polymeric matrix. A synergetic effect between CNTs and GNPs or CNTs and carbon black particles was observed [11,12] for epoxy and some thermoplastic matrices, such as styrene–butadiene, poly(ethersulfone), polyvinylidene fluoride, and poly(vinyl alcohol). However, it is doubtful that the synergetic effect can occur in every kind of composite, since Paszkiewicz et al. [13] did not find any evidence of it when investigating the electrical conductivity of polyethylene filled by CNTs or CNTs plus GNPs.

According to an alternative approach, surface laser treatments proved their suitability for locally improving the electrical conductivity of nanocarbon-filled polymers and, thus, for processing metal-free electrical circuits [14–16]. The laser beam causes the pyrolysis of the polymeric matrix on the composite surface with the formation of gaseous species that leave the material. In this manner, the filler/matrix ratio greatly increases, the percolation threshold is locally achieved, and conductive tracks form on the surface of an insulating composite with low mean filler content.

1.3. Polymer/Carbon Filler Piezoresistive Composites

The addition of carbon fillers shows potential not only for converting insulating polymers into conductive materials, but also for providing polymers of antistatic, electromagnetic absorption, and piezoresistive properties.

The piezoresistive behavior of polymers filled by CNTs, GNPs, thermally reduced graphene oxide, or carbon nano-blocks can be exploited for the fabrication of strain sensors [5,17–20]. The piezoresistive effect in these nanocomposites can be due to changes in the network formed by filler and variation of the piezoresistivity of fillers because of their own deformation. The stability and reproducibility of the electro-mechanical response of these nanocomposites is required for practical applications and should be further investigated. Systems with rather high filler content (from 5 to around 20 wt.% of filler) were studied [17,20], while efforts should be spent to keep the filler concentration and the material cost as low as possible.

1.4. Aim of the Work

In this paper, the laser processing of conductive tracks on several nanocomposites produced in a very similar manner was investigated, with the aim of demonstrating the viability of this approach for most composite systems, investigating the importance of laser writing parameters, and reducing the filler concentration needed for achieving locally electrical conductivity. The piezoresistive behavior of some nanocomposites and the effectiveness of laser treatment for improving the piezoresistive properties were also studied.

2. Materials and Methods

Several commercial thermoplastic polymers were used as composite matrices: high-density polyethylene (HDPE; Lupolen 4261 A IM, LyondellBasell, Houston, TX, USA), polypropylene–ethylene copolymer (PP; Hostacom CR 1171 G1, LyondellBasell), polycarbonate and acrylonitrile–butadiene–styrene blend (PC-ABS; Babyblend T65XF, Bayer Material Science-Covestro, Leverkusen, Germany), acrylonitrile–butadiene–styrene (ABS; Cycolac, Sabic, Riyadh, Saudi Arabia), and ethylene–propylene–diene monomer (EPDM; 719.A65 Forflex, SO.F.TER, Lebanon, TN, USA). These matrices were blended with carbonaceous micro and nanofillers: multiwall carbon nanotubes (MWCNTs; NC7000, Nanocyl, Sambreville, Belgium), graphene nanoplatelets (GNPs 25; AB304024 25 µm wide and 6–8 nm thick, ABCR Gute Chemie; GNPs grade 4, 1–2 µm wide and less than 4 nm thick, Cheaptube Inc., Cambridgeport, VT, USA), graphite flakes with a median size of 7 to 10 µm (Alfa Aesar, Ward Hill, MA, USA), and biochar OSR700 (UK Biochar Research Center, Edinburgh, Scotland). Also, masterbatches of composite materials were used to produce the samples under investigation. A masterbatch PP/15 wt.% CNTs (Nanocyl NC7000) was acquired from Hostacom, while other masterbatches (HDPE/6 wt.% CNTs, HDPE/12 wt.% GNPs 25, PC-ABS/2.75 wt.% CNTs, and HDPE/12 wt.% graphite) were produced in the laboratory by mixing the matrix and the filler. These masterbatches were processed by melt blending using a PlastiCorder Brabender W50E model. The material out of the mixer was then transferred to a pelletizing machine (Piovan RSP 15/15, Maria di Sala, Italy). Several composite samples with different compositions were finally prepared according to two possible paths: direct mixing of matrix and filler, or mixing of one or two masterbatches with the unfilled matrix. These composite samples were processed by mixing pellets of the starting materials using a twin-screw extruder (EuroLab 16mm XL 40:1 L/D Thermo Fisher Scientific, Waltham, MA, USA) with a final pelletizing unit (Thermo Haache Eurolab). The processing path for each material submitted to the investigation of electrical behavior is summarized in Table 1. Composite plates (67 mm × 12.8 mm × 2.9 mm) were fabricated by injection molding (Babyplast equipment of Cronoplast S.L.) of the composite pellets produced as described above. Larger plates (140 mm × 90 mm × 3 mm) were also produced (using a press Micro 65 series, Sandretto, Pont Canavese, Italy) with the aim of investigating the possible interaction between adjacent tracks.

Laser tracks were processed on the surface of these plates by using a CO_2 laser equipment (LASIT Towermark XL (Torre Annunziata, Italy), with a power of 100 W, a wavelength of 10.6 µm, and a spot size of 100 µm with 0 defocusing). Several parallel tracks were written at a distance of 1 cm on the plates in order to investigate their electrical resistance, as well as the inter-track resistance. A single track was produced in the middle of the plates in order to investigate the piezoelectric behavior. The laser treatment was performed under nitrogen atmosphere in order to avoid sample oxidation. The parameters adopted for the laser treatment were optimized for the different composites.

Table 1. Surface resistance per length unit measured for the laser tracks obtained under optimized conditions (P = power; S = scan rate; N = number of repetitions; F = frequency; D = defocus) on several polymer–carbon filler systems (* not good reproducibility).

Material	Filler	Production Process	Laser Parameters	Surface Resistance per Length Unit
HDPE/6 wt.% MWCNTs	MWCNTs Nanocyl NC7000	Masterbatch produced by melt compounding high-density polyethylene (HDPE) and multiwall carbon nanotubes (MWCNTs)	P = 10%, S = 100 mm/s, N = 25, F = 15 kHz, D = 50 mm	1.28 kΩ/cm
HDPE/4 wt.% MWCNTs	MWCNTs Nanocyl NC7000	Twin screw extrusion of masterbatch HDPE/MWCNTs and HDPE, pelletizing and injection molding	P = 10%, S = 100 mm/s, N = 25, F = 15 kHz, D = 50 mm	19.7 kΩ/cm
HDPE/4 wt.% MWCNTs/4 wt.%GNPs	MWCNTs Nanocyl NC7000; GNPs ABCR 25 μm 6–8 nm	Twin screw extrusion of masterbatch HDPE/MWCNTS and masterbatch HDPE/graphene-like nanoplatelets (GNPs), pelletizing and injection molding	P = 10%, S = 100 mm/s, N = 25, F = 15 kHz, D = 50 mm	46 kΩ/cm
HDPE/4 wt.% MWCNTs/4 wt.% graphite	MWCNTs Nanocyl NC7000; Graphite Alfa-Aesar 7–10 μm	Twin screw extrusion of masterbatch HDPE/MWCNTs and masterbatch HDPE/graphite, pelletizing and injection molding	P = 10%, S = 100 mm/s, N = 25, F = 15 kHz, D = 50 mm	7.01 kΩ/cm
PP/30 wt.% biochar	Biochar pellets OSR700 UK Biochar Research Center	Melt blending of PP and biochar, twin screw extrusion, pelletizing and injection molding	P = 15%, S = 50 mm/s, N = 7, F = 5 kHz, D = 30 mm	4 MΩ/cm (antistatic)
PP/2 wt.% CNTs	MWCNTs Nanocyl NC7000	Melt blending of masterbatch PP-MWCNTs and PP, pelletizing and injection molding	P = 20%, S = 50 mm/s, N = 25, F = 10 kHz, D = 200 mm	0.9 kΩ/cm
PP/1 wt.% CNTs	MWCNTs Nanocyl NC7000	Melt blending of masterbatch PP-MWCNTs and PP, pelletizing and injection molding	P = 20%, S = 200 mm/s, N = 25, F = 15 kHz, D = 100 mm	12.3 kΩ/cm
PC-ABS/1.0 wt.% CNTs	MWCNTs Nanocyl NC7000	Twin screw extrusion of masterbatch PC-ABS-MWCNTs and PC-ABS, pelletizing and injection molding	P = 5%, S = 300 mm/s, N = 30, F = 30 kHz, D = 0 mm	3.96 kΩ/cm
PC-ABS/0.75 wt.% CNTs	MWCNTs Nanocyl NC7000	Twin screw extrusion of masterbatch PC-ABS-MWCNTs and PC-ABS, pelletizing and injection molding	P = 5%, S = 100 mm/s, N = 20, F = 5 kHz, D = 0 mm	0.41 kΩ/cm
PC-ABS/0.5 wt.% CNTs	MWCNTs Nanocyl NC7000	Twin screw extrusion of masterbatch PC-ABS-MWCNTs and PC-ABS, pelletizing and injection molding	P = 10%, S = 100 mm/s, N = 20, F = 30 kHz, D = 0 mm	0.02 kΩ/cm
PP/5 wt.% GNPs	GNPs ABCR (1–2 μm)	Melt mixing, pelletizing and injection molding	P = 20%, S = 200 mm/s, N = 25, F = 15 kHz, D = 100 mm	≈5 * kΩ/cm
ABS/5 wt.% GNPs	GNPs ABCR (1–2 μm)	Melt mixing, pelletizing and injection molding	P = 20%, S = 200 mm/s, N = 25, F = 15 kHz, D = 100 mm	≈5 * kΩ/cm

The morphology of the tracks was investigated using a field-emission (FE)-SEM Zeiss MERLIN (Carl Zeiss AG, Oberkochen, Germany) and a Profilometer confocal microscope Leica DCM8 (Leica Microsystems Inc., Buffalo Grove, IL, USA).

Electrical resistance of the as-produced composites and tracks processed by laser writing was measured using a multimeter (Keitley 2700E, full scale value 120 MΩ, Keithley Instruments, Cleveland, OH, USA). Silver paint was deposited at the beginning and the end of the tracks with the aim of granting better contact with the steel probes of the multimeter.

For the investigation of the piezoelectric behavior, cyclic three-point bending tests were carried out using a dynamometer (Instron 5544, Norwood, MA, USA), contemporaneously measuring the displacement and the resistance variation by means of an extensometer and a Keithley 2700E multimeter. The software of the dynamometer (Bluehill3, Instron, Norwood, MA, USA) and that of the multimeter (Labview, version 2015, National Instruments, Austin, TX, USA) were interfaced with the data recording system.

3. Results

3.1. Laser Writing of Conductive Tracks on Carbon-Filled Polymers

Laser treatment of the surface of insulating composites can be exploited for locally changing their composition with the aim of increasing the carbon filler concentration until the percolation threshold for electrical conductivity is achieved. This approach offers the opportunity of creating conductive paths through the modification of the surface of composites that are not conductive and are not too expensive, because they contain rather low concentrations of expensive fillers, like CNTs and GNPs, or higher concentrations of cheap fillers.

In every case, the laser beam causes the pyrolysis of the polymeric matrix, which results in the formation of gaseous species that leave the material, thereby increasing the filler/matrix ratio. Laser irradiation was also exploited for improving the nanostructure of films made of CNTs deposited on different substrates [21–23]. This treatment resulted in the change of the film resistance owing to the decrease in impurity content in CNTs and defect healing/recrystallization. Therefore, in principle, laser irradiation could provide some conductivity improvement through filler modification. On the other hand, the treatment conditions we adopted for writing the conductive tracks seem very far from those suitable for modifying the structure of carbonaceous nanofillers (for instance, in this last case, the laser wavelength was one order of magnitude lower than that we used in the present investigation). For this reason, the change in filler concentration resulting from the polymeric matrix depletion can be considered as the main effect causing the enhancement of conductivity inside the tracks produced by laser writing.

The effect of the laser treatment is strictly related to the amount of energy given to the substrate, which in turn depends on the kind of laser and the processing parameters. Laser power and frequency, speed of movement of the laser beam on the surface, distance of the laser source from the treated surface (called defocusing hereafter), and numbers of laser runs on the same part of the surface (number of treatment repetitions) were the main parameters to be considered. The final electrical properties of the tracks that the laser beam writes on the surface also depend on the characteristics and the concentration of conductive filler, as well as on the tendency of the polymeric matrix to undergo thermal decomposition.

The concept of laser surface treatment and the morphology of the surface conductive track are shown in Figure 1. The laser writing on the composite surface locally causes a great enhancement of concentration of the carbonaceous filler or fillers, represented in Figure 1 as green rods and blue particles dispersed within the matrix. This effect is confined within the track and affects a surface layer of the material 1–2 mm wide and some hundreds of micrometers thick. The surface morphology after the laser action is also depicted at different magnifications in Figure 1, as shown in the middle of the track where the laser leaves a forest of nanotubes protruding from the polymeric substrate. For practical applications, the conductive tracks must be processed inside a non-conductive support in order to avoid short circuits.

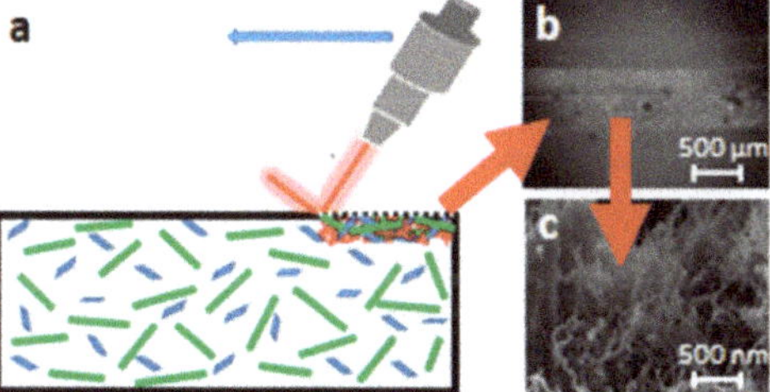

Figure 1. Effect of laser ablation on the surface (**a**); SEM micrographs at different magnifications of resulting tracks showing increased content of conductive filler (**b,c**).

In order to better understand the importance of the different processing parameters and their effect on the track conductivity, it is necessary to perform several tests, preferably using the design of experiment (DOE) approach, since it allows limiting the number of experiments [24]. As an example, the outcomes of the DOE approach in the case of laser treatment of polycarbonate/acrylonitrile butadiene styrene (PC/ABS) composite with 0.75 wt.% of CNTs are reported here. As the number of factors affecting the track conductivity was above four, a two-level fractional factorial design approach was adopted. Each laser trial differed from the others because two parameters out of five were changed. The lower and upper limit for the processing parameters were as follows: 5–30% of the maximum laser power, 100–600 mm/s for writing speed, 5–30 Hz for frequency, 0–50 for the number of repetition, and 0–50 mm for defocusing. The importance for conductivity enhancement of the different parameters and of their combinations is summarized by the Pareto plot in Figure 2. On the y-axis, each laser parameter or combinations of parameters possibly affecting the final conductivity of the tracks are reported. For each of these parameters or couple of parameters, the higher or lower importance of the effect on the final conductivity is represented by an index reported on the x-axis.

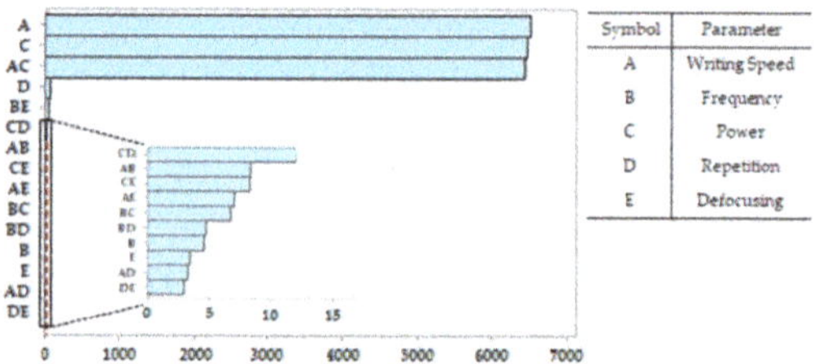

Figure 2. Pareto plot showing the relevance of different parameters and their combinations on the conductivity of tracks.

From this plot, it is clear that the resistance chiefly depends on power and writing speed, as well as from the combination of these to parameters, because they determine the amount of energy which is given to the substrate in the unit of time. The number of repetitions affects the final resistance less, while the other parameters exert only secondary effects. Nonetheless, many repetitions of the treatment along the same track result in the progressive depletion of the matrix and, as a consequence, in the increase of filler concentration and conductivity. The effect of the number of repetitions on the resistance of tracks processed on the surface of PP–matrix composites with different loads of CNTs (from 1 wt.% to 4 wt.%) is depicted in Figure 3. For each curve reported in this figure, the measure of resistance was repeated on the same track after progressively increasing the number of laser runs. The resistance decreased more or less quickly with the number of repetitions depending on the starting conductivity of the composite, which is initially controlled by the filler concentration.

The main importance of these three parameters (power, writing speed, and number of repetitions) was observed for many kinds of polymer–carbon composites submitted to laser functionalization, but there are also limits in the selection of both material and processing conditions that should be considered for practical applications. As a matter of fact, the conductive tracks can be obtained by laser-treating composites with filler concentrations below or above the percolation threshold. When the filler concentration is over the percolation threshold, the conductivity inside the tracks is very good, but to create conductive paths inside a conductive material has no practical relevance because, of course, short circuits form between the tracks. The amount of energy delivered by laser treatment by unit of surface and unit of time, which increases when power increases and writing speed decreases, greatly affects the conductivity and the morphology of the tracks. However, only a fraction of the maximum power of the laser can be used, because too much energy results in deep tracks, high temperature gradients, and high thermal stresses, which can even cause deformation of the sample. Similar effects can be observed when the writing speed is excessively low. Then, these two parameters can be changed only within limited ranges and not independently. Several repetitions of the laser passage can be adopted to deliver the energy over a longer period, thus reducing the risk of sample distortion. However, the repetition of the treatment has a noticeable effect on the morphology of the track, since it causes an increase in depth and width of the conductive path. The variation of track morphology with the number of repetitions can be assessed using profilometry, as shown in Figure 4a,b.

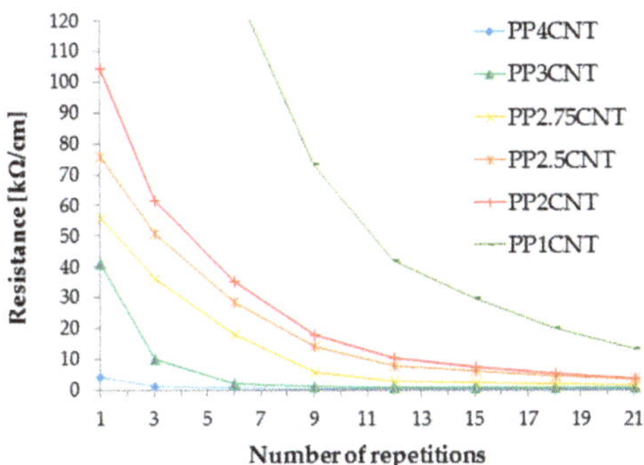

Figure 3. Decrease in track resistance with the number of laser treatments performed on polypropylene–ethylene copolymer (PP)/carbon nanotube (CNT) composites (P = 35%, scan rate = 200 mm/s, F = 15 kHz, defocus = 0 mm).

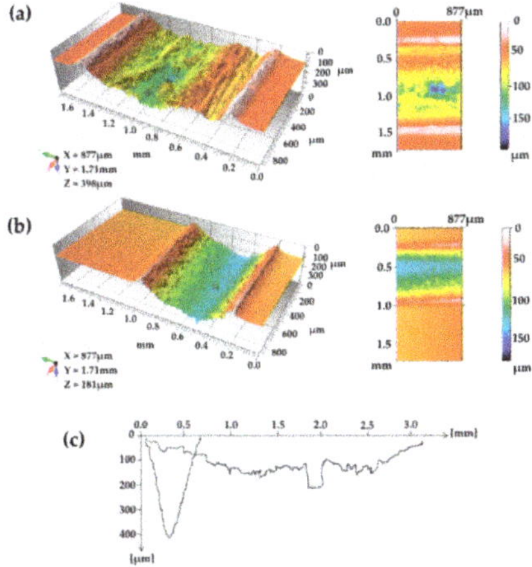

Figure 4. Profilometric measurements of tracks processed on polycarbonate and acrylonitrile–butadiene–styrene blend (PC-ABS)/0.5 wt.% CNT composite under different conditions: (**a**) wide and deep track resulting from repetitions (D = 0, P = 10, F = 5, S = 100, N = 20), (**b**) less severe treatment (D = 50, P = 50, F = 5, S = 600, N = 1), (**c**) comparison between narrow and deep track profile, obtained on PP/2 wt.% CNTs, resulting from low defocusing (D = 50, P = 15, F = 15, S = 200, N = 25) and wide track profile (D = 150, P = 40, F = 15, S = 200, N = 25).

Furthermore, defocusing is responsible for the track morphology. In Figure 4c, the profile of two tracks obtained on PP/2 wt.% CNT composites, with defocusing of 50 mm and 150 mm, are compared. The increase in defocusing (which must be coupled with the power increase, since the energy is spread on a larger surface) results in the track widening.

The widening of the tracks obtained on insulating composites can cause short circuits between adjacent tracks. This effect can be attributed to the presence of zones with increased filler concentration placed in between two tracks, resulting from an incomplete homogeneous dispersion of filler in the matrix.

Conclusively, the best processing parameters should be selected not only with the aim of improving the conductivity as much as possible and obtaining reproducible resistance values, but also taking into account that inter-track conductivity and material deformation must be avoided. Nevertheless, the choice of the most suitable processing parameters mainly depends on the characteristics of the filler (such as intrinsic electrical conductivity, size, and aspect ratio), the filler load, the homogeneity of filler distribution inside the matrix, and the response of the matrix to the laser action. For these reasons, the laser process should be tailored for each kind of composite. Processing parameters selected for maximizing conductivity are shown for several different matrix/filler systems in Table 1. Concentrations of fillers below the percolation threshold were generally adopted for the composite production, and the laser writing process was tailored to increase the track conductivity as much as possible without causing short circuits between the tracks. On the other hand, the percolation thresholds depend on the kind filler and matrix, and the effectiveness of the production process of the composite; therefore, very different CNT loads (from 0.5 wt.% to 4 wt.%) were used for different matrices. In addition, for composites with a filler load around (PP/2 wt.% CNTs, HDPE/4 wt.% CNTs) or even over (HDPE/6 wt.% CNTs), the percolation threshold [1] was treated in order to better understand the impact of the filler concentration on the conductivity of the paths produced by the laser action.

Laser power between 5% and 20% of the maximum power available and writing speed between 50 mm/s and 300 mm/s were adopted. Different combinations of these two parameters were used for different kinds of composites, which means that different amount of energy were required for the pyrolysis of different matrices, and that the concentration and kind of fillers can affect the energy adsorbed by the composite. The number of repetitions and the defocusing were tuned in order to obtain tracks of similar morphology on composites with different composition. Table 1 also shows that, in spite of the parameter optimization, the laser-writing process gives rise to conductive tracks showing very different resistance when different composite systems are treated. It is very hard to measure exactly the cross-section of each track owing to its irregular shape at a microscopic level. In addition, some conductive behavior of the material close to the track cannot be excluded. In fact, inside the thermally affected areas, some microstructure modification should also occur. For this reason, resistivity value was not calculated, but the measured resistance (which is not constant in every part of the track profile) was normalized with reference to the length unit of the track. Nonetheless, Table 1 shows that several polymeric matrices filled with CNTs or GNPs can be successfully submitted to laser ablation for the production of conductive tracks, and suggests some conclusions. When the CNT content inside each kind of matrix increased, the conductivity also generally increased. However, the opposite trend could be observed in some cases (see PC-ABS composites), very likely because CNTs can be more hardly dispersed in some matrices. When CNTs agglomerate and form bundles inside the matrix, the laser parameters must be changed in order to avoid inter-track conduction, and this can result in a worsening of the track conductivity. In addition, the variation of the filler load can require an adjustment of the processing parameters because the material response to the laser action depends on the filler/matrix ratio. The adoption of the same content of CNTs can result in different resistance of the tracks processed in the best manner on composites with different matrices. For instance, the track resistance was 12.3 kΩ/cm and 3.96 kΩ/cm for PP/1 wt.% CNTs and PC-ABS/1 wt.% CNTs, respectively. GNPs seemed less effective than CNTs when used alone or when they replaced part of the CNTs in the composite. The combination of a second filler with CNTs seemed more convenient when using graphite instead (see HDPE/4 wt.% MWCNTs/4 wt.% graphite system in comparison with HDPE/4 wt.% MWCNTs/4 wt.% GNP composite). A very high concentration of low-cost carbon particles was necessary to observe some conductivity in the tracks. This was the case of biochar, whose composite with 30 wt.% in a PP matrix showed laser tracks with poor conductivity. On the other hand, the resistance of these tracks was about six orders of magnitude lower than that of the unfilled matrix, which suggests that the laser treatment can also be exploited in this case to obtain antistatic properties.

3.2. Piezoresistive Behavior

The piezoresistive behavior of composites with a EPDM/PP matrix (60/40 weight ratio) and CNTs was investigated. Concentrations of CNTs from 1 wt.% to 5 wt.% were used. The percolation threshold was found around 3 wt.% CNTs (Figure 5). This percolation threshold found for the blend between the ethylene–propylene–diene monomer and polypropylene seems consistent with the literature [1,12,25], which reports that concentrations of CNTs of 2 wt.%, 7.5 wt.%, or 7 vol.% are necessary for achieving percolation when PP, PE, or commercial thermoplastic elastomer based on EPDM/PP, respectively, are used as composite matrices.

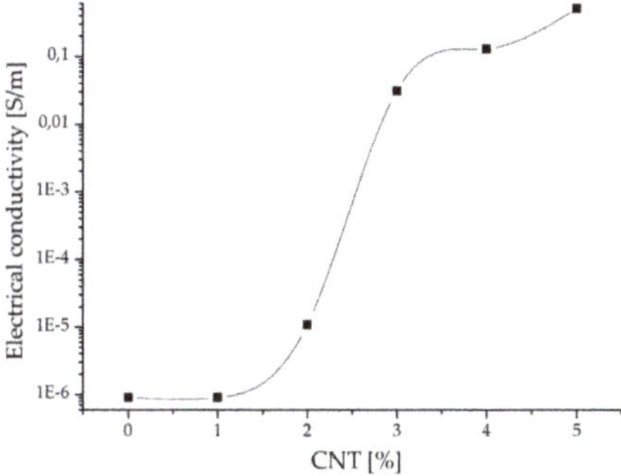

Figure 5. Percolation curve for ethylene–propylene–diene monomer (EPDM)/PP/CNT composites.

The piezoelectric behavior of such composites could be exploited for processing pressure sensors or switches based on the change in electrical resistivity occurring after deformation. The resistance of all composites with different filler loads progressively increased with the displacement occurring in the elastic and plastic fields when a flexural force was applied. However, only reversible deformations can be considered for practical applications, and then the displacement should be limited to the elastic field. The maximum displacement occurring in the elastic field was measured by a bending test. The stress/displacement curves obtained from three-point bending tests showed that the maximum elastic displacement of these composites increased with the increase in CNT concentration. On the other hand, only a little variation in resistance resulted from bending the material up to the elastic limit for samples containing 1 wt.% or 2 wt.% CNTs.

On the contrary, samples containing from 3 wt.% to 5 wt.% CNTs showed appreciable and almost linear resistance variation with displacement, but a good reproducibility of the piezoelectric effect was observed only for displacement over 1 mm. Figure 6 shows, for the composite with 4 wt.% CNTs, the typical variation of resistance with the cyclic change of displacement from 0 to 1.5 mm with a speed of 50 mm/min. It is possible to observe noise in the resistance signal when the deformation is recovered during each cycle, probably due to a rearrangement of CNTs inside the matrix occurring at the microscopic level.

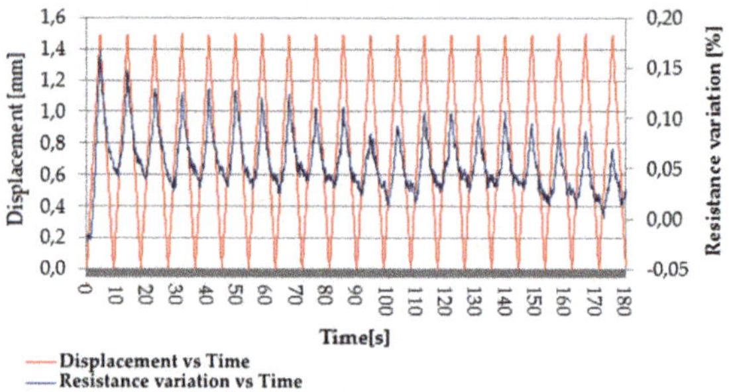

Figure 6. Piezoresistive behavior of EPDM/PP/4 wt.% CNT composite: cyclic resistance variation due to cyclic deformation from 0 to 1.5 mm.

However, the noise was greatly reduced when the maximum displacement increased to 2.5 mm (Figure 7).

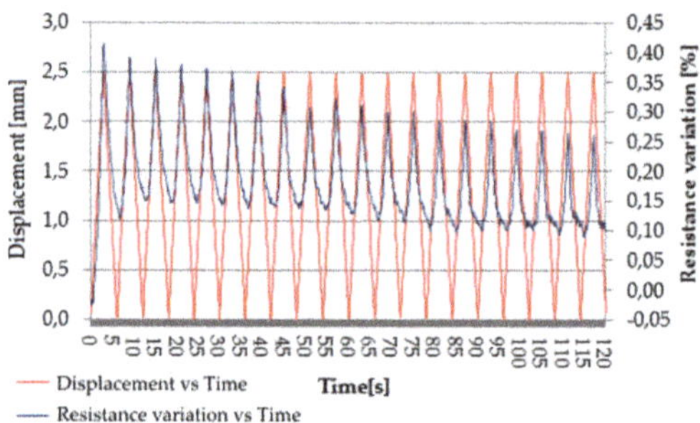

Figure 7. Piezoresistive behavior of EPDM/PP/4 wt.% CNT composite: cyclic resistance variation due to cyclic deformation from 0 to 2.5 mm (displacement speed: 50 mm/min).

The reproducibility of piezoresistive response to stress during long periods of cycling was also investigated; the resistance of unloaded material, as well as the maximum resistance change resulting from load application and consequent displacement, changed after a few hundreds of cycles; however, afterward, they remained constant with an increase in the number of cycles (Figure 8, Table 2). These outcomes prove that the composites under investigation can be exploited as pressure sensors, granting response to mechanical load for long periods.

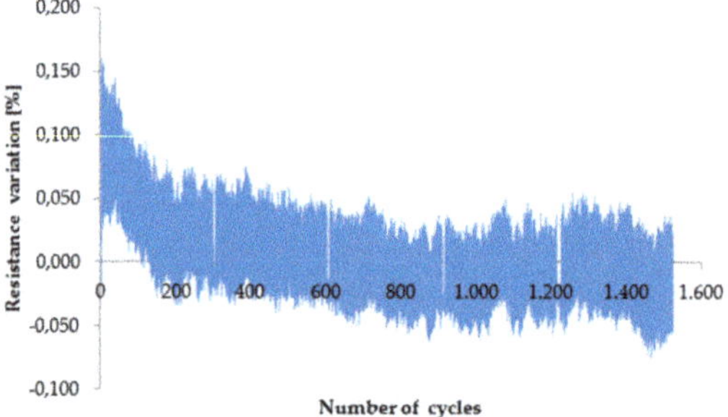

Figure 8. Testing of EPDM/PP/4 wt.% CNT composite up to 1500 cycles (displacement: 0–1.5 mm, speed: 50 mm/min).

Table 2. Resistance variation occurring during a single cycle of deformation and the average resistance value of a portion of the bar 40 mm long (displacement of up to 1.5 mm, displacement speed: 50 mm/min) after increasing the number of cycles.

Material (CNTs wt.%)	Resistance Variation (%)			Average Resistance (kΩ)		
	Cycle 1	After 300 Cycles	After 1000 Cycles	Cycle 1	After 300 Cycles	After 1000 Cycles
3	0.22	0.50	0.50	31.652	31.690	31.690
4	0.15	0.80	0.80	0.732	0.733	0.733
5	0.25	0.90	0.90	0.188	0.189	0.190

As laser surface treatment is able to greatly improve the conductibility of carbon-based composites, the piezoresistive behavior was also tested on specimens with conductive tracks processed on the surface. Samples with a very low concentration of CNTs in a PC-ABS matrix were used for this investigation. Conductive tracks were processed in the middle of composite bars which were then submitted to cyclic flexural deformation, while their deformation and the resistance of the laser track were measured. Samples with CNT loads of 0.5 wt.%, 0.75 wt.%, and 1.0 wt.% were tested. The parameters for laser treatment were selected with the purpose of not decreasing too much the electrical resistance. For instance, the following parameters were adopted for the functionalization of the PC-ABS/0.5 wt.% CNT sample: P = 5%, scan rate = 600 mm/s, F = 5 kHz, and defocus = 50 mm. The electrical resistance of the track was 464 kΩ/cm, 172 kΩ/cm, and 76 kΩ/cm for composites with 0.5 wt.%, 0.75 wt.%, and 1.0 wt.% CNTs, respectively. The specimen resistance variation was measured during 2000 cycles of deformation up to a maximum displacement of 0.5 mm. Typical change of resistance during cycling is depicted in Figures 9 and 10.

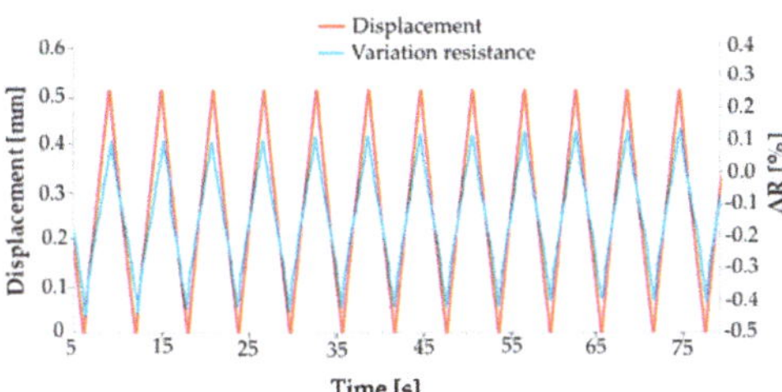

Figure 9. Displacement and resistance variation recorded during cycling of PC-ABS composite with 0.5 wt.% CNTs.

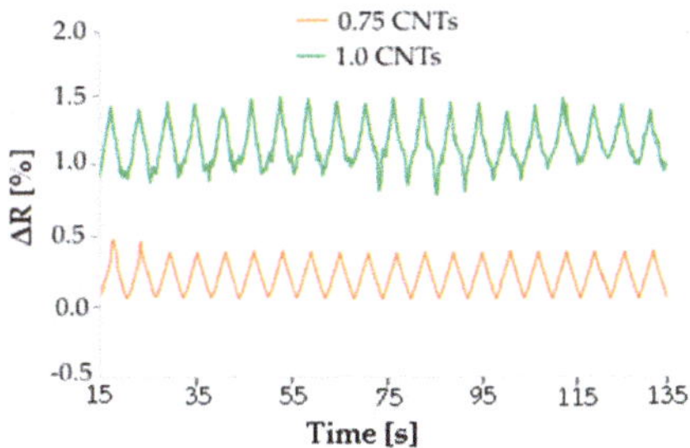

Figure 10. Resistance change during flexural fatigue cycles with maximum displacement of 0.5 mm for PC-ABS composites with 0.75 and 1.0 wt.% CNTs.

A resistance variation of 0.5% occurred during each cycle for the three kinds of composites, irrespective of the nominal concentration of filler. A precise correlation between displacement and resistance was always observed. Furthermore, the electrical signal was more sharp and regular when the filler concentration was 0.5 or 0.75 wt.%, while the composite with 1 wt.% CNTs gave more irregular resistance curves. Therefore, in the case of laser-functionalized piezoresistive composites, there is no reason to use samples showing enhanced filler content and conductivity, contrary to that which happened for non-laser-treated composites. In fact, piezoresistive behavior can also be achieved in the case of composites with low filler content by exploiting laser functionalization.

4. Conclusions

Laser treatment was successfully used to produce conductive tracks on the surface of several kinds of polymer–nanofiller composites with a filler content lower than the percolation threshold for electrical conduction. The effectiveness of this functionalization process for writing conductive tracks inside composites with low loads of conductive filler was proven. In general, the final resistance of these tracks depends on the kind of filler and matrix, and the laser processing parameters. Conductive tracks can be more easily obtained when using CNTs, while, to obtain the same result, a higher concentration of GNPs is required. Carbonaceous micro-fillers like graphite or biochar can also be used as fillers. The addition of a small quantity of graphite to the composite allows reducing the content of the CNTs because of a synergetic effect between the two fillers, thus making the material cheaper. When using biochar, even in rather high concentrations, only antistatic properties can be achieved with the laser treatment. Very different resistance values were observed when laser-treating the surface of polymer composites showing different matrices filled with the same or similar concentrations of CNTs.

Generally, the conductivity of laser tracks increased with the load of conductive filler, but there is not always a reason to increase the filler content, since its increase can cause agglomeration of filler particles, which results in a non-homogeneous filler distribution and, thus, a worse result of the laser treatment.

The processing parameters of laser treatment should be optimized for each kind of composite, depending on the composition, and the main parameters to be optimized are power, writing speed, and number of repetitions. The parameter optimization is limited by side effects such as distortion of the sample and excessive enlargement of the tracks, which causes short circuits. Moreover, other parameters that have little influence on the conductivity show a non-negligible effect on these side effects.

The addition of CNTs to a thermoplastic polymer also gives rise to piezoresistive behavior, which could be exploited for the fabrication of pressure sensors. The resistance variation with mechanical load and displacement depends on the CNT concentration. A good response to load was observed for EPDM/PP/CNT composites, but only for filler concentration exceeding 2 wt.%. Noise of the electrical signal was observed when the displacement was too low, but displacements over 1.5 mm were sufficient to overcome this drawback. The composite material with piezoresistive behavior can show some initial instability in terms of the range of resistance variation and average resistance value, probably due to a rearrangement of microstructure during the initial cycles of loading. However, the response of the materials soon stabilizes.

A surface laser treatment improved the piezoresistive behavior of PC-ABS/CNT composites. These composites with very small CNT concentrations (below the percolation threshold) showed good and stable piezoresistive behavior after the laser treatment. This characteristic offers potential for processing in situ sensors and switches using laser treatment of non-conductive composites with very low filler load.

Author Contributions: Conceptualization, C.B. and A.V.; methodology, C.B., A.C., A.V., and E.D.M.; validation, C.B. and A.V.; investigation, A.C., E.D.M., and E.P.; data curation, E.P. and M.P.; writing—original draft preparation, C.B.; writing—review and editing, C.B., A.V., and E.P.; supervision, C.B. and A.V.

Funding: This research received no external funding.

Acknowledgments: The authors wish to thank A. Tagliaferro for providing the biochar.

Conflicts of Interest: The authors declare no conflicts of interest.

References

1. Du, J.H.; Bai, J.; Cheng, H.M. The present status and key problems of carbon nanotube based polymer composites. *Express Polym. Lett.* **2007**, *5*, 253–273. [CrossRef]
2. Imtiaz, S.; Siddiq, M.; Kausar, A.; Muntha, S.T.; Ambreen, J.; Bibi, I. A Review Featuring Fabrication, Properties and Applications of Carbon Nanotubes (CNTs) Reinforced Polymer and Epoxy Nanocomposites. *Chin. J. Polym. Sci.* **2018**, *36*, 445–461. [CrossRef]
3. Mohan, V.B.; tak Lau, K.; Hui, D.; Bhattacharyya, D. Graphene-based materials and their composites: A review on production, applications and product limitations. *Compos. Part B Eng.* **2018**, *142*, 200–220. [CrossRef]
4. Kim, T.; Park, J.; Sohn, J.; Cho, D.; Jeon, S. Bioinspired, Highly Stretchable, and Conductive Dry Adhesives Based on 1D-2D Hybrid Carbon Nanocomposites for All-in-One ECG Electrodes. *ACS Nano* **2016**, *10*, 4770–4778. [CrossRef] [PubMed]
5. Alamusi; Hu, N.; Fukunaga, H.; Atobe, S.; Liu, Y.; Li, J. Piezoresistive strain sensors made from carbon nanotubes based polymer nanocomposites. *Sensors* **2011**, *11*, 10691–10723. [CrossRef] [PubMed]
6. Huang, C.L.; Wang, Y.J.; Fan, Y.C.; Hung, C.L.; Liu, Y.C. The effect of geometric factor of carbon nanofillers on the electrical conductivity and electromagnetic interference shielding properties of poly(trimethylene terephthalate) composites: A comparative study. *J. Mater. Sci.* **2017**, *52*, 2560–2580. [CrossRef]
7. Chiu, Y.C.; Huang, C.L.; Wang, C. Rheological and conductivity percolations of syndiotactic polystyrene composites filled with graphene nanosheets and carbon nanotubes: A comparative study. *Compos. Sci. Technol.* **2016**, *134*, 153–160. [CrossRef]
8. Dul, S.; Pegoretti, A.; Fambri, L. Effects of the Nanofillers on Physical Properties of Acrylonitrile-Butadiene-Styrene Nanocomposites: Comparison of Graphene Nanoplatelets and Multiwall Carbon Nanotubes. *Nanomaterials* **2018**, *8*, 674. [CrossRef]
9. Goh, P.S.; Ismail, A.F.; Ng, B.C. Directional alignment of carbon nanotubes in polymer matrices: Contemporary approaches and future advances. *Compos. Part A Appl. Sci. Manuf.* **2014**, *56*, 103–126. [CrossRef]
10. Jun, Y.S.; Um, J.G.; Jiang, G.; Yu, A. A study on the effects of graphene nano-platelets (GnPs) sheet sizes from a few to hundred microns on the thermal, mechanical, and electrical properties of polypropylene (PP)/GnPs composites. *Express Polym. Lett.* **2018**, *12*, 885–897. [CrossRef]
11. Chatterjee, S.; Nafezarefi, F.; Tai, N.H.; Schlagenhauf, L.; Nüesch, F.A.; Chu, B.T.T. Size and synergy effects of nanofiller hybrids including graphene nanoplatelets and carbon nanotubes in mechanical properties of epoxy composites. *Carbon N. Y.* **2012**, *50*, 5380–5386. [CrossRef]
12. Szeluga, U.; Kumanek, B.; Trzebicka, B. Synergy in hybrid polymer/nanocarbon composites. A review. *Compos. Part A Appl. Sci. Manuf.* **2015**, *73*, 204–231. [CrossRef]
13. Paszkiewicz, S.; Szymczyk, A.; Pawlikowska, D.; Subocz, J.; Zenker, M.; Masztak, R. Electrically and Thermally Conductive Low Density Polyethylene-Based Nanocomposites Reinforced by MWCNT or Hybrid MWCNT/Graphene Nanoplatelets with Improved Thermo-Oxidative Stability. *Nanomaterials* **2018**, *8*, 264. [CrossRef] [PubMed]
14. Zecchina, A.; Bardelli, F.; Bertarione, S.; Caputo, G.; Castelli, P.; Cesano, F.; Civera, P.; Demarchi, D.; Galli, R.; Innocenti, C.; et al. Process for Producing Conductive and/or Piezoresistive Traces on a Polymeric Substrate. Patent No. WO/2012/055934, 2 May 2012.
15. Liebscher, M.; Krause, B.; Pötschke, P.; Barz, A.; Bliedtner, J.; Möhwald, M.; Letzsch, A. Achieving electrical conductive tracks by laser treatment of non-conductive polypropylene/polycarbonate blends filled with MWCNTs. *Macromol. Mater. Eng.* **2014**, *299*, 869–877. [CrossRef]
16. Colucci, G.; Beltrame, C.; Giorcelli, M.; Veca, A.; Badini, C. A novel approach to obtain conductive tracks on PP/MWCNT nanocomposites by laser printing. *RSC Adv.* **2016**, *6*, 28522–28531. [CrossRef]
17. Cai, W.; Huang, Y.; Wang, D.; Liu, C.; Zhang, Y. Piezoresistive behavior of graphene nanoplatelets/carbon black/silicone rubber nanocomposite. *J. Appl. Polym. Sci.* **2014**, *131*, 1–6. [CrossRef]

18. Georgousis, G.; Pandis, C.; Kalamiotis, A.; Georgiopoulos, P.; Kyritsis, A.; Kontou, E.; Pissis, P.; Micusik, M.; Omastova, M. Strain sensing in polymer/carbon nanotube composites by electrical resistance measurement. In *Procedia Engineering*; Elsevier B.V.: Amsterdam, the Netherlands, 2012; Volume 47, pp. 774–777.

19. Palza, H.; Garzon, C.; Rojas, M. Elastomeric ethylene copolymers with carbon nanostructures having tailored strain sensor behavior and their interpretation based on the excluded volume theory. *Polym. Int.* **2016**, *65*, 1441–1448. [CrossRef]

20. Costa, P.; Nunes-Pereira, J.; Oliveira, J.; Silva, J.; Moreira, J.A.; Carabineiro, S.A.C.; Buijnsters, J.G.; Lanceros-Mendez, S. High-performance graphene-based carbon nanofiller/polymer composites for piezoresistive sensor applications. *Compos. Sci. Technol.* **2017**, *153*, 241–252. [CrossRef]

21. Bernal-Martínez, J.; Godínez-Fernández, R.; Roman-Aguirre, M.; Aguilar-Elguezabal, A. The electrical resistance of electrodes made of multi walled carbon nanotubes is modulated by nIR-laser. *Microelectron. Eng.* **2016**, *166*, 45–49. [CrossRef]

22. Souza, N.; Roble, M.; Diaz-Droguett, D.E.; Mücklich, F. Scaling up single-wall carbon nanotube laser annealing: Effect on electrical resistance and hydrogen adsorption. *RSC Adv.* **2017**, *7*, 5084–5092. [CrossRef]

23. Iakovlev, V.Y.; Sklyueva, Y.A.; Fedorov, F.S.; Rupasov, D.P.; Kondrashov, V.A.; Grebenko, A.K.; Mikheev, K.G.; Gilmutdinov, F.Z.; Anisimov, A.S.; Mikheev, G.M.; et al. Improvement of optoelectronic properties of single-walled carbon nanotube films by laser treatment. *Diam. Relat. Mater.* **2018**, *88*, 144–150. [CrossRef]

24. Ballantine, L.B. *An Introduction to Design of Experiments: A Simplified Approach*; ASQ Quality Press: Milwaukee, WI, USA, 1999; ISBN 0-87389-444-8.

25. Dang, Z.M.; Shehzad, K.; Zha, J.W.; Mujahid, A.; Hussain, T.; Nie, J.; Shi, C.Y. Complementary percolation characteristics of carbon fillers based electrically percolative thermoplastic elastomer composites. *Compos. Sci. Technol.* **2011**, *72*, 28–35. [CrossRef]

 *micromachines*

Article

Pressure Sensitivity Enhancement of Porous Carbon Electrode and Its Application in Self-Powered Mechanical Sensors

Keren Dai [1], Xiaofeng Wang [2,*], Zheng You [2] and He Zhang [1,*]

[1] ZNDY of Ministerial Key Laboratory, School of Mechanical Engineering, Nanjing University of Science and Technology, Nanjing 210094, China; dkr@njust.edu.cn
[2] Beijing Innovation Center for Future Chips, Department of Precision Instrument, Tsinghua University, Beijing 100084, China; yz-dpi@tsinghua.edu.cn
* Correspondence: xfw@tsinghua.edu.cn (X.W.); hezhangz@njust.edu.cn (H.Z.); Tel.: +86-010-6277-6000 (X.W.)

Received: 26 December 2018; Accepted: 10 January 2019; Published: 16 January 2019

Abstract: Microsystems with limited power supplies, such as electronic skin and smart fuzes, have a strong demand for self-powered pressure and impact sensors. In recent years, new self-powered mechanical sensors based on the piezoresistive characteristics of porous electrodes have been rapidly developed, and have unique advantages compared to conventional piezoelectric sensors. In this paper, in order to optimize the mechanical sensitivity of porous electrodes, a material preparation process that can enhance the piezoresistive characteristics is proposed. A flexible porous electrode with superior piezoresistive characteristics and elasticity was prepared by modifying the microstructure of the porous electrode material and adding an elastic rubber component. Furthermore, based on the porous electrode, a self-powered pressure sensor and an impact sensor were fabricated. Through experimental results, the response signals of the sensors present a voltage peak under such mechanical effects and the sensitive signal has less clutter, making it easy to identify the features of the mechanical effects.

Keywords: porous electrode; pressure sensitivity; self-powered sensors; mechanical impact

---

1. Introduction

In recent years, self-powered mechanical sensors have attracted wide attention for their applications in wearable smart devices and intelligent microsystems with limited energy supply [1–5]. For example, a self-powered pressure sensor can achieve skin-like tactile sensing without a power source, which benefits energy saving and promotes practical applications such as electronic skin [6–8]. A self-powered impact sensor can detect the characteristics of a target hit by an artillery shell, thus promoting the intelligence of weapon microsystems (such as a smart fuze) with limited volume and power supply [9].

Traditional self-powered mechanical sensors applied in measurements of static pressure and dynamic impact are mainly piezoelectric sensors [10–12], some of which benefit from porous materials [13,14]. However, due to the inherent physical properties of piezoelectric crystals, high frequency oscillations in the sensitive signals under mechanical effects are unavoidable, and complex signal processing must be applied to obtain the features of the target (e.g., the number of impacts during a continuous impact process) [15,16]. These insurmountable shortcomings make it urgent to develop mechanical sensors based on new principles.

In recent years, the dual functions of energy storage and piezoresistive sensitivity in porous electrodes, such as activated carbon and carbon nanotubes, have been investigated [17], and have become a new hotspot in research of self-powered mechanical sensors. For static pressure sensing,

Zhang et al. realized a fusion of supercapacitors and self-powered pressure sensors using a carbon nanotube–polydimethylsiloxane film [18]. For dynamic impact sensing, the previous research of our team realized a novel device combing a supercapacitor and impact sensor based on an activated carbon film [19,20]. The sensitive signal has fewer high frequency oscillations, making it easier to identify target features. This unique characteristic is referred to as "self-filtering", and a theoretical model was proposed to explain it in our former research [21].

For these self-powered mechanical sensors, the material properties of porous electrodes have a decisive influence on their energy storage density and mechanical sensitivity [18,20,21]. However, the methods for enhancing the mechanical sensitivity of porous electrodes have not been systematically studied, so that the optimal design and practical application of these self-powered mechanical sensors are limited for better performance.

In this paper, an enhancement method for pressure sensitivity of porous electrodes is proposed. An elastic polymer is added to the porous energy storage material, which forms a spatial network skeleton structure at the microscopic level by a shearing process. On the one hand, the micro-elastic skeleton structure can improve the elasticity of the electrode, so that the electrode can withstand greater stress; on the other hand, the deformation effect of the elastic polymer skeleton under mechanical effects results in a more significant change in the contact state between the conductive particles, and thus enhances the mechanical sensitivity of the electrode. Finally, based on the proposed high elastic piezoresistive film, a self-powered mechanical sensor was fabricated, and its superior sensing performance was experimentally verified. In general, the preparation process of the porous electrode proposed in this paper achieves effective enhancement of pressure sensitivity, and promotes practical applications of the self-powered impact sensor, based on the piezoresistive effect of the porous electrode.

2. Materials and Methods

As shown in Figure 1a, traditional mechanical sensors need an external power source, which can be much larger than the sensor itself. In this paper, a self-powered mechanical sensor that is composed of two porous electrodes separated by a membrane is proposed, and it simultaneously achieves superior performance for both energy storage and mechanical sensitivity. This novel self-powered mechanical sensor has broad application prospects in the fields of electronic skin, intelligent fuzes, and so on. With the material modification in this paper, the flexibility of the sensor can be significantly enhanced. So, it can be used as a pressure sensor for electronic skin, realizing a tactile sensing for a robot. Furthermore, with a compact package, it can also serve as a high acceleration impact sensor to detect the structural characteristics of an attack target for an intelligent fuze, realizing intelligent ignition control of a projectile [22].

On the one hand, the porous electrode enables energy storage and provides energy supply during impact, as shown in Figure 1b. The porous electrode is composed of activated carbon and carbon nanotubes (or other porous materials); there are a lot of microscopic pores inside, which are filled with electrolyte (e.g., sulfuric acid). When the device is charged, an electrochemical double-layer structure forms at the microscopic solid–liquid interface due to the adsorption force between positive and negative ions, which realizes energy storage [23]. After being fully charged, the device can be discharged with certain current similarly to a power source.

On the other hand, the mechanical sensitivity of the device is achieved due to the piezoresistive effect of the porous electrode. As shown in Figure 1c, the conductivity of the porous electrode is determined by the number of micro-conductive chains. Under mechanical effects, the electrode will be deformed due to its loose microstructure [20,21], and the conductive particles are more likely to be closer to each other, increasing the number of conductive chains and greatly enhancing the conductivity of the electrode. The change in conductivity is reflected in the change of output voltage, thus realizing the perception of the mechanical effect.

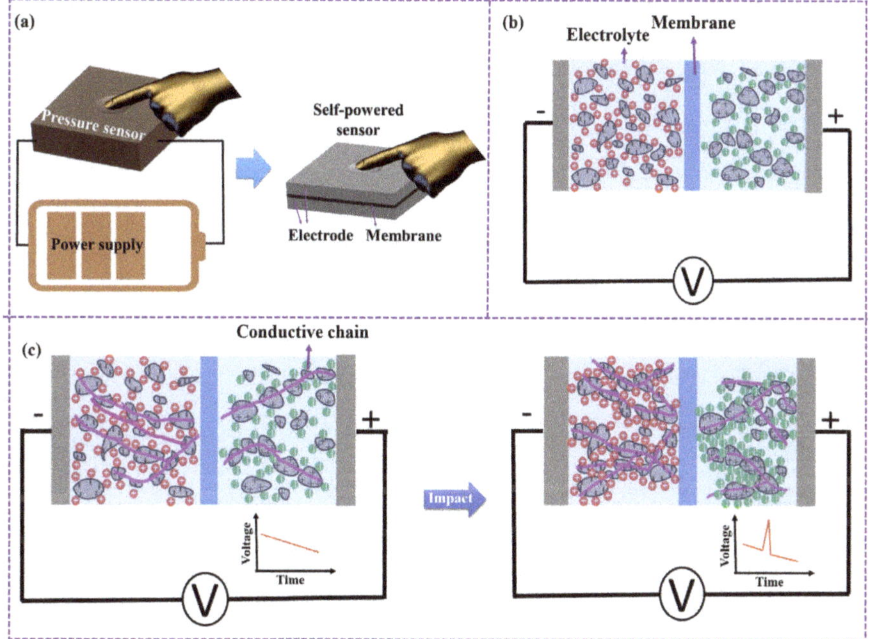

Figure 1. Diagram and mechanism of self-powered mechanical sensors. (**a**) A self-powered mechanical sensor works without a power supply. (**b**) Based on a porous electrode, energy storage is realized by the electric double-layer effect. (**c**) Based on a porous electrode, mechanical sensitivity is realized by the piezoresistive effect.

Therefore, enhancing the piezoresistive performance of porous electrodes is the key to realizing self-powered mechanical sensors, and the enhancement can be achieved through the optimization of the micro-morphology of the porous electrode. Previous literature [20] has studied the relationship between electrode conductivity and porosity according to the general equivalent medium equation [24]:

$$\sigma_m = \begin{cases} \left(\dfrac{\Phi - \Phi_c}{1 - \Phi_c}\right)^{t_m} \sigma_h, & \Phi_c \leq \Phi \leq 1 \\ 0, & 0 < \Phi < \Phi_c \end{cases}, \tag{1}$$

where σ_m represents the conductivity of the porous electrode, σ_h represents the conductivity of the carbon particles, Φ represents the volume ratio of the carbon particles, and Φ_c represents the critical volume ratio, t_m is a coefficient.

According to Equation (1), a larger porosity is critical to improve the piezoresistive performance of the porous electrode, so an electrode with loose microstructure is preferred. However, for normal carbon electrodes, the large porosity also results in weak mechanical strength of the electrode, making it easy to be broken under mechanical effects. In order to solve this contradiction, a novel preparation process is proposed, forming an elastic micro-networked structure with polytetrafluoroethylene and achieving superior piezoresistive performance and elasticity simultaneously. The specific steps of the film preparation process are as follows (Figure 2a):

1. Mix activated carbon powder and a binder powder (e.g., polytetrafluoroethylene and rubber) evenly by concussion and grinding. The mass ratio of the activated carbon powder and the binder powder is 85:15.
2. The mixed powder is lightly pressed under high temperature heating (150–180 °C) to form a film.

3. A shearing force is applied along the surface of the film, and the elastic binder and rubber component in the film are stretched. A micro-networked structure is formed, which has an envelope effect on the activated carbon micro-particles.

The porous electrode prepared by the above process has good elasticity, as shown in Figure 2b, and thus can be applied to flexible electronic devices such as electronic skin. In the following, in-depth tests and analyses of the piezoresistive characteristics of the prepared porous electrode are carried out. The experimental platform is composed of a mechanical press/tensile machine and an electrochemical workstation, as shown in Figure 2c,d. The experimental platform for the piezoresistive characteristics of the electrode film is shown in Figure 2c. Pressure is applied to the electrode film by a press machine, and the resistance change in the electrode is recorded by the electrochemical workstation. The experimental platform for the elasticity of the electrode film is shown in Figure 2d. The electrode film is stretched by a tensile machine until it breaks. The amount of tension and the length at which the film is stretched (i.e., the distance that the clamp moves) is recorded by the tensile machine.

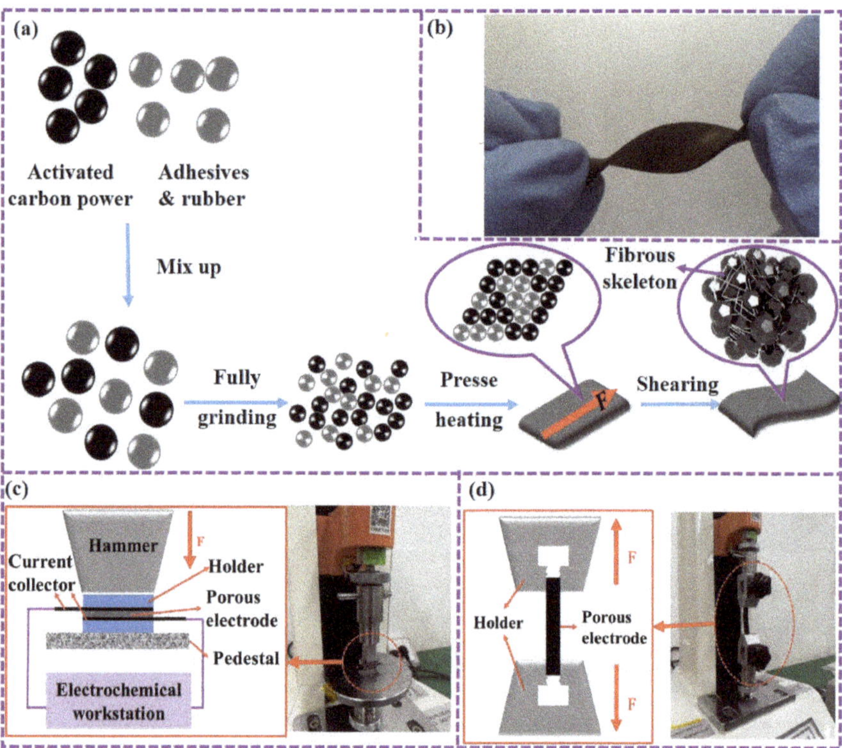

Figure 2. Preparation and testing of the porous electrodes. (**a**) Preparation process for microstructure formation. (**b**) Flexibility of the electrode under extreme torsion. (**c**) Experimental platform for piezoresistive testing. (**d**) Experimental platform for tensile testing.

3. Results

3.1. Pressure Sensitivity Enhancement of the Porous Electrode by Shearing

Four consecutive processes of pressurization and depressurization were applied to the electrode film. As shown in Figure 3a, when the pressure increases from 0 MPa to 0.3 MPa, the resistance of the electrode (10 mm × 10 mm × 1 mm) decreases from more than 100 Ω to less than 1 Ω. When the pressure is removed, the resistance almost returns to the original value. Figure 3b further illustrates that the resistance–pressure curves corresponding to the four consecutive pressurization–depressurization processes are almost coincident, verifying the reliability and stability of the electrode's application for mechanical sensing based on its piezoresistive characteristic.

In the electrode preparation process in Figure 2a, the shearing process along the surface of the film is the key to modifying the material for the electrode. As shown in Figure 3c, the electrode with the shearing process has much better piezoresistive sensitivity and a wider sensitive range than the electrode without the shearing process. When the pressure is increased from 0 MPa to 0.3 MPa, the former's resistance drops by 162.40 Ω, and its resistance decreases with the increasing pressure, until 0.3 MPa. However, the latter's resistance changes by only 26.34 Ω, and stops decreasing around 0.15 MPa. In addition, the elasticity characteristics of the two are completely different. As shown in Figure 3d, the electrode film (50 mm × 10 mm × 1 mm) with the shearing process can withstand a tensile force up to nearly 0.2 kgF, and fracture occurs when the length of the stretch exceeds 25% of the original length. On the other hand, the electrode film without the shearing process can only withstand a tensile force of less than 0.02 kgF, and fracture occurs when the length of the stretch is only about 2% of the original length.

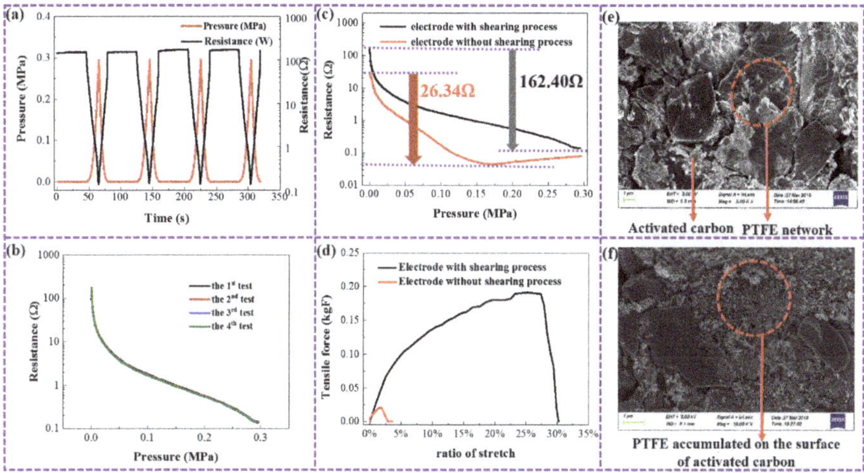

Figure 3. Changes in the microstructure and macroscopic properties of the porous electrodes by a shearing process. (**a**) Resistance change of the electrode during four consecutive pressurization–depressurization processes. (**b**) Repeatability of electrode piezoresistive characteristics during four consecutive pressurization–depressurization processes. (**c**) Piezoresistive characteristics of the electrodes with and without the shearing process. (**d**) Elasticity of the electrodes with and without the shearing process. (**e**) Electron micrograph of the electrodes with the shearing process where PTFE is the abbreviation of polytetrafluoroethylene. (**f**) Electron micrograph of the electrodes without the shearing process.

This tremendous improvement in piezoresistive sensitivity and elasticity is due to the changes in the microstructure of the porous electrode materials by the shearing process. As shown in Figure 3e, a loose micro-networked skeleton structure is formed in the electrode with the shearing process, which has an envelope effect on the particles distributed throughout the material. Energy spectrum tests show that the content of fluorine in the network skeleton structure is significantly higher than that in the granular particles, and it can be presumed that the network skeleton structure is mainly formed by the polytetrafluoroethylene binder under the shearing force. Since the polytetrafluoroethylene has good elasticity and its network skeleton structure has a loose microstructure, significant micro-deformation and resistance changes can be realized under pressure. At the same time, the superior elasticity property of the polytetrafluoroethylene network skeleton also enables the electrode to withstand larger stress and deformation. On the contrary, for the electrode without the shearing process, there are fewer microscopic network skeleton structures, but some cluster structures instead, as shown in Figure 3f. Energy spectrum tests show that the content of fluorine in the cluster structure is significantly higher than that of other regions, indicating that the binder has not been sufficiently dispersed during the preparation of the electrode film, but is clustered in some areas. As a result, the amount of binder in other areas was insufficient, making it easier to fracture under mechanical effects.

3.2. Pressure Sensitivity Enhancement of the Porous Electrode by the Rubber Component

Further experimental results reveal the more significant influence of rubber additives on the piezoresistive characteristics of the porous electrode. As shown in Figure 4a, as the mass fraction of Ethylene-Propylene-Diene Monomer (EPDM) rubber was increased from 0% to 20%, the magnitude of the resistance drop in the porous electrode, when the pressure increased from 0.05 MPa to 0.3 MPa, increases significantly. In addition, the sensitivity range of the electrode in response to the pressure is also broadened. For the electrode without rubber, the resistance remains at a small value without change when the pressure is increased to 0.3 MPa; while for the electrode with rubber mass fractions of 10% and 20%, when the pressure increases from 0.3 MPa to 0.4 MPa, the resistance continues to drop by 1.55 Ω and 12.8 Ω, respectively. This remarkable modulation of the piezoresistive effect also resulted from the changes in the microstructure of the porous electrode. As shown in Figure 4b, the electron micrograph of the porous electrode with rubber has two different network skeleton structures: one that is similar to the polytetrafluoroethylene skeleton in Figure 3e, and another network skeleton that is obviously more robust. Through energy spectrum tests, the fluorine content in the robust skeleton structure is significantly lower than the polytetrafluoroethylene skeleton structure in Figure 3e. So, it can be presumed that these robust network skeletons are formed by the rubber stretched under the effect of the shearing force. Since the elasticity of the network skeleton structure formed by rubber is much stronger than the polytetrafluoroethylene network skeleton, it has a more significant influence on the piezoresistive characteristics of the porous electrode.

Although the increase in the rubber component can improve the piezoresistive characteristics of the porous electrode, it also causes a loss of its energy storage characteristics. As shown in Figure 4c, as the mass ratio of rubber increases, the volume ratio of the microscopic pores of the electrode decreases remarkably, resulting in a loss of specific surface area, as shown in Figure 4d. The loss of specific surface area means that the micro-interface of the electrochemical double-layer energy storage structure is reduced [25]:

$$i_{\mathrm{DL}} = a_{\mathrm{v}} C_{\mathrm{dl}} \frac{\partial \left(\Psi_{\mathrm{s}} - \Psi_{\mathrm{l}} \right)}{\partial t}, \tag{2}$$

where i_{DL} represents the current density resulting from electric double-layer effect, a_{v} represents the specific surface area of the electrodes, and C_{dl} represents the capacitance associated with the electric double-layer effect. Ψ_{s} and Ψ_{l} represents the potential of the electrode phase and electrolyte phase, respectively.

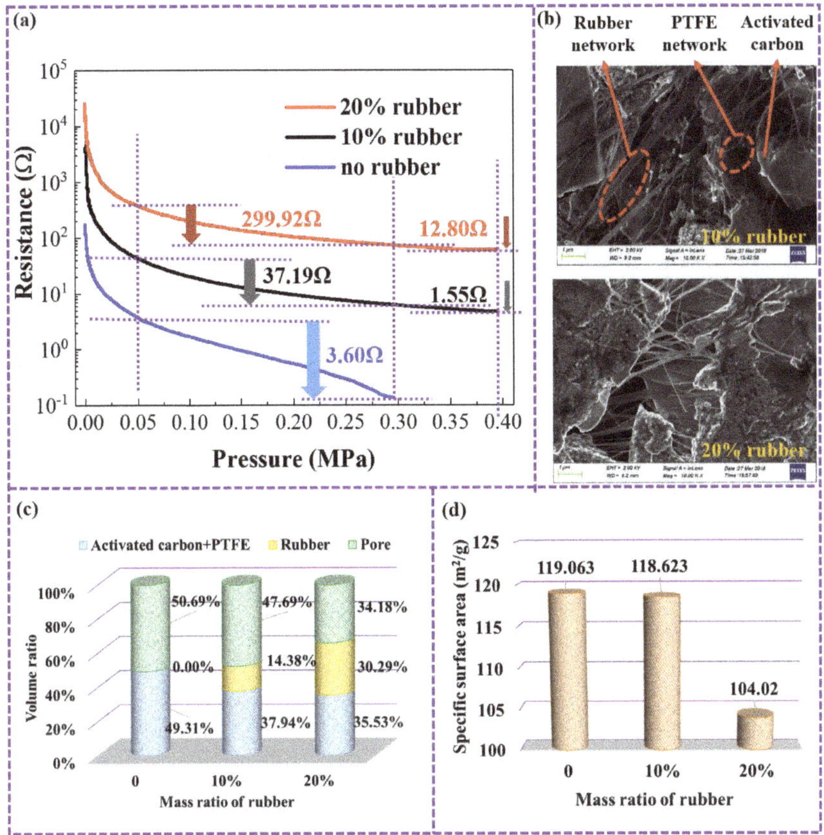

Figure 4. Changes in the microstructure and macroscopic properties of the porous electrodes by adding elastic rubber. (**a**) Piezoresistive characteristics of electrodes with different mass ratios of rubber. (**b**) Electron micrograph of the electrodes with different mass ratios of rubber. (**c**) Volume ratio of electrodes with different mass ratios of rubber. (**d**) Specific surface area of electrodes with different mass ratios of rubber.

So, as the mass ratio of the rubber increases, the energy storage performance of the porous electrode decreases, which is not conducive to self-powered mechanical sensors. Therefore, a moderate amount of rubber component is preferred to keep a balance between the energy storage characteristics and the piezoresistive characteristics. In the following section, with a prototype assembly and performance test, a porous electrode with a rubber mass ratio of 10% is chosen.

3.3. Performance of Applications in Self-Powered Mechanical Sensors

With flexible and compact packages, a self-powered pressure sensor and a self-powered impact sensor, based on the porous electrodes with piezoresistive sensitivity enhancement, are both realized.

A picture of the self-powered pressure sensor is shown in Figure 5a. With superior flexibility, this sensor can be applied to flexible electronic devices such as electronic skin. Figure 5b shows the sensitive signal of the sensor under continuous pressing. It is obvious that the voltage of the sensor increases and forms a peak with almost no cluttering data during each pressing process, and the continuous pressing process can be clearly identified. The amplitude of the sensitive voltage peak signal reaches about 10 mV, which is enough for signal acquisition and processing, verifying that the piezoresistive enhancement proposed in this paper is effective for applications.

A picture of the self-powered impact sensor is shown in Figure 5c. Its impact sensitivity characteristics are tested by a Machete Hammer system which utilizes the potential energy of the counterweight to produce an ultra-high impact of up to 30,000 g, as shown in Figure 5d. The experimental results are shown in Figure 5e, The response signal of the sensor has a clear voltage peak when it is subjected to a high-*g* impact, with a wide range from 7800 g to 23,000 g. Considering that the impact process in practical applications is about 1 millisecond [26], the sampling rate for signal acquisition of the sensor is set as 50 kHz. The voltage peak has almost no clutter even at such a high sample rate, showing superior impact feature recognition ability over traditional piezoelectric sensors.

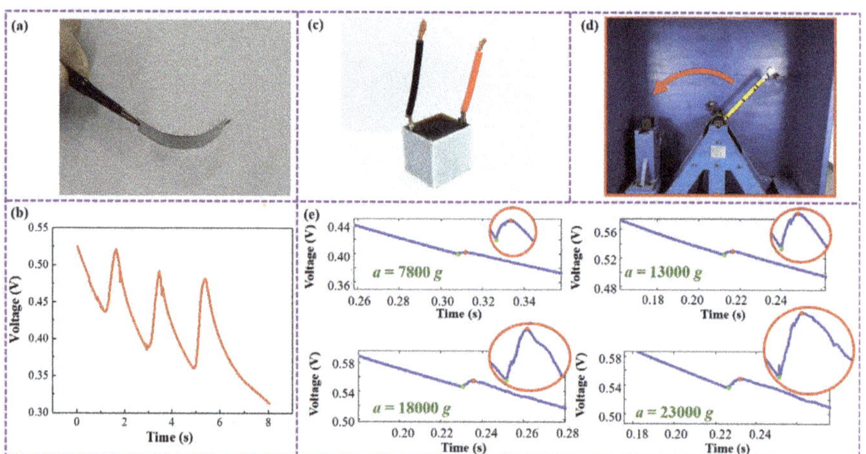

Figure 5. Applications as self-powered mechanical sensors. (**a**) The flexible pressure sensor device. (**b**) Response signal of the pressure sensor under continuous pressing. (**c**) The impact sensor device. (**d**) The Machete Hammer system. (**e**) Response signal of the impact sensor under different accelerations of impact.

The fewer high frequency components in the response signal of the self-powered sensor is due to its essential 'self-filtering' mechanism, which has been revealed in a previous study [21]. The response of the porous electrode can be approximated by the vibration of the electrode in the electrolyte under an external force:

$$f(t) - kx(t) - \mu\frac{\mathrm{d}x(t)}{\mathrm{d}t} = m\frac{\mathrm{d}^2x(t)}{\mathrm{d}t^2},$$ (3)

where $f(t)$ is the external force as a function of time, k is the equivalent stiffness coefficient of the electrode, $x(t)$ is the displacement as a function of time, μ is the equivalent damping coefficient of the electrolyte, and m represents the equivalent mass of the electrode.

To investigate the frequency components of $x(t)$, a Fourier transformation is applied to both sides of Equation (3):

$$F(\omega) - kX(\omega) - \mu j\omega X(\omega) = -m\omega^2 X(\omega),$$ (4)

where $F(\omega)$ and $X(\omega)$ are the Fourier transformations of $f(t)$ and $x(t)$, respectively. Thus, the spectrum of displacement is:

$$X(\omega) = \frac{F(\omega)}{k - m\omega^2 + j\mu\omega} = \frac{|F(\omega)|}{\sqrt{(m\omega^2 - k)^2 + (\mu\omega)^2}}e^{j\Phi(\omega)},$$ (5)

where $\Phi(\omega)$ represents the phase-frequency characteristics of $X(\omega)$.

The porous carbon electrode has a small equivalent stiffness coefficient and a large viscous damping coefficient (compared with a silicon beam structure and air damping in traditional impact sensors), which can effectively suppress high frequency components in the response signal according to Equation (5).

4. Conclusions

This paper presents a preparation process that can enhance the piezoresistive characteristics of porous electrodes. The experimental results show that by applying a shearing force along the surface of the electrode film, the binder is stretched to form a microscopic network skeleton structure. This has an envelop effect on porous particles such as activated carbon, and significantly enhances the piezoresistive characteristics and elasticity of the electrode. In addition, by adding a highly elastic rubber component, another more robust microscopic network skeleton structure is formed by the rubber under the shearing process, and a flexible porous electrode with superior piezoresistive characteristics and elasticity can be more effectively obtained. Based on such porous electrodes, both a self-powered flexible pressure sensor and a self-powered impact sensor were realized, with response signals of voltage peaks under pressure and high acceleration impact stimuli. Moreover, the sensors have superior ability to recognize features of mechanical processes, such as the number of consecutive pressing processes. These superior self-powered sensing characteristics provide new technological approaches for microsystems with limited power sources, such as electronic skin and smart fuzes.

Author Contributions: K.D. and X.W. conceived and designed the experiments; K.D. performed the experiments; K.D. and Z.Y. analyzed the data; H.Z. contributed analysis tools; K.D. wrote the paper.

Funding: This work was supported by the Central University Special Funding for Basic Scientific Research (Grant No. 309171B8805).

References

1. Wang, Z.L. Self-powered nanosensors and nanosystems. *Adv. Mater.* **2012**, *24*, 280–285. [CrossRef] [PubMed]
2. Lin, L.; Xie, Y.; Wang, S.; Wu, W.; Niu, S.; Wen, X.; Wang, Z.L. Triboelectric active sensor array for self-powered static and dynamic pressure detection and tactile imaging. *ACS Nano* **2013**, *7*, 8266–8274. [CrossRef] [PubMed]
3. Lee, J.H.; Yoon, H.J.; Kim, T.Y.; Gupta, M.K.; Lee, J.H.; Seung, W.; Ryu, H.; Kim, S.W. Micropatterned P (VDF-TrFE) Film-Based Piezoelectric Nanogenerators for Highly Sensitive Self-Powered Pressure Sensors. *Adv. Funct. Mater.* **2015**, *25*, 3203–3209. [CrossRef]
4. Dai, K.; Wang, X.; Yi, F.; Jiang, C.; Li, R.; You, Z. Triboelectric nanogenerators as self-powered acceleration sensor under high-g impact. *Nano Energy* **2018**, *45*, 84–93. [CrossRef]
5. Wang, S.; Lin, L.; Wang, Z.L. Triboelectric nanogenerators as self-powered active sensors. *Nano Energy* **2015**, *11*, 436–462. [CrossRef]
6. Ma, M.; Zhang, Z.; Liao, Q.; Yi, F.; Han, L.; Zhang, G.; Liu, S.; Liao, X.; Zhang, Y. Self-powered artificial electronic skin for high-resolution pressure sensing. *Nano Energy* **2017**, *32*, 389–396. [CrossRef]
7. Zhou, Y.; He, J.; Wang, H.; Qi, K.; Nan, N.; You, X.; Shao, W.; Wang, L.; Ding, B.; Cui, S. Highly sensitive, self-powered and wearable electronic skin based on pressure-sensitive nanofiber woven fabric sensor. *Sci. Rep.* **2017**, *7*, 12949. [CrossRef]
8. Wang, S.; Xu, J.; Wang, W.; Wang, G.J.N.; Rastak, R.; Molinalopez, F.; Chung, J.W.; Niu, S.; Feig, V.R.; Lopez, J. Skin electronics from scalable fabrication of an intrinsically stretchable transistor array. *Nature* **2018**, *555*. [CrossRef]
9. Narasimhan, V.; Li, H.; Jianmin, M. Micromachined high-g accelerometers: A review. *J. Micromech. Microeng.* **2015**, *25*, 033001. [CrossRef]

10. Akiyama, M.; Morofuji, Y.; Kamohara, T.; Nishikubo, K.; Tsubai, M.; Fukuda, O.; Ueno, N. Flexible piezoelectric pressure sensors using oriented aluminum nitride thin films prepared on polyethylene terephthalate films. *J. Appl. Phys.* **2006**, *100*, 114318. [CrossRef]

11. Mortet, V.; Petersen, R.; Haenen, K.; D'Olieslaeger, M. Wide range pressure sensor based on a piezoelectric bimorph microcantilever. *Appl. Phys. Lett.* **2006**, *88*, 133511. [CrossRef]

12. de Reus, R.; Gulløv, J.O.; Scheeper, P.R. Fabrication and characterization of a piezoelectric accelerometer. *J. Micromech. Microeng.* **1999**, *9*, 123. [CrossRef]

13. Zhang, Y.; Sun, H.; Jeong, C.K. Biomimetic Porifera Skeletal Structure of Lead-Free Piezocomposite Energy Harvesters. *ACS Appl. Mater. Interfaces* **2018**, *10*, 35539–35546. [CrossRef] [PubMed]

14. Chun, J.; Kim, J.W.; Jung, W.S.; Kang, C.Y.; Kim, S.W.; Wang, Z.L.; Baik, J.M. Mesoporous pores impregnated with Au nanoparticles as effective dielectrics for enhancing triboelectric nanogenerator performance in harsh environments. *ACS Appl. Mater. Interfaces* **2015**, *8*, 3006–3012. [CrossRef]

15. Zhang, B.; Shi, G. Mechanical Filtering for Target Recognition of Hard Target Penetration. *J. Detect. Control* **2010**, *32*, 25–29.

16. Wang, J.; Rong, L.I.; Huang, H. Layer Penetrating Adhesion Signal Characteristic Extractting Based on Wavelet Coefficients. *J. Detect. Control* **2016**, *38*, 13–17+23.

17. Kang, I.; Schulz, M.J.; Kim, J.H.; Shanov, V.; Shi, D. A carbon nanotube strain sensor for structural health monitoring. *Smart Mater. Struct.* **2006**, *15*, 737. [CrossRef]

18. Song, Y.; Chen, H.; Su, Z.; Chen, X.; Miao, L.; Zhang, J.; Cheng, X.; Zhang, H. Highly Compressible Integrated Supercapacitor–Piezoresistance-Sensor System with CNT–PDMS Sponge for Health Monitoring. *Small* **2017**, *13*, 1702091. [CrossRef]

19. Dai, K.; Wang, X.; Yin, Y.; Hao, C.; You, Z. Voltage Fluctuation in a Supercapacitor During a High-g Impact. *Sci. Rep.* **2016**, *6*, 38794. [CrossRef]

20. Dai, K.; Wang, X.; Yi, F.; Yin, Y.; Jiang, C.; Niu, S.; Li, Q.; You, Z. Discharge voltage behavior of electric double-layer capacitors during high-g impact and their application to autonomously sensing high-g accelerometers. *Nano Res.* **2018**, *11*, 1146–1156. [CrossRef]

21. Dai, K.; Wang, X.; Li, R.; Zhang, H.; You, Z. Theoretical study and applications of self-sensing supercapacitors under extreme mechanical effects. *Extreme Mech. Lett.* **2019**, *26*, 53–60. [CrossRef]

22. Hou, C.; Liu, Y.T.; Yang, X. Penetrating Hard Target Weapons and the Key Technology Research of Its Smart Fuse. *Aero Weapon.* **2012**, *2*, 44–48.

23. Wang, H.; Pilon, L. Accurate Simulations of Electric Double Layer Capacitance of Ultramicroelectrodes. *J. Phys. Chem. C* **2011**, *115*, 16711–16719. [CrossRef]

24. McLachlan, D.S.; Blaszkiewicz, M.; Newnham, R.E. Electrical Resistivity of Composites. *J. Am. Ceram. Soc.* **1990**, *73*, 2187–2203. [CrossRef]

25. Lin, C.; Ritter, J.A.; Popov, B.N.; White, R.E. A Mathematical Model of an Electrochemical Capacitor with Double-Layer and Faradaic Processes. *J. Electrochem. Soc.* **1999**, *146*, 3168–3175. [CrossRef]

26. Dong, L. Study on Signal Acquisition Echnology of Multi-Layer Penetration Overloads. Master's Thesis, North University of China, Taiyuan, China, 2013.

 *micromachines*

Article

Spray-On Liquid-Metal Electrodes for Graphene Field-Effect Transistors

Jordan L. Melcher [1,*], Kareem S. Elassy [1], Richard C. Ordonez [2], Cody Hayashi [2], Aaron T. Ohta [1] and David Garmire [1]

[1] Department of Electrical Engineering, University of Hawaii at Manoa, Honolulu, HI 96822, USA; elassy@hawaii.edu (K.S.E.); aohta@hawaii.edu (A.T.O.); david.garmire@gmail.com (D.G.)
[2] Space and Naval Warfare Systems Center Pacific, Pearl City, HI 96782, USA; richard.c.ordonez@navy.mil (R.C.O.); cody.hayashi@navy.mil (C.H.)
* Correspondence: melcherj@hawaii.edu

Received: 1 December 2018; Accepted: 8 January 2019; Published: 14 January 2019

Abstract: Advancements in flexible circuit interconnects are critical for widespread adoption of flexible electronics. Non-toxic liquid-metals offer a viable solution for flexible electrodes due to deformability and low bulk resistivity. However, fabrication processes utilizing liquid-metals suffer from high complexity, low throughput, and significant production cost. Our team utilized an inexpensive spray-on stencil technique to deposit liquid-metal Galinstan electrodes in top-gated graphene field-effect transistors (GFETs). The electrode stencils were patterned using an automated vinyl cutter and positioned directly onto chemical vapor deposition (CVD) graphene transferred to polyethylene terephthalate (PET) substrates. Our spray-on method exhibited a throughput of 28 transistors in under five minutes on the same graphene sample, with a 96% yield for all devices down to a channel length of 50 μm. The fabricated transistors possess hole and electron mobilities of 663.5 cm^2/(V·s) and 689.9 cm^2/(V·s), respectively, and support a simple and effective method of developing high-yield flexible electronics.

Keywords: graphene; Galinstan; Liquid-Metal; spray-on; aerosol; honey; mobility; contact resistance; TLM; I-V characteristics

---

1. Introduction

Since the 1960's, integrated circuit electrodes have been fabricated from traditional copper, nickel, silver, and gold metals [1]. Although these traditional metals are used widely in practically all consumer electronics today, they are susceptible to degradation under repeated stress and strain [2]. As the commercial sector demands nanomaterial-based flexible device applications, there is an immediate need for novel and inexpensive fabrication methods that can provide reliable flexible electrodes that resist damage after repeated deformation.

Liquid-Metal (LM) Galinstan is a commercially available eutectic alloy comprising of 68% gallium, 22% indium, and 10% tin that exhibits a conductivity of 2.30 × 10^6 S/m, a desirable vapor pressure (<1 × 10^{-6} Pa at 500 °C) compared with mercury (0.1713 Pa at 20 °C), and a stable liquid state across a broad temperature range (−19 °C to 1300 °C) [3]. Galinstan has been studied as a candidate for flexible integrated-circuit electrodes due to its deformability and non-toxic nature. Although there are several metals that have higher conductivity compared to LM, such as copper, the low contact resistance of LM in contact with graphene is a desirable benefit over the high contact resistance of metals, such as copper, in contact with graphene [4,5].

To deposit liquid-metal on electronic devices, mask deposition and microcontact printing have been adopted to improve manufacturing yield to over 95%, while simultaneously decreasing feature size down to 1 μm [6–12]. Such techniques are convenient due to design simplicity, reliability, and high throughput. However, when drop-casting LM with mask deposition, there are immediate drawbacks. The thickness of the patterned LM vary due to the curvature created by the LM surface tension and wetting to the mask and substrate during lift-off. In addition, microcontact printing LM on 2D nanomaterials is not suitable because of the risk of damage to the nanomaterial substrate as the contact head moves or drags across the print surface. A technique known as atomization is desirable for LM deposition on 2D nanomaterials due to the ability to reduce bulk liquids to a fine mist to create thin traces with homogenous thickness. However, the cost of commercial atomization systems is high, making it difficult to adopt such techniques in small labs and standard academic environments [13].

In this article, we demonstrate the use of a low-cost novel spray-on deposition technique for LM using off-the-shelf equipment that improves upon contemporary deposition techniques. We demonstrate the utility of the spray-on deposition of liquid-metal with a proven flexible material combination, consisting of Galinstan source and drain electrodes, an electrolytic gate comprised of honey, and a graphene channel that forms a flexible graphene field-effect transistor [14]. Galinstan was integrated with graphene not only as a solution to enhance flexibility and robustness of the fabricated device, but also as a means to overcome the undesirable high contact resistance of graphene in contact with standard electrode materials copper, gold, and silver [15]. We will aim to convince the reader that our stencil and deposition technique improves yield and simplicity, and decreases cost flexible nanomaterial-based electronics.

2. Materials and Methods

Figure 1a–e illustrates the process to fabricate graphene field effect transistors with our low-cost rapid prototyping LM spray-on technique. First, the stencil used to pattern LM electrodes was designed using CAD software (Silhouette Studio, Lindon, UT, USA) and cut into a Polyethylene terephthalate (PET) substrate using a vinyl cutter (Silhouette Portrait, Lindon, UT, USA). The stencil was adhered firmly at the edges with scotch tape onto a commercially bought Chemical Vapor Deposition (CVD) monolayer graphene sample transferred to a PET substrate (Graphene Platform, Shibuya-Ku, Tokyo) (Figure 1a). LM was than loaded into a commercially bought paint-gun reservoir, with the air compressor pressure set to 110 psi. The paint gun was mounted and positioned at a 90-degree angle with the exit aperture facing towards the target surface with an optomechanical stage and sprayed for ~4 s or until a homogenous LM thickness was deposited on the target surface (Figure 1b,c). The stencil was then removed slowly to reveal the desired LM pattern (Figure 1d, Figure 2). This process was repeated to create LM electrode pairs with differing channel lengths. There were four pairs of each of the following channel lengths: 1 mm, 500 μm, 400 μm, 300 μm, 200 μm, 100 μm, and 50 μm (28 electrode pairs in total), on a single 2 inch × 1 inch CVD graphene sample. Our methods are potentially compatible with complex shapes on the order of several hundreds of microns [16]. Out of the 28 electrode pairs, a single 50 μm electrode pair could not be measured. This fabrication error is due to the vinyl cutter reaching its minimum resolution limit. Despite the error, preparation of the LM mask took less than five minutes, and the liquid-metal spray duration took less than 10 s to pattern several pairs of LM electrodes. The total cost of materials, including the graphene sample, is under $200.

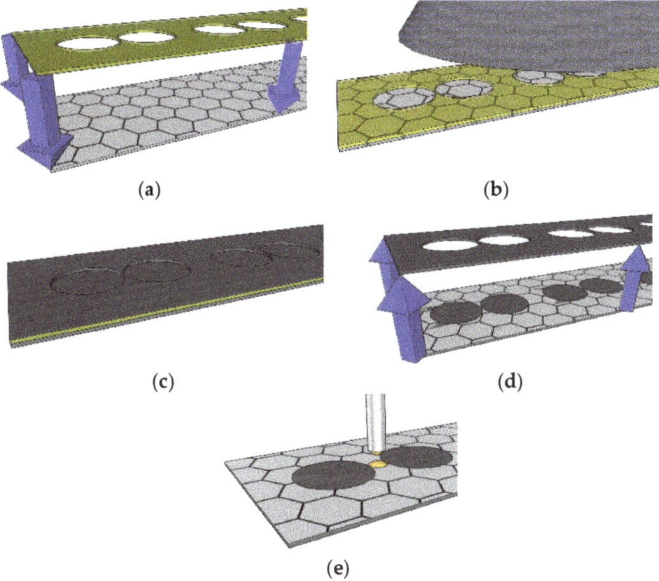

(a)   (b)

(c)   (d)

(e)

Figure 1. Fabrication process for graphene field-effect transistors with spray-on liquid-metal electrodes: (**a**) electrode patterns are cut into a Polyethylene Terephthalate (PET) substrate and placed flush on graphene surface to act as stencil; (**b**,**c**) liquid-metal is sprayed on the graphene surface with a paint gun; (**d**) the stencil is carefully removed, resulting in patterned electrodes on graphene; (**e**) an electrolytic top-gate material (honey) is drop-casted between the electrode pair to complete the device.

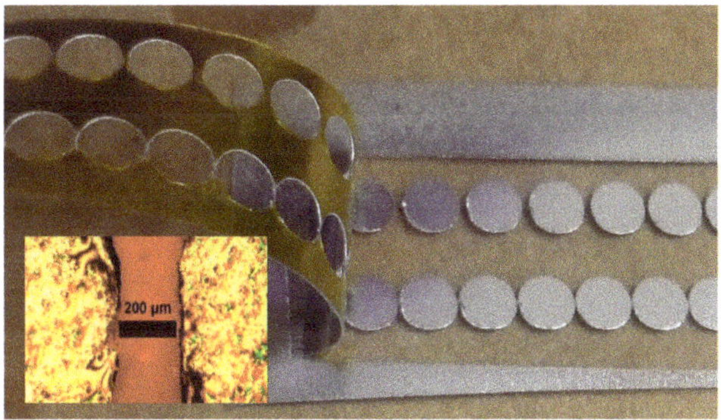

Figure 2. Spray-on Liquid-Metal electrode stencil removal process step. Inset: Magnified view of a liquid-metal electrode pair with a channel width of 200 μm.

It is important to note the quality of monolayer graphene produced commercially has been an issue when fabricating graphene devices, and verification of quality is mandatory before experimentation [17]. The Raman spectrum for the CVD graphene on PET (Graphene Platform) is illustrated in Figure 3. Raman measurements were taken at three different sites and the ratios between the second order overtone (2D) peak and in-plane vibrational mode (G) peak were computed to identify the disorder in graphene [18]. Due to the strong vibrational modes of polymeric PET, the PET Raman spectrum was subtracted from the graphene/PET Raman spectrum, leaving only the Raman spectrum due to graphene. Overall, the samples used for experimentation were of monolayer graphene, with minor disorder due to $I_{2D}/I_G > 2$ for all three spots imaged. There did exist a defect site D, which may be due to the structural disorder caused when transferring graphene to PET. Only electrical measurements can determine the effect the defect site D has on the quality of graphene.

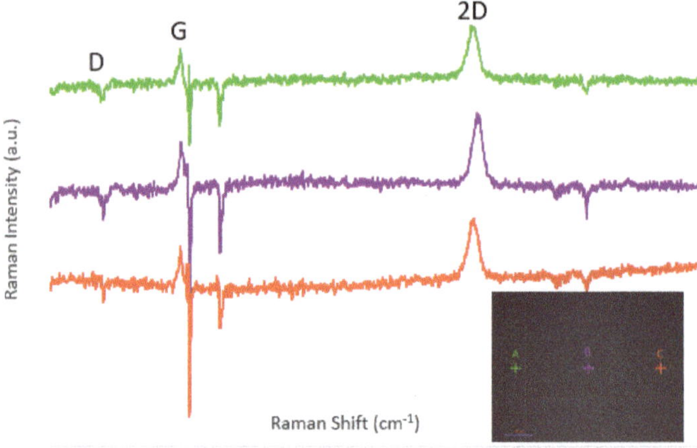

Figure 3. Raman Spectrum of graphene with in-plane vibrational mode (G), second order overtone (2D), and defect site (D) identified. Each Raman measurement was taken with a 532 nm, 2 mW Laser, with an exposure time of 5 s over an area of 25 μm. Inset: Image of Chemical Vapor Deposition (CVD) graphene on Polyethylene terephthalate.

To complete the three-terminal graphene field-effect transistors (GFET) device, honey was loaded into a syringe and drop-casted between each electrode pair to act as an electrolytic gate dielectric, as seen in Figure 4. Honey was chosen as an electrolytic gate dielectric due to its conformability, low-cost, and ease of accessibility [14]. Standard oxides, such as aluminum and silicon oxide, can be used, but are not the focus of this manuscript due to fabrication complexity [19].

Figure 4. (**a**) Illustration and (**b**) picture of graphene field-effect transistor with spray-on liquid-metal electrodes and honey gate dielectric.

3. Results and Discussion

To demonstrate the utility of our spray-on LM technique, graphene charge transport characteristics for several GFET devices with varying channel lengths (1 mm, 500 μm, 400 μm, 300 μm, 200 μm, 100 μm, and 50 μm) were extracted from Current-Voltage (I-V) measurements taken via an Agilent 4155C Semiconductor Parameter Analyzer (Santa Clara, CA, USA) and probe station. Each GFET device was subjected to a gate voltage (V_g) sweep from ± 5 V with a drain voltage (V_d) of 10 mV and the charge transport characteristics were plotted in Figure 5a,b for comparison. The on:off ratio and gate-leakage current density are also illustrated in Figures 5c and 4d, respectively, for comparison.

Tungsten micromanipulator probes were used to make electrical contact to the LM electrodes. Tungsten was chosen as the micromanipulator probe material because tungsten is one of the few materials that does not amalgamate with LM Galinstan [20]. A third micromanipulator probe was used to contact the electrolytic gate dielectric comprised of honey [14] (Figure 4a). Honey was adopted as an electrolytic gate dielectric in order to provide a rapid minimalistic method to actuate the graphene charge transport characteristics. Honey is a polar organic molecule, and in contact with a charged metal surface will form an electric double layer (EDL), as the charged ions that comprise honey diffuse to the graphene/honey interface. Applying either positive of negative potential to the honey via a third micromanipulator probe will enable electron or hole transport in the graphene channel. The authors implore the readers to implement this simple and rapid LM-patterning technique with alternative dielectrics to optimize performance characteristics and tradeoffs for their particular application.

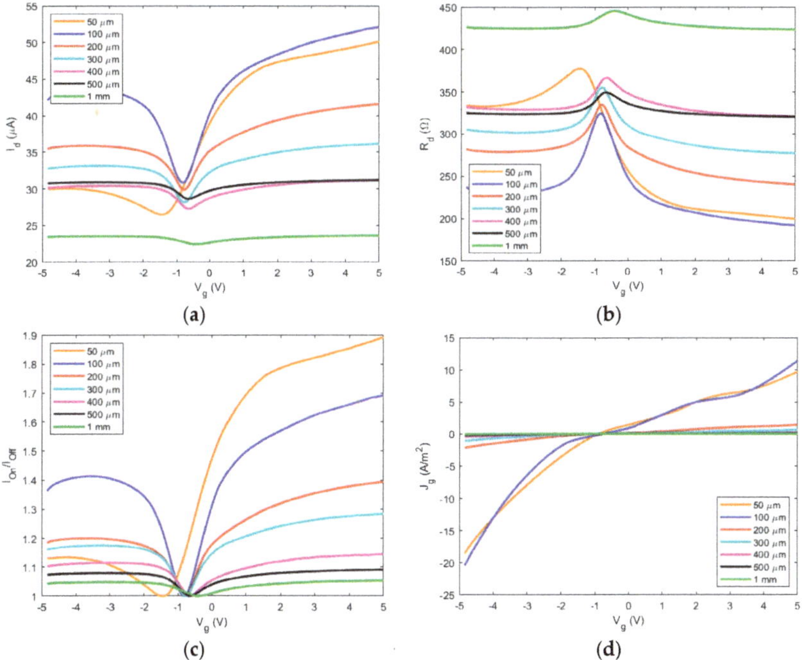

Figure 5. Graphene charge transport characteristics for several graphene field-effect transistors with varying channel-length: (**a**,**b**) illustrates the drain current (I_d) and drain resistance (R_d) as a function of top-gate voltage (V_g); (**c**) details the on:off ratio (I_{on}/I_{off}) and (**d**) illustrates the gate-leakage current density.

The ambipolar nature of all graphene field-effect transistors is made clear with the V-shape in Figure 5a; I_{ds} vs. V_g curve. The dual polarity allows the device to operate in either electron or hole conduction mode, which is beneficial for applications such as digital or analog circuit modulation [21]. Notice the minimum drain current, also known as the Dirac peak, for each device is at a negative V_g value, illustrating there is an overall intrinsic n-type doping characteristic that may be brought upon by conduction of electrons through the honey top-gate dielectric. It is commonly known that graphene exhibits p-type behavior in contact with atmospheric oxygen [22]. Therefore, there is reason to believe that the honey allows intrinsic n-type doping behavior and may be due to the composition of sucrose, glucose, fructose, and ash content in the honey [14]. To our benefit, the dirac shift is rather small with respect to $V_g = 0$ and is rather convenient for low-power devices. This, in part, is due to the nanoscale charge accumulation at the graphene-honey interface, also known as electric double layer (EDL), that can be actuated by altering the gate voltage [23–25].

Figure 5c illustrates the on:off ratio for the test devices. With exception to the 100 μm and 50 μm devices, there is a clear trend: as the channel length of each GFET decreases, the on:off ratio increases. This is an expected result based on well-reported improvement of semiconductor devices as miniaturization occurs [26]. In addition to the anomaly in the on:off ratio trend with the 100 μm and 50 μm devices, these devices also exhibit a noticeable asymmetry in electron and hole conduction branches. This asymmetry may be, in part, due to partial charge pinning due to low-resistivity graphene-metal contacts, as has been documented before [27]. However, an observation of the gate current density in Figure 5d shows there is reason to believe the primary cause of asymmetry, and the low on:off ratio, is due to the gate leakage current in short-channel devices. The 100 μm and 50 μm devices possess significant gate current densities in comparison to the other devices. Correspondingly, the 100 μm and 50 μm devices seem to possess the largest asymmetry between electron and hole branch, and exhibit anomalous on:off ratios.

The high gate-leakage current density at the smaller channel lengths may be due to the reduced electrode separation distance between the graphene-dielectric interface. It is noted that in the aforementioned process, the PET stencil did not form an ideal, airtight contact with the graphene in all devices during fabrication. Therefore, there were stray microscale LM-spray residues deposited under the masked areas, and in some cases shorted the source and drain electrodes. In addition, for ease of fabrication, the honey dielectric was drop-casted by hand and applied unnecessary pressure to the LM electrodes causing the LM to move, hence the failure of our methods to produce repeatable working devices at the 50 μm channel length size. Due to these circumstances, there was additional stray conductance between the source/drain electrodes and gate electrode, and the additional gate-leakage current density at smaller channel lengths is to be expected. However, the readers are encouraged to optimize this spray-on process with more ideal stencil-mask materials that provide stronger adhesion to graphene, and produce minimal LM residue. Additionally, other liquid dielectrics, such as lower-conductivity ionic gels, can be utilized in place of honey. Lastly, traditional back-gated graphene field effect transistor topologies can be utilized to eliminate stray conductance produced from overlap of the top-gate dielectric.

The transconductance (g_m) of a 1 mm GFET, Figure 6a, and electron and hole mobilities of the same 1 mm GFET, Figure 6b, were extracted from the following relationships:

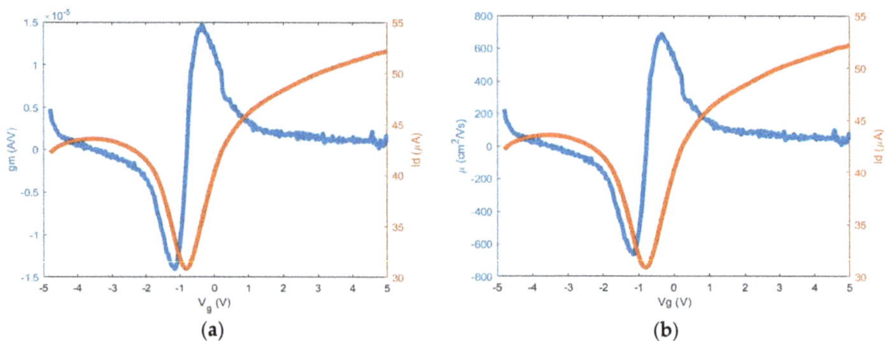

Figure 6. (**a**) Transconductance and (**b**) mobility vs. gate voltage.

Transconductance:

$$g_m = \frac{\frac{\partial I_D}{\partial t}}{\frac{\partial V_G}{\partial t}} \ (\mathrm{A/V}) \tag{1}$$

Electron and Hole Mobility:

$$\mu_e, \; -\mu_p = \frac{L g_m}{W C_{ox} V_D} \; (\text{cm}^2/(\text{V·s}))$$ (2)

The extracted hole and electron mobilities measured for the GFET devices are shown in Table 1 and are comparable to several reported devices with similar methods [14,28–32]. It is important to note the hole and electron mobilities increase as the channel length decreases and may be due, in part, to the lower probability of defects within the graphene channel at shorter channel lengths. However, there was an exception in the 50 µm channel length case and this may be due, in part, to a large dominating gate leakage effect. Other factors may include the quality of the graphene or honey between the LM source and drain electrodes. It is well documented that commercially grown graphene is non-uniform across a given area [33].

Table 1. This table details the electron and hole mobility of each channel length device and is compared to those in other works.

Device	Hole Mobility (cm^2/(V·s))	Electron Mobility (cm^2/(V·s))
1 mm	60.17	40.47
500 µm	179.5	105.3
400 µm	241	151
300 µm	371	346
200 µm	393.8	438.5
100 µm	663.5	689.9
50 µm	123	543
Ordonez [14]	213	166
Lu [28]	300	230
Kim [29]	203	91
Wang [30]	N/A	0.04
Kam [31]	154	154.6
Lee [32]	1188	422

4. Conclusions

It has been demonstrated that our spray-on LM method is a viable technique to fabricate nanomaterial devices. We have shown that top-gated graphene field-effect transistor devices with feature sizes down to 100 µm can be fabricated and exhibit standard charge transport characteristics with minimal effort. All devices were produced in a time frame of 30 s per transistor with a reliability of 96%. Our methods can be improved with fully-automated spray and drop-casting techniques, or even with back-gated devices, to improve the speed and reliability of this method. Furthermore, a strong-adhesion stencil mask and higher-resolution stencil-cutter can be used to significantly improve the minimum channel length achieved in this first attempt. This technique, with further modification, can be used inexpensively to mass-produce nanomaterial devices.

Author Contributions: Conceptualization, R.C.O. and D.G.; data curation, J.L.M. and K.S.E.; formal analysis, J.L.M.; funding acquisition, R.C.O. and C.H.; investigation, J.L.M.; methodology, J.L.M. and K.S.E.; project administration, R.C.O., A.T.O., and D.G.; resources, A.T.O.; software, J.L.M. and C.H.; supervision, A.T.O. and D.G.; validation, R.C.O., C.H., A.T.O., and D.G.; visualization, J.L.M.; writing—original draft, J.L.M.; writing—review and editing, K.S.E., R.C.O., C.H., A.T.O., and D.G.

Funding: This research was funded by the University of Hawaii's Hawaii Nano-Device Laboratory and by the Space and Naval Warfare Systems Center Pacific Naval Innovative Science and Engineering (NISE) program, grant number N66001-18-P-0009.

Conflicts of Interest: The authors declare no conflict of interest.

References

1. Alferov, Z.I. The history and future of semiconductor heterostructures. *Semiconductors* **1998**, *32*, 1–14. [CrossRef]
2. Greer, J.R.; Oliver, W.C.; Nix, W.D. Size dependence of mechanical properties of gold at the micron scale in the absence of strain gradients. *Acta Mater.* **2005**, *53*, 1821–1830. [CrossRef]
3. Liu, T.; Prosenjit Sen, P.; Kim, C.J. Characterization of nontoxic liquid-metal alloy galinstan for applications in microdevices. *J. Microelectromech. Syst.* **2012**, *21*, 443–450. [CrossRef]
4. Secor, E.B.; Cook, A.B.; Tabor, C.E.; Hersam, M.C. Wiring up liquid metal: Stable and robust electrical contacts enabled by printable graphene inks. *Adv. Electron. Mater.* **2018**, *4*, 1700483. [CrossRef]
5. Giubileo, F.; Di Bartolomeo, A. The role of contact resistance in graphene field-effect devices. *Prog. Surf. Sci.* **2017**, *92*, 143–175. [CrossRef]
6. Roberts, P.; Damian, D.D.; Shan, W.; Lu, T.; Majidi, C. Soft-matter capacitive sensor for measuring shear and pressure deformation. In Proceedings of the 2013 IEEE International Conference on Robotics and Automation, Karlsruhe, Germany, 6–10 May 2013; pp. 3529–3534.
7. Jeong, S.H.; Hagman, A.; Hjort, K.; Jobs, M.; Sundqvist, J.; Wu, Z. Liquid alloy printing of microfluidic stretchable electronics. *Lab Chip* **2012**, *12*, 4657–4664. [CrossRef]
8. Tabatabai, A.; Fassler, A.; Usiak, C.; Majidi, C. Liquid-phase gallium–indium alloy electronics with microcontact printing Langmuir. *Langmuir* **2013**, *29*, 6194–6200. [CrossRef] [PubMed]
9. Kramer, R.K.; Majidi, C.; Wood, R.J. Masked deposition of gallium–indium alloys for liquid-embedded elastomer conductors. *Adv. Funct. Mater.* **2013**, *23*, 5292–5296. [CrossRef]
10. Gozen, B.A.; Tabatabai, A.; Ozdoganlar, O.B.; Majidi, C. High-density soft-matter electronics with micron-scale line width. *Adv. Mater.* **2014**, *26*, 5211–5216.
11. Tang, S.; Zhu, J.; Sivan, V.; Gol, B.; Soffe, R.; Zhang, W.; Mitchell, A.; Khoshmanesh, K. Creation of liquid metal 3D microstructures using dielectrophoresis. *Adv. Funct. Mater.* **2015**, *25*, 4445–4452. [CrossRef]
12. Ladd, C.; So, J.H.; Muth, J.; Dickey, M.D. 3D printing of free standing liquid metal microstructures. *Adv. Mater.* **2013**, *25*, 5081–5085. [CrossRef]
13. Jeong, S.H.; Hjort, K.; Wu, Z. Tape transfer atomization patterning of liquid alloys for microfluidic stretchable wireless power transfer. *Sci. Rep.* **2015**, *5*, 8419.
14. Ordonez, R.; Hayashi, C.; Torres, C.; Melcher, J.; Kamin, N.; Severa, G.; Garmire, D. Rapid fabrication of graphene field-effect transistors with liquid-metal interconnects and electrolytic gate dielectric made of honey. *Sci. Rep.* **2017**, *7*. [CrossRef] [PubMed]
15. Ordonez, R.C.; Hayashi, C.K.; Torres, C.M.; Hafner, N.; Adleman, J.R.; Acosta, N.M.; Melcher, J.; Kamin, N.M.; Garmire, D. Conformal liquid-metal electrodes for flexible graphene device interconnects. *IEEE Trans. Electron Devices* **2016**, *63*, 4018–4023. [CrossRef]
16. Khondoker, M.A.H.; Sameoto, D. Fabrication methods and applications of microstructured gallium based liquid metal alloys. *Smart Mater. Struct.* **2016**, *25*, 093001. [CrossRef]
17. Graf, D.; Molitor, F.; Ensslin, K.; Stampfer, C.; Jungen, A.; Hierold, C.; Wirtz, L. Spatially resolved Raman spectroscopy of single-and few-layer graphene. *Nano Lett.* **2007**, *7*, 238–242. [CrossRef] [PubMed]
18. Childres, I.; Jauregui, L.A.; Park, W.; Cao, H.; Chen, Y.P. Raman spectroscopy of graphene and related materials. *New Dev. Photon Mater. Res.* **2013**, *1*.
19. Nistor, R.A.; Newns, D.M.; Martyna, G.J. The role of chemistry in graphene doping for carbon-based electronics. *ACS Nano* **2011**, *5*, 3096–3103. [CrossRef] [PubMed]
20. Lyon, R.N. *Liquid-Metal Handbook*, 2nd ed.; Atomic Energy Commission, Department of the Navy: Washington, DC, USA, 1952; pp. 170–171.
21. Lin, Y.F.; Xu, Y.; Wang, S.T.; Li, S.L.; Yamamoto, M.; Aparecido-Ferreira, A.; Li, W.; Sun, H.; Nakaharai, S.; Jian, W.B. Ambipolar $MoTe_2$ transistors and their applications in logic circuits. *Adv. Mater.* **2014**, *26*, 3263–3269. [CrossRef] [PubMed]
22. Ryu, S.; Liu, L.; Berciaud, S.; Yu, Y.J.; Liu, H.; Kim, P.; Flynn, G.W.; Brus, L.E. Atmospheric oxygen binding and hole doping in deformed graphene on a SiO_2 substrate. *Nano Lett.* **2010**, *10*, 4944–4951. [CrossRef] [PubMed]
23. Stern, O. The theory of the electrolytic double layer. *Z. Elektrochem. Angew. Phys. Chem.* **1924**, *30*, 508–516.
24. Booth, F. The dielectric constant of water and the saturation effect. *J. Chem. Phys.* **1951**, *19*, 391–394. [CrossRef]

25. Schmickler, W. Electronic effects in the electric double layer. *Chem. Rev.* **1996**, *96*, 3177–3200. [CrossRef]
26. Dragoman, M.; Dinescu, A.; Dragoman, D. Solving the graphene electronics conundrum: High mobility and high on-off ratio in graphene nanopatterned transistors. *Phys. E Low-Dimens. Syst. Nanostruct.* **2018**, *97*, 296–301. [CrossRef]
27. Di Bartolomeo, A.; Giubileo, F.; Santandrea, S.; Romeo, F.; Citro, R.; Schroeder, T.; Lupina, G. Charge transfer and partial pinning at the contacts as the origin of a double dip in the transfer characteristics of graphene-based field-effect transistors. *Nanotechnology* **2012**, *22*, 275702. [CrossRef] [PubMed]
28. Lu, C.; Lin, Y.; Yeh, C.; Huang, J.; Chiu, P. High mobility flexible graphene field-effect transistors with self-healing gate dielectrics. *ACS Nano* **2012**, *6*, 4469–4474. [CrossRef]
29. Kim, B.; Jang, H.; Lee, S.; Hongm, B.; Ahn, J.; Cho, J. High-performance flexible graphene field effect transistors with ion gel gate dielectrics. *Nano Lett.* **2010**, *10*, 3464–3466. [CrossRef] [PubMed]
30. Wang, D.; Noël, V.; Piro, B. Electrolytic gated organic field-effect transistors for application in biosensors—A review. *Electronics* **2016**, *5*, 9. [CrossRef]
31. Kam, K.; Tengan, B.; Hayashi, C.; Ordonez, R.; Garmire, D. Polar organic gate dielectrics for graphene field-effect transistor-based sensor technology. *Sensors* **2018**, *18*, 2774. [CrossRef] [PubMed]
32. Lee, S.K. Stretchable graphene transistors with printed dielectrics and gate electrodes. *Nano Lett.* **2011**, *11*, 4642–4646. [CrossRef] [PubMed]
33. Lili, L.; Qing, M.; Wang, Y.; Chen, S. Defects in graphene: Generation, healing, and their effects on the properties of graphene: A review. *J. Mater. Sci. Technol.* **2015**, *31*, 599–606.

 *micromachines*

Article

Plasma Surface Functionalization of Carbon Nanofibres with Silver, Palladium and Platinum Nanoparticles for Cost-Effective and High-Performance Supercapacitors

Zelun Li, Shaojun Qi, Yana Liang, Zhenxue Zhang, Xiaoying Li and Hanshan Dong *

School of Metallurgy and Materials, University of Birmingham, Birmingham B15 2TT, UK;
zl381@cam.ac.uk (Z.L.); S.Qi@bham.ac.uk (S.Q.); YXL452@student.bham.ac.uk (Y.L.);
z.zhang.1@bham.ac.uk (Z.Z.); X.li.1@bham.ac.uk (X.L.)
* Correspondence: h.dong.20@bham.ac.uk; Tel.: +44-121-414-7105

Received: 5 October 2018; Accepted: 19 December 2018; Published: 21 December 2018

Abstract: Due to their relatively low cost, large surface area and good chemical and physical properties, carbon nanofibers (CNFs) are attractive for the fabrication of electrodes for supercapacitors (SCs). However, their relatively low electrical conductivity has impeded their practical application. To this end, a novel active-screen plasma activation and deposition technology has been developed to deposit silver, platinum and palladium nanoparticles on activated CNFs surfaces to increase their specific surface area and electrical conductivity, thus improving the specific capacitance. The functionalised CNFs were fully characterised using scanning electron microscope (SEM), energy dispersive X-ray analysis (EDX) and X-ray diffraction (XRD) and their electrochemical properties were evaluated using cyclic voltammetry and electrochemical impedance spectroscopy. The results showed a significant improvement in specific capacitance, as well as electrochemical impedance over the untreated CNFs. The functionalisation of CNFs via environmental-friendly active-screen plasma technology provides a promising future for cost-effective supercapacitors with high power and energy density.

Keywords: carbon nanofibres (CNFs); active-screen plasma sputtering (ASPS) technology; supercapacitors (SCs); silver (Ag); platinum (Pt) and palladium (Pd) nanoparticles

---

1. Introduction

The enormous growth of electric vehicles and portable electronic devices has boosted the needs for energy storage devices simultaneously with high power and energy density. The most common energy storage devices are batteries, fuel cells and supercapacitors (SCs). The latter have attracted significant attention and research progress due to their advantageous properties over the formers. SCs exhibit greater power density satisfying the increasing demand for high power electrical appliances and fast charging, and they can withstand larger numbers of charge-discharge cycles [1,2]. However, SCs suffer from lower energy densities, high costs of raw materials and manufacturing which limit their widespread use and commercialisation [3–11]. One of the most crucial and greatest challenges to achieve these targets is indeed control of nanoscale materials and structures used for the SCs [12–14]. A large quantity of processing methods have been presented to fabricate SC electrodes based on carbon materials [15], especially carbon nanoparticles, graphene and carbon nanotubes (CNTs) [16,17]. In the past few years, two-dimensional (2D) nanomaterials including graphene, graphene-like materials, such as MXenes and transition-metal dichalcogenide (TMDs) have been explored to develop supercapacitors with enhanced electrochemical performance [18–25].

The energy storage in SCs is based on the electrostatic forces in the formation of an electrochemical double layer (EDL) between electrons and ions as well as fast redox reaction. Theoretically, the energy density of a supercapacitor is proportional to the specific capacitance and the operating cell voltage, given below [1,26,27]

$$E_s = \frac{1}{2}C_s\Delta V^2 \tag{1}$$

where E_s is the energy density, C_s is the specific capacitance, and ΔV is the operating cell voltage. Therefore, improved energy density can be achieved through increasing specific capacitance and extending operating cell voltage. According to Equation (2), specific capacitance of an electrode is dependent of the surface area of the electrode [1], therefore materials with a high surface area are preferable for electrode fabrication.

$$C_s = \frac{\varepsilon A}{md} \tag{2}$$

where ε is the electrolyte dielectric constant, A is the surface area of the electrode material, d is the effective thickness of the EDL, and m is the mass of the electrode material. Also, it is crucial to retain high power density when improving energy density. Power density of supercapacitors is associated with the rate capability of electrodes which reflects how fast charge/discharge cycling can be when good capacitive behaviour is maintained. Rate capability of electrode materials is related with their electrochemical impedance, and hence low electrochemical impedance is also demanded for electrode materials [28].

Carbon materials such as carbon nanotubes (CNTs) and carbon nanofibres (CNFs) are considered as good anode materials for SCs due to their accessibility, chemical inertness in different solutions, easy processability and good temperature tolerance [29–32]. Also, many physical and chemical active methods allow for production of the material with an improved surface area and a controllable pore structure [29,32,33]. Although CNTs show unique tubular porous structures and prominent electrical properties [15], their applications are limited by the production costs [32]. Instead, inexpensive CNFs can be easily manufactured using different methods such as electrospinning [34–36] or vapor growth [37,38]. Also, CNFs are easy to disperse, process and functionalise achieving high conductivity and improved surface area. Therefore, CNFs have attracted great interest for making cost-effective supercapacitor electrodes. It was found that electrical conductivity of CNT films could be increased following the deposition of metallic nanoparticles such as gold, silver, platinum and palladium on CNTs by plasma sputtering [18]. Active-screen plasma sputtering technology offers a green and easy route to achieve the deposition of metallic nanoparticles [39–42], compared to electrochemical [43,44] and electroless deposition [45,46] because the former is a physical deposition method which does not involve disposal of polluting wastes or chemicals.

This work aimed at functionalising the inexpensive CNFs by coating with Ag, Pt and Pd nanoparticles via the environmentally-friendly active-screen plasma sputtering and hence improving their electrochemical properties. The results showed that the specific capacitance and electrochemical impedance of CNFs were improved significantly and good cyclability was achieved.

2. Materials and Methods

2.1. Sample Preparation

CNFs were supplied by Sigma-Aldrich Company Ltd. (Dorset, UK) and they were pyrolytically stripped. The carbon content is more than 98% with trace amounts of sulphur, calcium, silicon, nickel, chromium, sodium, magnesium and iron. Commercial CNF powders (Sigma-Aldrich) (5 g) were dispersed in isopropanol (20 mL), provided by Struers (Catcliffe, UK), to form a suspension which was then sonicated at a power of 200 W for 1 h to disperse the CNFs. The suspension was distributed into glass Petri dishes (54 mm dimeter) and left to dry overnight, allowing the CNFs attaching to the Petri dishes.

2.2. Active-Screen Plasma Sputtering (ASPS)

The setup of active-screen plasma sputtering (ASPS) is demonstrated in Figure 1. The Petri dishes including CNFs were introduced in the vacuum chamber of a laboratory scale modified DC furnace (Klöckner Ionon GMBH, Bergisch Gladbach, Germany) with different targets of silver, palladium and platinum (specification: 80 mm × 80 mm × 0.2 mm; purity: 99.99%; Birmingham Metal Co. Ltd., Birmingham, UK), respectively. The distance between target plates and Petri dishes was 33 mm. ASPS was carried out at 320 °C at a heating rate of 500°/h in an atmosphere containing 75% hydrogen and 25% argon at a pressure of 0.75 mbar. The treatment time ranged from 0.1 to 1.0 h. The sample codes and treatment details are listed in Table 1.

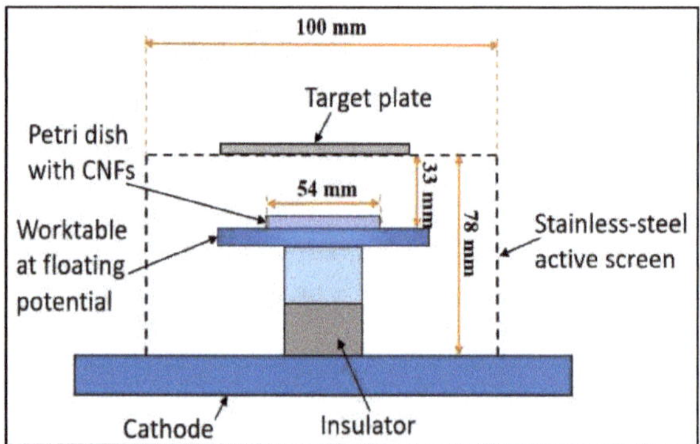

Figure 1. Schematic diagram of the active-screen plasma sputtering (ASPS) setup inside the plasma furnace.

Table 1. Experimental conditions of active-screen plasma sputtering (ASPS).

Sample Code	Target Material	Gas	Pressure (mbar)	Temperature °C	Time (h)
CNFs/Ag-0.1h	Silver				0.1
CNFs/Ag-0.5h	Silver				0.5
CNFs/Ag-1.0h	Silver				1.0
CNFs/Pt-0.1h	Platinum				0.1
CNFs/Pt-0.2h	Platinum	25% Ar + 75% H_2	0.75	320	0.2
CNFs/Pt-0.5h	Platinum				0.5
CNFs/Pd-0.1h	Palladium				0.1
CNFs/Pd-0.5h	Palladium				0.5

2.3. Fabrication of Electrodes

The structure of the home-made working electrode is shown in Figure 2. A copper wire was cold mounted in a tube using mixture of epoxy resin and epoxy resin hardener. The resin was then ground down until the end of the copper wire was fully exposed. A disc of conductive double-sided copper tape was attached to the end. Before the electrochemical experiment, different CNFs were attached to different conductive tapes, respectively. The weight of the attached CNFs (m) was measured and used to calculate their specific capacitance.

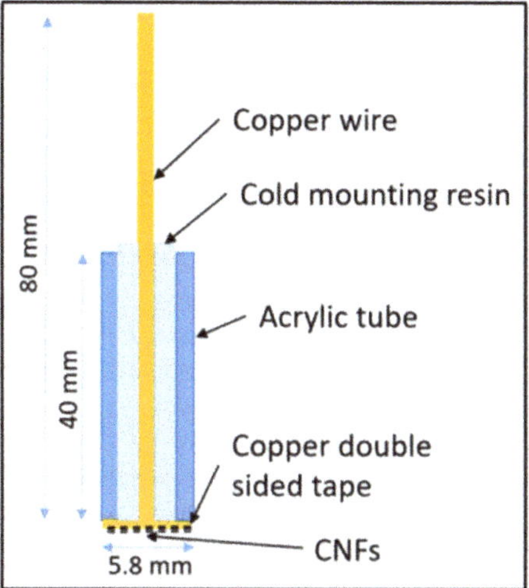

Figure 2. Schematic diagram of the electrodes.

2.4. Microstructure Characterisation

Joel 7000 scanning electron microscope (SEM) (Joel, Tokyo, Japan) equipped with an Oxford Inca energy dispersive X-ray spectroscopy (EDX) (Oxford Instruments, Abingdon, UK), was utilised to observe the morphology and collect the elemental information of deposited nanoparticles on CNFs. The phases of nanoparticles were identified using Bruker D8 ADVANCE diffractometer.

2.5. Electrochemical Test

Home-made three-electrode cell consisting of a working electrode (CNFs), a saturated calomel electrode (SCE) and a platinum inert counter electrode was designed for half-cell measurement. The solution was Na_2SO_4 (anhydrous, 99%) in de-ionized water at concentration of 1 M. The tests were carried out at room temperature. Cyclic voltammetry (CV) cycled between -0.2 and -0.7 V with respect to the SCE at different charge/discharge linear scan rates (i.e. 10, 25, 50, 100 and 200 mV/s) unfolded the charging and discharging performance of the working electrode. Cyclic voltammograms were obtained to calculate the specific capacitance of the untreated and functionalised CNFs by Equation (3) [47]. The first few cycles were neglected due to unstable initial conditions caused by electrode activation [48].

$$C_s = \frac{Q}{2m\Delta V} = \frac{\frac{\int I dV}{v}}{2m\Delta V} = \frac{\frac{\int \frac{I}{m} dV}{v}}{2\Delta V} \tag{3}$$

where C_s and Q are the specific capacitance of material and total charge in one cycle under a specific potential scan rate (v), respectively, I is the measured current through the circuit, V is the applied potential (vs. SCE), $\int I dV$ is the integral area of a CV curve integrating the forward and backward scans in the cyclic voltammogram, ΔV is the operating cell voltage, J is the measured current density.

Electrochemical impedance spectroscopy (Nyquist diagrams) over a frequency range of 0.05 to 100,000 Hz around the open circuit potential with an alternating current (AC) perturbation of 10 mV revealed the relationship between imaginary and real impedances ($Z_{|imag|}$ and $Z_{|real|}$). CNFs/Pd-0.5h, CNFs/Pt-0.5h and CNFs/Ag-1.0h were then charged and discharged at a constant current density of 2, 0.8 and 0.8 A/g, respectively, for 2000 cycles to test their cyclability in terms of capacitance retention.

3. Results

3.1. Microstructure

Figure 3a–i show the typical morphology of the untreated CNFs and functionalised CNFs. Typical diameters of the CNFs range from 150 to 450 nm. The untreated CNFs have a smooth surface, while the functionalised CNFs have relatively rough surfaces. The EDX results display the chemical composition of the nanoparticles deposited on CNFs. The existence of silver, palladium and platinum elements was detected in functionalised CNFs, respectively. Traces of copper were observed due to the copper tape for SEM sample preparation, and traces of iron and chromium were also identified, which mainly come from the sputtering effect of the active-screen due to plasma bombardment [40,49]. Figure 3b–f show that the size of Ag and Pd nanoparticles increased with the sputtering time. The size of the Pd nanoparticles was generally greater than that of Ag nanoparticles after the same sputtering times (0.1 h and 0.5 h, respectively). The particle size of Pd reached up to 250 nm after sputtering for 0.5 h, while that of Ag was around 100 nm after sputtering for 1.0 h. The morphology of the nanoparticles appeared to change from spherical to more nodular with increasing sputtering time. The surface morphology of CNFs/Pd became inhomogeneous after sputtering for 0.5 h, whereas the distribution of the Ag nanoparticles was relatively uniform even after sputtering for 1.0 h. In comparison, the surface morphology of the CNFs/Pt was quite different (Figure 3g–i). The size of the platinum nanoparticles deposited on CNFs was very small and almost remained constant with the sputtering time. The CNFs/Pt achieved a smooth surface finish and a homogeneous as well as fine particle distribution. In addition, there were valleys between the nanoparticles, and the size of these valleys differed with nanoparticle types and sputtering time. For CNFs/Ag and CNFs/Pd, the size of valleys increased with increasing sputtering time, with the valleys becoming large as tens of nanometres after 1.0 and 0.5 h, respectively, whereas the CNFs/Pt-0.5h displayed valleys of negligible sizes and remained constant. XRD characterisation revealed equilibrium phase of the deposited element and one of the samples is shown in Figure 3m. It can be seen that the major peaks of pure silver were identified, indicating the existence of pure silver instead of silver compounds. The peaks of copper were due to the copper tape background.

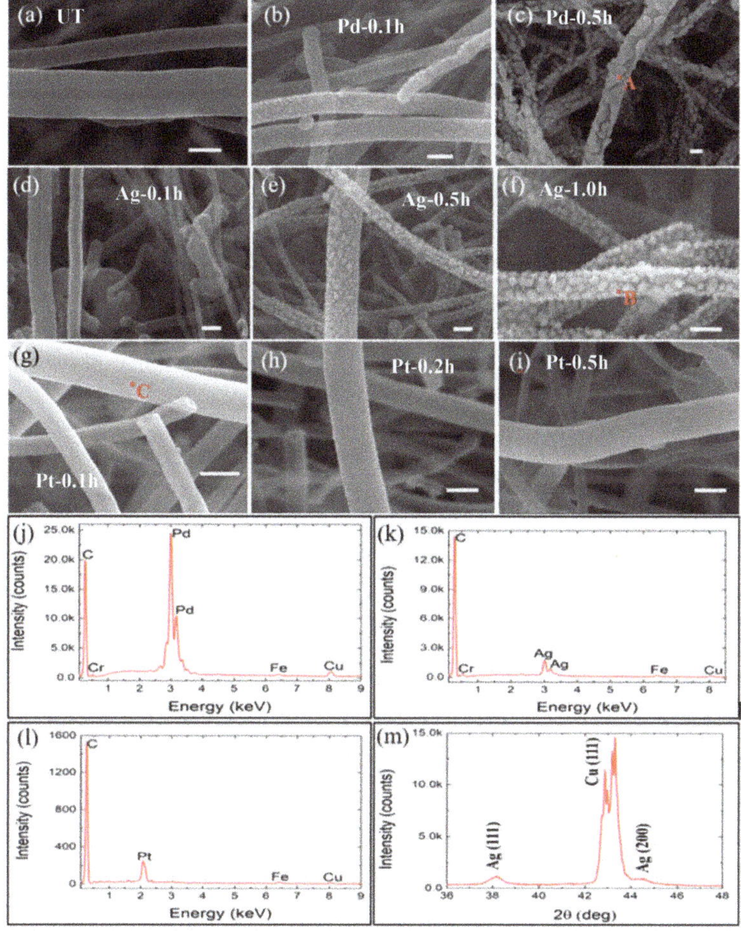

Figure 3. SEM images of: (**a**) untreated CNFs; (**b**) CNFs/Pd-0.1h; (**c**) CNFs/Pd-0.5h; (**d**) CNFs/Ag-0.1h; (**e**) CNFs/Ag-0.5h; (**f**) CNFs/Ag-1.0h; (**g**) CNFs/Pt-0.1h; (**h**) CNFs/Pt-0.2h; (**i**) CNFs/Pt-0.5h. All scale bars are representing 300 nm. EDX spectrum of: (**j**) CNFs/Pd-0.5h taken from spot 'A', denoted in (**c**); (**k**) CNFs/Ag-1.0h taken from spot 'B', denoted in (**f**); (**l**) CNFs/Pt-0.1h taken from spot 'C', denoted in (**g**). (**m**) XRD pattern indicating the phase of CNFs/Ag-0.5h.

3.2. Electrochemical Performance

Figure 4a–d show typical CV curves of untreated CNFs and functionalized CNFs. The data was taken after 10 cycles. At low scan rates (10 and 25 mV/s), all CV curves exhibited quasi-rectangular shapes without obvious redox peaks, demonstrating an appropriate capacitive behavior [50]. However, the distortion from a rectangular shape became more pronounced with increasing scan rates. At high scan rates (100 and 200 mV/s), the CV curves of all samples except for CNFs/Ag-1.0h and CNFs/Pd-0.5h exhibited considerably huge distortion, indicating poor rate capability of the system [51]. This might attribute to the limited ion-penetration diffusion into the materials, where only the outer surface was in use for charge storage [52]. In comparison, CNFs/Ag-1.0h and CNFs/Pd-0.5h exhibited good rate capability [51].

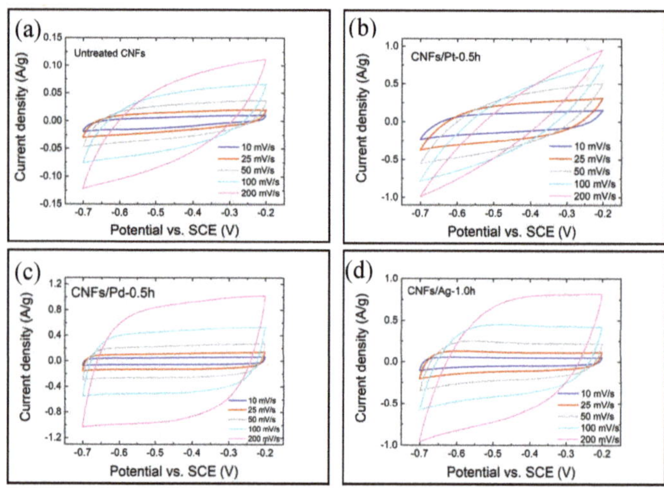

Figure 4. Typical cyclic voltammograms for electrodes made of: (**a**) untreated CNF; (**b**) CNFs/Pt-0.5h; (**c**) CNFs/Pd-0.5h and (**d**) CNFs/Ag-1.0h.

Figure 5a–c show that at same scan rates, the specific capacitances of functionalised CNFs were all greater than that of untreated CNFs, and they increased with the sputtering time. The specific capacitances decreased as the scan rate increased, with CNFs/Pt showing the greatest shrinkage. CNFs/Pt-0.5h, CNFs/Pd-0.5h and CNFs/Ag-1.0h exhibited the best performance within their own groups, and their values were 12.9, 5.7 and 5.6 times larger than that of untreated CNFs at 10 mV/s, respectively, and 3.0, 13.0 and 10.9 times larger at 200 mV/s. Figure 5d illustrates that the specific capacitance of CNFs/Pt-0.5h declined dramatically with scan rates, losing approximately 90% of its capacity at 200 mV/s compared with that at 10 mV/s, whilst CNFs/Ag-1.0h and CNFs/Pd-0.5h experienced less reductions in capacity as the scan rate increased. The specific capacitance for CNFs/Pd-0.5h only decreased by 27%, indicating its good rate capability [53].

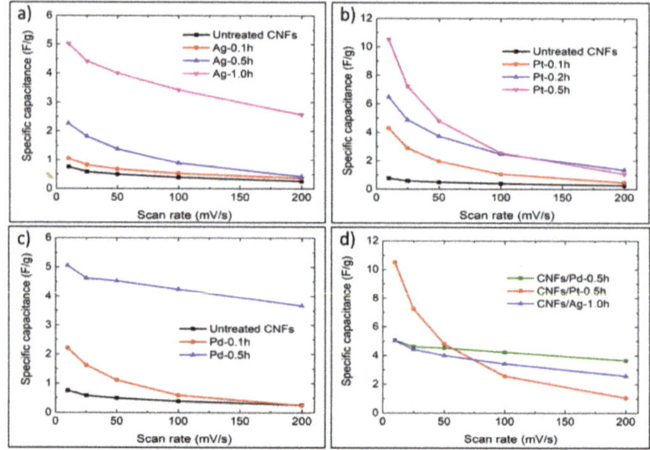

Figure 5. Specific capacitances for: (**a**) CNFs/Ag series; (**b**) CNFs/Pt series; (**c**) CNFs/Pd series; (**d**) CNFs/Ag-1.0h, CNFs/Pt-0.5h and CNFs/Pd-0.5h, at a series of scan rates (10 to 200 mV/s).

In Figure 6, the capacity of CNFs/Pt-0.5h and CNFs/Pd-0.5h decreased by 15.8% and 16.9% of the initial values, respectively, over 2000 cycles. CNFs/Ag-1.0h fluctuates around the initial value. The capacity of qualified SCs commonly falls by less than 20% of the initial value within their lifetime [54]. The average charging time is: 1.30 s for Ag-1.0h at current density of 0.8 A/g, 1.40 s for Pt-0.5h at 0.8 A/g, 0.77 s for Pd-0.5h at 2 A/g. Therefore, the results demonstrated that CNFs/Ag-1.0h, CNFs/Pt-0.5h and CNFs/Pd-0.5h all exhibited good cycle stability for use in at least five years if they were charged and discharged once a day, and the former had the best performance.

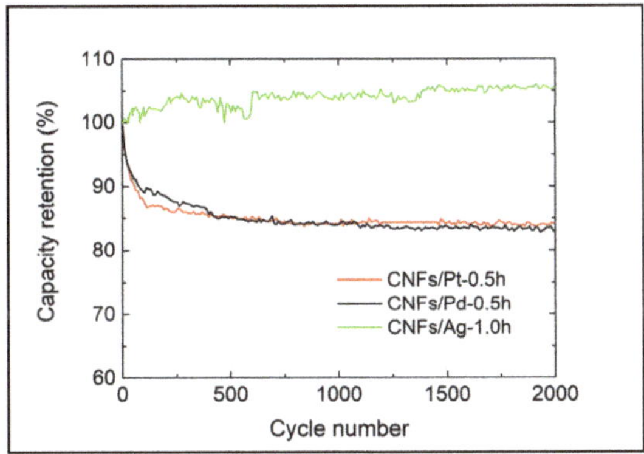

Figure 6. Capacitance retention of CNFs/Pd-0.5h, CNFs/Pt-0.5h and CNFs/Ag-1.0h.

In Nyquist plots, the intersection of the projected dash lines and the x-axis indicates the equivalent series resistance (ESR) of the electrodes, which is associated with the electrode porous structure and determines their rate capability [5,55]. The intersections between the semicircle and the x-axis at lower and higher $Z_{|real|}$ (e.g., in Figure 7b) represent the equivalent internal resistance (Rs) and charge transfer resistance (Rct) of the electrode material, respectively [34]. Rs involves the resistance of the electrode material and electrolyte, as well as the contact resistance at the interface between current collector and electrode material. Rct are related with the ion and electron transfer process during the formation of EDL. Figure 7a shows that all functionalised CNFs were shifted to the left and achieved smaller semicircle diameters than untreated CNFs, indicating lower charge transfer resistance of the functionalised CNFs [34]. Also, the slopes of functionalised CNFs were steeper than that of the untreated CNFs, indicating that functionalised CNFs can achieve faster EDL formation which leads to higher rate capability and hence higher power density [35,36]. Figure 7b–d show that the Rct and ESR for each kind of materials both decreased with increasing sputtering time, indicating low electron-and-charge transfer resistance and high rate capability, respectively [28]. Figure 7e compares the best material in each group. CNFs/Pd-0.5h exhibited the smallest Rct and ESR, as well as the steepest slope of the diffusion tail, indicating the fastest ion diffusion and penetration in the formation of EDL [35,36]. Therefore, it can be inferred that CNFs/Pd-0.5h had the best rate capability, which is consistent with the CV results.

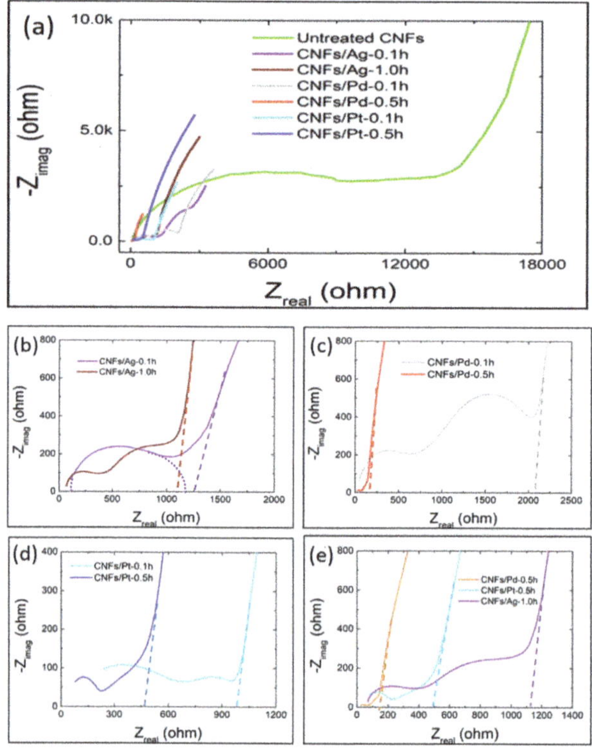

Figure 7. Nyquist plots of: (**a**) untreated and functionalised CNFs; (**b**) CNFs/Ag series; (**c**) CNFs/Pd series and (**d**) CNFs/Pt series and (**e**) CNFs/Ag-1.0h, CNFs/Pd-0.5h and CNFs/Pt-0.5h.

4. Discussion

4.1. Effect of the Nanoparticle Deposition on the Electrochemical Performance of CNFs

The electrochemical performance for the functionalised CNFs displayed a significant improvement possibly due to the corresponding surface structures [28,55,56]. The increase of specific capacitance for functionalised CNFs may have attributed to the porous structures. In the SEM images, the valleys existing between the nanoparticles can be considered as porous structures contributing to increased surface area for ion adsorption, therefore improving the specific capacitance. Pd and Ag nanoparticles tended to agglomerate to form larger nanoparticles, whereas Pt nanoparticles piled up evenly on the previous nanoparticles. Consequently, the pore sizes of CNFs/Ag-1.0h and CNFs/Pd-0.5h were in the range of tens of nanometres while that of CNFs/Pt-0.5h was only a few nanometres. The desirable electrode materials present a hierarchical porous structure including macropores (>50 nm) for accommodating ions, mesopores (2–50 nm) for facilitating the ion transport, and nanopores (<2 nm) for boosting the charge storage [57,58]. The accessibility of ions from electrolyte into nanopores (<2 nm) depends on both the ion size and the pore dimension. It is believed that maximum EDL capacitance is yielded when the pore size is approximate to the ion size, and both larger and smaller pores cause a rapid drop in capacitance [59]. The EDL formed at CNFs is mainly due to the adsorption of hydrated Na^+ ions with an average size of 0.358 nm [60,61]. The low specific capacitance of untreated CNFs may result from unfavourable ion adsorption due to its surface structures. In contrast, a high capacitance was obtained on CNFs/Pt-0.5h at small scan rates possibly because CNFs/Pt-0.5h achieved a large number of nanopores close to the size of Na^+ ions. Accordingly, Na^+ ions could access this kind of nanopore cavities and form strong interactions within the cavities, resulting in high adsorption and enhanced charge storage [62]. However, the pore size remained constant with sputtering time of up to 0.5 h, and hence, mesopores might be negligible in CNFs/Pt. The fine structure of CNFs/Pt might lead to longer paths for ion diffusion and thus higher charge transfer resistance, hindering fast formation of the EDL which in turn led to poor rate capability [36,42,43]. On the contrary, perhaps owing to the large numbers of pores equivalent to mesopores existing in their surface structures, CNFs/Pd and CNFs/Ag obtained improved rate capability with the increasing sputtering time. The relatively simple structure of mesopores could shorten the ion diffusion paths and lower the charge transfer resistance effectively, therefore promoting the ion transport and the adsorption/desorption process [58,63,64]. Their capacitances were smaller than that of CNFs/Pt-0.5h at low scan rates possibly due to the lack of nanopores [65]. The present results showed that the ASPS technology may provide a way to achieve controllable porous structures and surface properties, enabling the manipulation of electrochemical performance of SCs.

It is noted that the performance of the supercapacitors in our research was not as good as the other literature results. CNFs/Pt-0.5h showed the highest specific capacitance (~10.5 F/g) at 10 mV/s, but it was lower than the state-of-art literature results which are normally hundreds of F/g. Nonetheless, this feasibility study shows the potential of the active screen plasma treatments for the energy storage applications, and follow-on work has been planned to optimise the process and improve the performance of the final products.

4.2. Potential Application of the Functionalised CNFs

From the CV and EIS analysis, CNFs/Pt-0.5h exhibited the highest specific capacitance only at low scan rates, while CNFs/Pd-0.5h displayed the lowest ESR, excellent rate capability and highest specific capacitance at high scan rates. A small ESR corresponding to fast EDL formation is the key for high rate capability and high power density for SCs [1,28]. CNFs/Ag-1.0h showed lower specific capacitance and rate capability compared to CNFs/Pd-0.5h, but better cycle stability than CNFs/Pd-0.5h, which may contribute to a more consistent performance. Above all, CNFs/Pd-0.5h can be a suitable SC electrode material for short-term use, while CNFs/Ag-1.0h might be appropriate for long-term use.

5. Conclusions

The deposition of Ag, Pd and Pt nanoparticles on CNFs via ASPS was successful. The surface morphology of functionalised CNFs depended on target materials and sputtering time. The sizes of Ag and Pd nanoparticles both increased with sputtering time; their shapes changed from spherical to nodular, and their distribution became less uniform, whereas Pt nanoparticles remained constantly small in size. The cyclic voltammetry performance of CNFs was significantly improved after functionalisation. The specific capacitance of functionalised CNFs increased with sputtering time for each target material. CNFs/Pt-0.5h showed the highest specific capacitance (~10.5 F/g) at 10 mV/s but poor rate capability, while CNFs/Pd-0.5h and CNFs/Ag-1.0h exhibited competitive values at high scan rates. CNFs/Pd-0.5h exhibited the lowest ESR in the EIS analysis, suggesting excellent rate capability, while the excellent life-cycle was achieved on CNFs/Ag-1.0h.

Author Contributions: H.D. and X.L. conceived the research concept, provided all research facilities, supervised the research project, and finalised the paper. Z.L. performed the experimental work, collected data, and drafted the manuscript. Y.L. and Z.Z. contributed to some of the experimental work. S.Q. and X.L. helped to interpret the results and improved the manuscript.

Funding: The research received funding from the European Union's Horizon 2020 programme under agreement no. GA685844.

Acknowledgments: Zelun Li would like to thank Santiago Corujeira Gallo (Deakin University, Australia) for his very helpful discussion on the experimental set-up and data analysis.

Conflicts of Interest: The authors declare no conflict of interest.

References

1. Wang, G.; Zhang, L.; Zhang, J. A review of electrode materials for electrochemical supercapacitors. *Chem. Soc. Rev.* **2012**, *41*, 797–828. [CrossRef]
2. Wang, H.; Xu, Z.; Yi, H.; Wei, H.; Guo, Z.; Wang, X. One-step preparation of single-crystalline Fe_2O_3 particles/graphene composite hydrogels as high performance anode materials for supercapacitors. *Nano Energy* **2014**, *7*, 86. [CrossRef]
3. Erwin, W.R.; Zarick, H.F.; Talbert, E.M.; Bardhan, R. Light trapping in mesoporous solar cells with plasmonic nanostructures. *Energy Environ. Sci.* **2016**, *9*, 1577. [CrossRef]
4. Malgras, V.; Ji, Q.; Kamachi, Y.; Mori, T.; Shieh, F.-K.; Wu, K.C.-W.; Ariga, K.; Yamauchi, Y. Templated synthesis for nanoarchitectured porous materials. *Bull. Chem. Soc. Jpn.* **2015**, *88*, 1171. [CrossRef]
5. Liu, H.; Huang, Z.; Wei, S.; Zheng, L.; Xiao, L.; Gong, Q. Nano-structured electron transporting materials for perovskite solar cells. *Nanoscale* **2016**, *8*, 6209. [CrossRef] [PubMed]
6. Bairi, P.; Minami, K.; Hill, J.P.; Nakanishi, W.; Shrestha, L.K.; Liu, C.; Harano, K.; Nakamura, E.; Ariga, K. Supramolecular differentiation for construction of anisotropic fullerene nanostructures by time-programmed control of interfacial growth. *ACS Nano* **2016**, *10*, 8796. [CrossRef] [PubMed]
7. Sakaushi, K.; Antonietti, M. Carbon- and nitrogen-based porous solids: A recently emerging class of materials. *Bull. Chem. Soc. Jpn.* **2015**, *88*, 386. [CrossRef]
8. Huang, Z.; Che, S. Fabrication of mesostructured silica materials through Co-structure-directing route. *Bull. Chem. Soc. Jpn.* **2015**, *88*, 617. [CrossRef]
9. Yamamoto, E.; Kuroda, K. Colloidal mesoporous silica nanoparticles. *Bull. Chem. Soc. Jpn.* **2016**, *89*, 501. [CrossRef]
10. Tabuchi, H.; Urita, K.; Moriguchi, I. Effect of carbon nanospace on charge–discharge properties of Si and SiOx nanoparticles-embedded nanoporous carbons. *Bull. Chem. Soc. Jpn.* **2015**, *88*, 1378. [CrossRef]
11. Wu, K.; Lian, T. Quantum confined colloidal nanorod heterostructures for solar-to-fuel conversion. *Chem. Soc. Rev.* **2016**, *45*, 3781. [CrossRef] [PubMed]
12. Lewis, N.S. Research opportunities to advance solar energy utilization. *Science* **2016**, *351*, aad1920. [CrossRef] [PubMed]
13. Xiao, Z.; Lei, H.; Zhang, X.; Zhou, Y.; Hosono, H.; Kamiya, T. Ligand-hole in [SnI_6] unit and origin of band gap in photovoltaic perovskite variant Cs_2SnI_6. *Bull. Chem. Soc. Jpn.* **2015**, *88*, 1250. [CrossRef]
14. Shibayama, N.; Ozawa, H.; Ooyama, Y.; Arakawa, H. Highly efficient cosensitized plastic-substrate dye-sensitized solar cells with black dye and pyridine-anchor organic dye. *Bull. Chem. Soc. Jpn.* **2015**, *88*, 366. [CrossRef]
15. Feng, D.; Lv, Y.; Wu, Z.; Dou, Y.; Han, L.; Sun, Z.; Xia, Y.; Zheng, G.; Zhao, D. Free-standing mesoporous carbon thin films with highly ordered pore architectures for nanodevices. *J. Am. Chem. Soc.* **2011**, *133*, 15148–15156. [CrossRef] [PubMed]
16. Zhang, L.L.; Zhao, X.S. Carbon-based materials as supercapacitor electrodes. *Chem. Soc. Rev.* **2009**, *38*, 2520–2531. [CrossRef] [PubMed]
17. Xiao, K.; Ding, L.X.; Liu, G.; Chen, H.; Wang, S.; Wang, H. Freestanding, hydrophilic nitrogen-doped carbon foams for highly compressible all solid-state supercapacitors. *Adv. Mater.* **2016**, *28*, 5997–6002. [CrossRef]
18. Pushparaj, V.L.; Shaijumon, M.M.; Kumar, A.; Murugesan, S.; Ci, L.; Vajtai, R.; Linhardt, R.J.; Nalamasu, O.; Ajayan, P.M. Flexible energy storage devices based on nanocomposite paper. *Proc. Natl. Acad. Sci. USA* **2007**, *104*, 13574–13577. [CrossRef] [PubMed]
19. Zhu, Y.; Murali, S.; Stoller, M.D.; Ganesh, K.J.; Cai, W.; Ferreira, P.J.; Pirkle, A.; Wallace, R.M.; Cychosz, K.A.; Thommes, M.; et al. Carbon-based supercapacitors produced by activation of graphene. *Science* **2011**, *332*, 1537. [CrossRef] [PubMed]
20. Peng, X.; Peng, L.; Wu, C.; Xie, Y. Two dimensional nanomaterials for flexible supercapacitors. *Chem. Soc. Rev.* **2014**, *43*, 3303. [CrossRef]
21. Peng, L.; Zhu, Y.; Li, H.; Yu, G. Chemically integrated inorganic-graphene two-dimensional hybrid materials for flexible energy storage devices. *Small* **2016**, *12*, 6183. [CrossRef] [PubMed]
22. Peng, L.; Zhu, Y.; Chen, D.; Ruoff, R.S.; Yu, G. Two-dimensional materials for beyond-lithium-ion batteries. *Adv. Energy Mater.* **2016**, *6*, 1600025. [CrossRef]

23. Khan, A.H.; Ghosh, S.; Pradhan, B.; Dalui, A.; Shrestha, L.K.; Acharya, S.; Ariga, K. Bull. Two-dimensional (2D) nanomaterials towards electrochemical nanoarchitectonics in energy-related applications. *Bull. Chem. Soc. Jpn.* **2017**, *90*, 627–648. [CrossRef]

24. Sengottaiyan, C.; Jayavel, R.; Bairi, P. Cobalt oxide/reduced graphene oxide composite with enhanced electrochemical supercapacitance performance. *Bull. Chem. Soc. Jpn.* **2017**, *90*, 955–962. [CrossRef]

25. Dubal, D.P.; Chodankar, N.R.; Kim, D.-H.; Gomez-Romero, P. Towards flexible solid-state supercapacitors for smart and wearable electronics. *Chem. Soc. Rev.* **2018**, *47*, 2065–2129. [CrossRef] [PubMed]

26. Yang, C.; Zhou, M.; Xu, Q. Three-dimensional ordered macroporous MnO_2/carbon nanocomposites as high-performance electrodes for asymmetric supercapacitors. *Phys. Chem. Chem. Phys.* **2013**, *15*, 19730–19740. [CrossRef]

27. Zhong, C.; Deng, Y.; Hu, W.; Qiao, J.; Zhang, L.; Zhang, J. A review of electrolyte materials and compositions for electrochemical supercapacitors. *Chem. Soc. Rev.* **2015**, *44*, 7484–7539. [CrossRef]

28. Yang, L.; Cheng, S.; Ding, Y.; Zhu, X.; Wang, Z.L.; Liu, M. Hierarchical network architectures of carbon fiber paper supported cobalt oxide nanonet for high-capacity pseudocapacitors. *Nano Lett.* **2012**, *12*, 321–325. [CrossRef]

29. Frackowiak, E.; Béguin, F. Carbon materials for the electrochemical storage of energy in capacitors. *Carbon* **2001**, *39*, 937–950. [CrossRef]

30. Fan, Z.; Yan, J.; Wei, T.; Zhi, L.; Ning, G.; Li, T.; Wei, F. Asymmetric supercapacitors based on graphene/MnO_2 and activated carbon nanofiber electrodes with high power and energy density. *Adv. Funct. Mater.* **2011**, *21*, 2366–2375. [CrossRef]

31. Wang, D.-W.; Li, F.; Liu, M.; Lu, G.Q.; Cheng, H.-M. 3D aperiodic hierarchical porous graphitic carbon material for high-rate electrochemical capacitive energy storage. *Angew. Chem.* **2008**, *120*, 379–382. [CrossRef]

32. Zhai, Y.; Dou, Y.; Zhao, D.; Fulvio, P.F.; Mayes, R.T.; Dai, S. Carbon materials for chemical capacitive energy storage. *Adv. Mater.* **2011**, *23*, 4828–4850. [CrossRef] [PubMed]

33. Raymundo-Piñero, E.; Leroux, F.; Béguin, F. A high-performance carbon for supercapacitors obtained by carbonization of a seaweed biopolymer. *Adv. Mater.* **2006**, *18*, 1877–1882. [CrossRef]

34. Kim, C.; Ngoc, B.T.N.; Yang, K.S.; Kojima, M.; Kim, Y.A.; Kim, Y.J.; Endo, M.; Yang, S.C. Self-sustained thin webs consisting of porous carbon nanofibers for supercapacitors via the electrospinning of polyacrylonitrile solutions containing zinc chloride. *Adv. Mater.* **2007**, *19*, 2341–2346. [CrossRef]

35. Kim, C.; Yang, K.S.; Kojima, M.; Yoshida, K.; Kim, Y.J.; Kim, Y.A.; Endo, M. Fabrication of electrospinning-derived carbon nanofiber webs for the anode material of lithium-ion secondary batteries. *Adv. Funct. Mater.* **2006**, *16*, 2393–2397. [CrossRef]

36. Wang, G.; Pan, C.; Wang, L.; Dong, Q.; Yu, C.; Zhao, Z.; Qiu, J. Activated carbon nanofiber webs made by electrospinning for capacitive deionization. *Electrochim. Acta* **2012**, *69*, 65–70. [CrossRef]

37. Al-Saleh, M.H.; Sundararaj, U. A review of vapor grown carbon nanofiber/polymer conductive composites. *Carbon* **2009**, *47*, 2–22. [CrossRef]

38. Tibbetts, G.; Lake, M.; Strong, K.; Rice, B. A review of the fabrication and properties of vapor-grown carbon nanofiber/polymer composites. *Compos. Sci. Technol.* **2007**, *67*, 1709–1718. [CrossRef]

39. Janas, D.; Koziol, K.K.K. The influence of metal nanoparticles on electrical properties of carbon nanotubes. *Appl. Surf. Sci.* **2016**, *376*, 74–78. [CrossRef]

40. Li, C.; Bell, T.; Dong, H. A study of active screen plasma nitriding. *Surf. Eng.* **2013**, *18*, 174–181. [CrossRef]

41. Soin, N.; Roy, S.S.; Karlsson, L.; McLaughlin, J.A. Sputter deposition of highly dispersed platinum nanoparticles on carbon nanotube arrays for fuel cell electrode material. *Diam. Relat. Mater.* **2010**, *19*, 595–598. [CrossRef]

42. Zacharia, R.; Rather, S.-U.; Hwang, S.W.; Nahm, K.S. Spillover of physisorbed hydrogen from sputter-deposited arrays of platinum nanoparticles to multi-walled carbon nanotubes. *Chem. Phys. Lett.* **2007**, *434*, 286–291. [CrossRef]

43. He, Z.; Chen, J.; Liu, D.; Zhou, H.; Kuang, Y. Electrodeposition of Pt–Ru nanoparticles on carbon nanotubes and their electrocatalytic properties for methanol electrooxidation. *Diam. Relat. Mater.* **2004**, *13*, 1764–1770. [CrossRef]

44. Lin, Z.; Ji, L.; Zhang, X. Electrodeposition of platinum nanoparticles onto carbon nanofibers for electrocatalytic oxidation of methanol. *Mater. Lett.* **2009**, *63*, 2115–2118. [CrossRef]

45. Chen, X.; Xia, J.; Peng, J.; Li, W.; Xie, S. Carbon-nanotube metal-matrix composites prepared by electroless plating. *Compos. Sci. Technol.* **2000**, *60*, 301–306. [CrossRef]

46. Faraji, S.; Ani, F.N. The development supercapacitor from activated carbon by electroless plating—A review. *Renew. Sustain. Energy Rev.* **2015**, *42*, 823–834. [CrossRef]

47. Arulepp, M.; Permann, L.; Leis, J.; Perkson, A.; Rumma, K.; Jänes, A.; Lust, E. Influence of the solvent properties on the characteristics of a double layer capacitor. *J. Power Sources* **2004**, *133*, 320–328. [CrossRef]

48. Allagui, A.; Freeborn, T.J.; Elwakil, A.S.; Maundy, B.J. Reevaluation of performance of electric double-layer capacitors from constant-current charge/discharge and cyclic voltammetry. *Sci. Rep.* **2016**, *6*, 38568. [CrossRef]

49. Corujeira-Gallo, S.; Dong, H. On the fundamental mechanisms of acitve screen plasma nitriding. *Vacuum* **2010**, *84*, 321–325. [CrossRef]

50. Numao, S.; Judai, K.; Nishijo, J.; Mizuuchi, K.; Nishi, N. Synthesis and characterization of mesoporous carbon nano-dendrites with graphitic ultra-thin walls and their application to supercapacitor electrodes. *Carbon* **2009**, *47*, 306–312. [CrossRef]

51. Hsia, B.; Marschewski, J.; Wang, S.; In, J.B.; Carraro, C.; Poulikakos, D.; Grigoropoulos, C.P.; Maboudian, R. Highly flexible, all solid-state micro-supercapacitors from vertically aligned carbon nanotubes. *Nanotechnology* **2014**, *25*, 055401. [CrossRef] [PubMed]

52. Bao, L.; Zang, J.; Li, X. Flexible Zn_2SnO_4/MnO_2 core/shell nanocable-carbon microfiber hybrid composites for high-performance supercapacitor electrodes. *Nano Lett.* **2011**, *11*, 1215–1220. [CrossRef] [PubMed]

53. Yan, J.; Wei, T.; Shao, B.; Ma, F.; Fan, Z.; Zhang, M.; Zheng, C.; Shang, Y.; Qian, W.; Wei, F. Electrochemical properties of graphene nanosheet/carbon black composites as electrodes for supercapacitors. *Carbon* **2010**, *48*, 1731–1737. [CrossRef]

54. Testing Electrochemical Capacitors: Cyclic Charge-Discharge-Stacks. Available online: https://www.gamry.com/application-notes/battery-research/electrochemical-capacitors-cyclic-charge-discharge-and-stacks (accessed on 9 June 2018).

55. Gamby, J.; Taberna, P.L.; Simon, P.; Fauvarque, J.F.; Chesneau, M. Studies and characterisations of various activated carbons used for carbon/carbon supercapacitors. *J. Power Sources* **2001**, *101*, 109–116. [CrossRef]

56. Gui, Z.; Zhu, H.; Gillette, E.; Han, X.; Rubloff, G.W.; Hu, L.; Lee, S.B. Natural cellulose fiber as substrate for supercapacitor. *ACS Nano* **2013**, *7*, 6037–6046. [CrossRef] [PubMed]

57. Cai, J.; Niu, H.; Wang, H.; Shao, H.; Fang, J.; He, J.; Xiong, H.; Ma, C.; Lin, T. High-performance supercapacitor electrode from cellulose-derived, inter-bonded carbon nanofibers. *J. Power Sources* **2016**, *324*, 302–308. [CrossRef]

58. Liu, C.; Yu, Z.; Neff, D.; Zhamu, A.; Jang, B.Z. Graphene-based supercapacitor with an ultrahigh energy density. *Nano Lett.* **2010**, *10*, 4863–4868. [CrossRef] [PubMed]

59. Liu, C.; Li, F.; Ma, L.P.; Cheng, H.M. Advanced materials for energy storage. *Adv. Mater.* **2010**, *22*, E28–E62. [CrossRef]

60. Largeot, C.; Portet, C.; Chmiola, J.; Taberna, P.-L.; Gogotsi, Y.; Simon, P. Relation between the ion size and pore size for an electric double-layer capacitor. *JACS* **2008**, *130*, 2730–2731. [CrossRef]

61. Volkov, A.G.; Paula, S.; Deamer, D.W. Two mechanisms of permeation of small neutral molecules and hydrated ions across phospholipid bilayers. *Bioelectrochem. Bioenerg.* **1997**, *42*, 153–160. [CrossRef]

62. Nightingale, E.R. Phenomenological theory of ion solvation. Effective radii of hydrated ions. *J. Phys. Chem.* **1959**, *63*, 1381–1387. [CrossRef]

63. Yuan, L.; Xiao, X.; Ding, T.; Zhong, J.; Zhang, X.; Shen, Y.; Hu, B.; Huang, Y.; Zhou, J.; Wang, Z.L. Paper-based supercapacitors for self-powered nanosystems. *Angew. Chem. Int. Ed. Engl.* **2012**, *51*, 4934–4938. [CrossRef] [PubMed]

64. Zhi, M.; Xiang, C.; Li, J.; Li, M.; Wu, N. Nanostructured carbon-metal oxide composite electrodes for supercapacitors: A review. *Nanoscale* **2013**, *5*, 72–88. [CrossRef]

65. Yamada, H.; Nakamura, H.; Nakahara, F.; Moriguchi, I.; Kudo, T. Electrochemical study of high electrochemical double layer capacitance of ordered porous carbons with both meso/macropores and micropores. *J. Phys. Chem. C* **2007**, *111*, 227–233. [CrossRef]

Article

White-Light Photosensors Based on Ag Nanoparticle-Reduced Graphene Oxide Hybrid Materials

Wei-Chen Tu [1,*], Xiang-Sheng Liu [1], Shih-Lun Chen [1], Ming-Yi Lin [2], Wu-Yih Uen [1], Yu-Cheng Chen [3] and Yu-Chiang Chao [4]

[1]   Department of Electronic Engineering, Chung Yuan Christian University, Taoyuan 32023, Taiwan; a2001751@gmail.com (X.-S.L.); chrischen@cycu.edu.tw (S.-L.C.); wuyih@cycu.edu.tw (W.-Y.U.)
[2]   Department of Electrical Engineering, National United University, Miaoli 36003, Taiwan; mylin@nuu.edu.tw
[3]   School of Electrical and Electronic Engineering, Nanyang Technological University, Singapore 639798, Singapore; yucchen@ntu.edu.sg
[4]   Department of Physics, National Taiwan Normal University, Taipei 10610, Taiwan; ycchao@ntnu.edu.tw
*   Correspondence: wctu@cycu.edu.tw; Tel.: +88-632-654617

Received: 17 October 2018; Accepted: 7 December 2018; Published: 11 December 2018

Abstract: The unique and outstanding electrical and optical properties of graphene make it a potential material to be used in the construction of high-performance photosensors. However, the fabrication process of a graphene photosensor is usually complicated and the size of the device also is restricted to micrometer scale. In this work, we report large-area photosensors based on reduced graphene oxide (rGO) implemented with Ag nanoparticles (AgNPs) via a simple and cost-effective method. To further optimize the performance of photosensors, the absorbance and distribution of the electrical field intensity of graphene with AgNPs was simulated using the finite-difference time-domain (FDTD) method through use of the surface plasmon resonance effect. Based on the simulated results, we constructed photosensors using rGO with 60–80 nm AgNPs and analyzed the characteristics at room temperature under white-light illumination for outdoor environment applications. The on/off ratio of the photosensor with AgNPs was improved from 1.166 to 9.699 at the bias voltage of -1.5 V, which was compared as a sample without AgNPs. The proposed photosensor affords a new strategy to construct cost-effective and large-area graphene films which raises opportunities in the field of next-generation optoelectronic devices operated in an outdoor environment.

Keywords: photosensor; reduced graphene oxide; Ag nanoparticles; solution process; finite-difference time-domain

1. Introduction

Two-dimensional material graphene with unique properties distinguished from bulk materials has been one of the most celebrated inventions [1,2] in the applications of photosensors and therefore the development of high-efficiency photosensors has become an emerging area of research [3–5]. The interesting properties of two-dimensional graphene materials include high flexibility, high mobility, a large surface to volume ratio, broadband absorption which results from its structure, symmetrical conduction/valence bands and the linear dispersion of Dirac massless electrons [6–8]. However, photosensors based on single-layer graphene usually exhibit low responsivity and short carrier lifetime due to their zero bandgap. Moreover, graphene-based photosensors are typically fabricated under high-temperature and high-vacuum growth conditions. Besides, single-layer graphene may be broken during the transferring process, therefore the size of graphene flakes and devices is limited, which has become the bottleneck for the future development of graphene [9,10]. In recent years, a solution-processed reduced graphene oxide (rGO) with a tunable bandgap that can range from

1.00 eV to 1.69 eV through modifying its oxygen content has gained much attention, owing to its cost advantage. Consequently, the development of rGO devices and rGO nanocomposites has been widely explored [11–15]. To further improve the performance of graphene-based photosensors, several approaches can be applied. For example, authors formed a vertical built-in field in the graphene channel to trap the photoinduced electrons [16] or fabricated a gated multilayer photodetector, integrated on a photonic waveguide to reduce dark current and increase responsivity [17–20]. However, these structures were conventionally obtained by using the time-consuming e-beam deposition method or a complicated photolithography process.

Recently, photosensors with plasmonic structures have been proposed as an effective way to improve the absorption in the active layer which would contribute to the enhanced photocurrent [21–24]. The physics of surface plasmon resonance lie in the oscillation of conducting electrons at the interface of materials with different permitivities, such as the metal and the non-conducting media. By modifying the shape, structure or surrounding of the metal and the non-conducting media, the characteristics of surface plasmon resonance can be tailored, which is of interest to their potential use in optoelectronic devices [25,26]. Based on this theory, the performance of photosensors can be improved by decorating rGO with plasmonic metal nanoparticles. This not only preserves the outstanding properties of the original material but also provides the improved characteristics. For example, the Ag nanoparticles have a superior ability to enhance the photocurrent through light scattering and the surface plasmon resonance effect [27–29]. As a result, graphene with plasmonic nanoparticles can construct a hybrid system that benefits modified and enhanced light absorption [30]. While Ag nanoparticle-graphene nanocomposites have demonstrated enhanced carrier separation, very little research has been performed for the Ag nanoparticle-rGO hybrid system constructed on large-area photosensors and operated under white-light illumination for outdoor environment applications.

Here, we provide the fabrication method of rGO films on SiO_2/Si substrates to construct large-area photosensors via a cost-effective solution process. To further optimize the characteristics of rGO films and improve the performance of photosensors, the absorbance and distribution of the electrical field intensity of graphene with different size plasmonic Ag nanoparticles was simulated through the finite-difference time-domain (FDTD) method. Then, 60–80 nm Ag nanoparticles combined with rGO were spin-coated on SiO_2/Si substrates to form the plasmonic photosensors and the characteristics of the photosensors were measured under white-light illumination using a solar simulator for outdoor environment applications. It was found that the photocurrent of photosensors based on rGO with Ag nanoparticles was enhanced and the on/off ratio was therefore improved from 1.166 to 9.699, comparing the photosensors without Ag nanoparticles. This can be attributed to the effective surface plasmon resonance and light scattering effect. Our fabrication procedure and measured results offer a valuable guide to improving the performance of large-area photosensors or other devices via a simple way which can be applied for future wearable and outdoor devices.

2. Materials and Methods

Figure 1a shows the schematic structure of a solution-processed rGO photosensor on a SiO_2/Si substrate. First, Si substrates were cleaned using hydrofluoric acid and then rinsed using deionized (DI) water to remove the native oxide on the Si. Then, 300 nm-thick of SiO_2 was deposited on the Si by the sputtering system. The absorbing layer, rGO, was obtained by reducing the graphite oxide (GO). To synthesize rGO, GO in DI water was sonicated followed by adding hydrobromic acid into the GO colloids. Then the mixtures were refluxed in an oil bath at 110 °C for 24 h. The rGO flakes were obtained after the processes of filtration, washing and desiccation. To construct the active layer of photosensors, rGO flakes in DI water were spin-coated on SiO_2/Si substrates at a spin rate of 3000 rpm. Finally, a 120 nm Ag film which served as a contact electrode was sputtered on the rGO/SiO_2/Si. The size of the photosensor was 1.5 cm in width and 1.5 cm in length; the rGO detective channel of the photosensor was 0.2 cm in width and 1.5 cm in length. To further improve the performance of photosensors, 60–80 nm Ag nanoparticles (commercial products produced by Golden Innovation Business Company) mixed with rGO solution were spin-coated on SiO_2/Si at a spin rate of 3000 rpm. The shape and the diameter of the Ag nanoparticles were spherical and 60–80 nm, respectively. The structure and fabrication processes of the rGO and Ag nanoparticle-rGO photosensor were the same besides the adding of Ag nanoparticles. The SEM image of the rGO with Ag nanoparticles photosensor on the SiO_2/Si substrate is illustrated in Figure 1b, indicating that the rGO was sheet-like and the rGO flakes can construct a continuous film. Moreover, the plasmonic Ag nanoparticles were successfully coated on the rGO flakes.

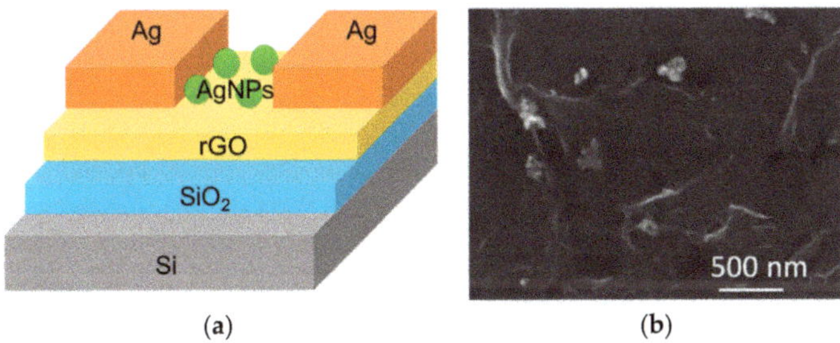

Figure 1. (**a**) Structure of a Ag nanoparticle-reduced graphene oxide (AgNPs-rGO) photosensor. The Ag nanoparticles mixed with reduced graphene oxide (rGO) were spin-coated on SiO_2/Si to improve the performance of photosensors; (**b**) SEM image of rGO with 60–80 nm Ag nanoparticles (AgNPs).

3. Results and Discussion

Raman spectroscopy is a useful and informative technique used to investigate disorder in sp^2 carbon materials. As shown in Figure 2, characteristic D, G, and 2D peaks of rGO film are observed at 1334, 1591, and 2730 cm^{-1}, respectively. The D peak is resulted from the existence of dislocations or vacancies in the graphene and the G peak is originated from the in-phase stretching vibration of symmetric sp^2 C-C bonds [31,32]. Based on the peaks in the measured Raman spectra, we can confirm that the material we used was rGO. As the Ag nanoparticle combined rGO film is formed, the intensity of Raman spectra exhibits an obvious enhancement due to enhanced Raman scattering induced by the intense local electromagnetic fields of the plasmonic Ag nanoparticles.

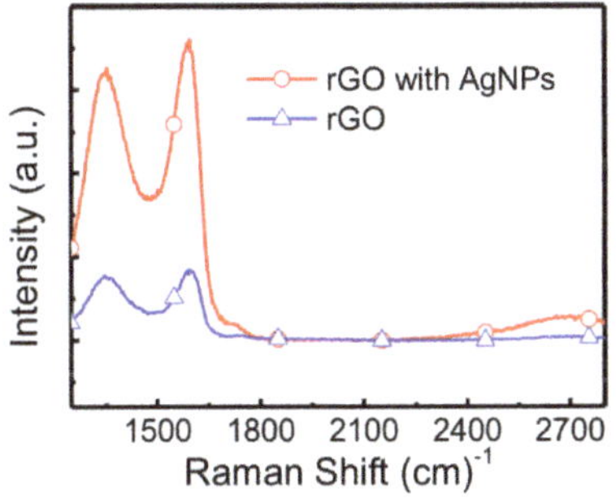

Figure 2. Raman spectra of rGO and Ag nanoparticles-rGO hybrid films. The Raman intensity of rGO with Ag nanoparticles is enhanced owing to the surface plasmon resonance effect.

The oscillation of carriers generated by the surface plasmon resonance can radiate electromagnetic energy at the same frequency as that of the surface plasmon resonance, which leads to elastic/Rayleigh scattering. Meanwhile, the surface plasmon resonance is dependent on the size and shape of particles as well as the dielectric constant of the surrounding materials [33,34]. To further investigate the surface plasmon resonance effect and the morphology of graphene on the absorbance and distribution of the electric field intensity, we executed simulated absorbance spectra of flat graphene and rough graphene with Ag nanoparticles using the FDTD method. The flat structure was constructed as Ag nanoparticles/flat and few-layer graphene/SiO_2/Si; the rough structure was designed as Ag nanoparticles/rough and multilayer graphene/SiO_2/Si. The roughness of graphene was around tens nanometers. Figure 3a,b displays the simulated absorbance spectra of 20 nm, 40 nm, 60 nm and 80 nm Ag nanoparticles on flat graphene/SiO_2/Si and rough graphene/SiO_2/Si, respectively. Based on the full Mie equation, for larger nanoparticles (diameter >20 nm), the extinction cross section is dependent on higher-order multipole modes and the extinction characteristics are dominated by quadrupole, octopole absorption and scattering [33]. These higher oscillation modes are related to the particle size, in addition, the maximum absorption may shift to a longer wavelength and the bandwidth will increase with increasing particle size. As displayed in Figure 3a, the absorbance curves of flat graphene with any size of Ag nanoparticles shift toward longer wavelengths, and the larger size of Ag nanoparticles, the larger the shift. In the case of the absorbance spectra of rough graphene with Ag nanoparticles as shown in Figure 3b, the curves also exhibit a shift, however, the shifted wavelengths are not linearly dependent on the particle's size because the surface plasmon resonance and the light scattering phenomena are more complex. Nevertheless, the frequency of the surface plasmon resonance can be modified and the absorbance of Ag nanoparticles/graphene/SiO_2/Si is increased in several parts of the wavelength range.

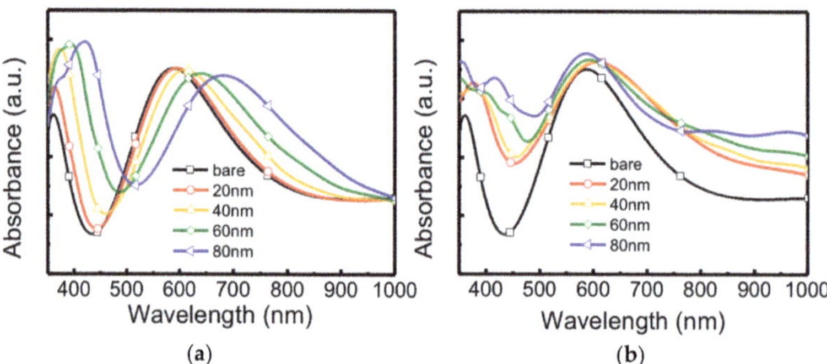

Figure 3. Simulated absorbance spectra of bare graphene and graphene with different size of Ag nanoparticles on (**a**) flat and few-layer graphene/SiO_2/Si and (**b**) rough and multi-layer graphene/SiO_2/Si.

To explore the effect of graphene morphology on the surface plasmon resonance, the electric field intensity distributions of 60 nm Ag nanoparticles on flat and rough graphene were both simulated. Figure 4a shows the electric field intensity distribution of Ag nanoparticles on flat and few layer graphene/SiO_2/Si at wavelengths of 410 nm, 510 nm and 650 nm, respectively. The electric field intensities around the Ag nanoparticles at wavelengths 410 nm and 510 nm show the highest and lowest values, respectively, which agrees with the simulated absorption spectra at the wavelengths of 410 nm, 510 nm and 650 nm. Figure 4b exhibits the electric field intensity distribution of Ag nanoparticles on rough and multilayer graphene/SiO_2/Si at wavelengths of 410 nm, 510 nm and 650 nm, respectively. The roughness of multilayer graphene is about 20 nm. Similarly, the electric field intensity around the Ag nanoparticles at the wavelength of 410 nm is stronger than those of other wavelengths. From these simulated results, it is revealed that the electrical field intensity distributions can be enhanced through the addition of Ag nanoparticles on both flat and rough graphene surfaces.

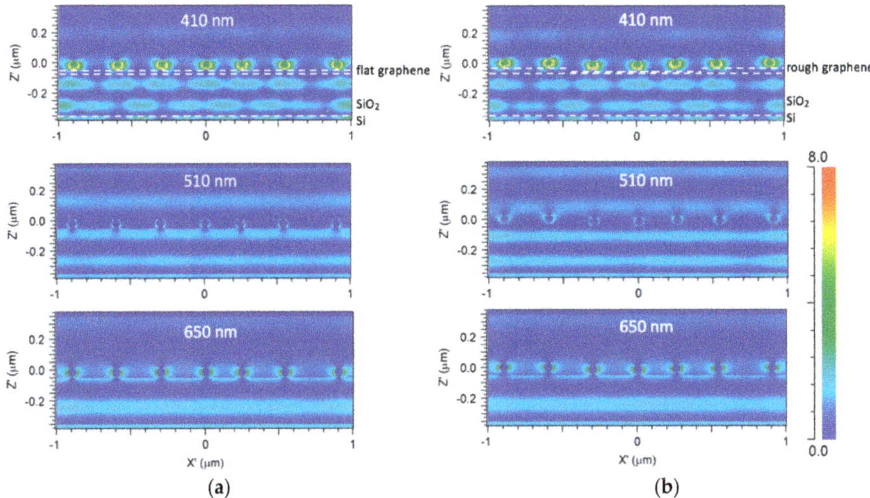

Figure 4. (**a**) Simulated electric field intensity distributions of 60 nm Ag nanoparticles/flat and few-layer graphene/SiO$_2$/Si at wavelengths of 410 nm, 510 nm and 650 nm, respectively; (**b**) simulated electric field intensity distributions of 60 nm Ag nanoparticles/rough and multilayer graphene/SiO$_2$/Si at wavelengths of 410 nm, 510 nm and 650 nm, respectively.

To investigate the difference in the surface plasmon resonance effect between the simulated and experimental results, the optical absorbance spectra of 60–80 nm Ag nanoparticle-rGO on SiO$_2$/Si with different spin rates, including 1500 rpm, 2000 rpm, 2500 rpm, 3000 rpm and 3500 rpm were measured through ultraviolet-visible-near infrared (UV-vis-NIR) spectroscopy in a wavelength range from 350 nm to 1000 nm, as shown in Figure 5. When several metal nanoparticles are close to other nanoparticles, the coupling effect between particles becomes very important. By controlling the spin rate of Ag nanoparticles/rGO solution on substrates, the density of Ag nanoparticles and the thickness of rGO can be tuned. Figure 5 displays that the absorbance curves of Ag nanoparticle-rGO with different spin rates is shifted about 10 nm which is beneficial to specific optoelectronic devices.

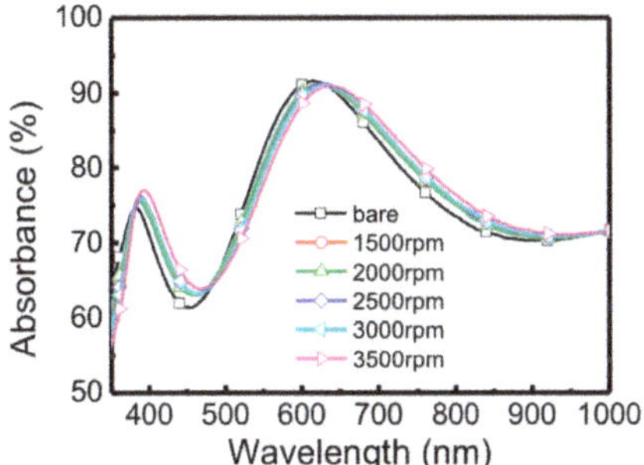

Figure 5. Absorbance spectra of rGO with a spin rate of 3000 rpm and 60–80 nm Ag nanoparticle-rGO with spin rates of 1500 nm, 2000 nm, 2500 nm, 3000 nm and 3500 nm, respectively.

To design photosensors operated in an outdoor environment, typical current-voltage (I-V) characteristics of rGO and Ag nanoparticle-rGO photosensors under white-light illumination by the light source of an AM 1.5 G solar simulator with power density of 1.03 kW/m^2 were measured. Figure 6a shows the structure of a rGO-based photosensor with electrical connections to detect the

I-V characteristics. The applied voltage of back gate (V_g) was set as 0 V for low-power applications and one of the Ag electrodes served as a drain and the other electrode as a source. The voltage between the drain and source (V_{ds}) was swept from −3 V to +3 V. When the rGO absorbs incident light, excitons (electron-hole pairs) are obtained at the Schottky-like metal-rGO interface. In addition, defects in the rGO film can help dissociate excitons into free carriers and some of them have sufficient energy to overcome the Schottky barrier. In Figure 6b, I-V curves of a rGO photosensor without Ag nanoparticles are displayed. The light and dark currents are distinguishable even at a low bias because the photosensor collects more photocurrent, owing to large absorbing area. I-V characteristics of a photosensor with 60 nm Ag nanoparticles are displayed in Figure 6b. A further improvement of the on/off ratio is achieved by the surface plasmon resonance effect caused by the Ag nanoparticles. At the bias of −1.5 V, the dark current, light current and on/off ratio of the photosensor without Ag nanoparticles was 0.241 µA, 0.281 µA and 1.166, respectively; the dark current, light current and on/off ratio of the photosensor with 60–80 nm Ag nanoparticless was 0.163 µA, 15.809 µA and 9.699, respectively. The responsivities of photosensors without and with Ag nanoparticles at bias voltage of −1.5 V were 0.52 µA/W and 202.53 µA/W, respectively. As incident light with proper frequency illuminates on the plasmonic structure, the oscillation of conducting electrons at the interface of materials arises and the electrical field intensity surrounding the plasmonic structure is increased, leading to the improved photocurrent. As a result, the on/off ratio of rGO with Ag nanoparticles can be enhanced, which is attributed to the plasmon-generated carriers from the AgNPs into the surrounding rGO and the plasmon-enhanced direct carrier excitation of rGO electrons. Additionally, the dark current of the photosensor with Ag nanoparticles shows a higher shunt resistance than that of the photosensor without Ag nanoparticles, resulting in the improved on/off ratio and a lower dark current.

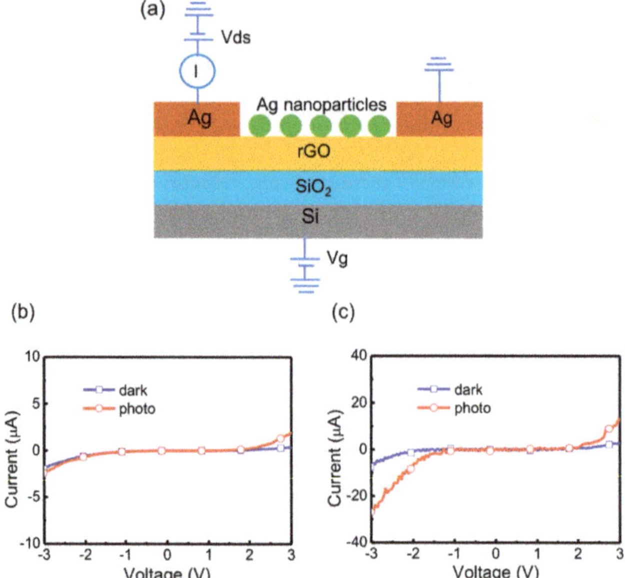

Figure 6. (**a**) Cross-sectional view of the rGO photosensor with electrical connections for current-voltage (I-V) measurement. The substrate is served as a back gate; one of the Ag electrodes sets as drain and the other is the source. Current-voltage curves of (**b**) a rGO photosensor without Ag nanoparticles and (**c**) a rGO photosensor with 60–80 nm Ag nanoparticles.

In Figure 7a, light currents of photosensors with and without Ag nanoparticles operated at the bias −2 V were measured under different levels of light density illumination. For the photosensor without Ag nanoparticles, more photons were absorbed and they generated more carriers as the intensity of light was increased. For the photosensor with Ag nanoparticles, the photocurrent increased with the increase of light density when the light power density was lower than 0.9 kW/m^2 and the light current was saturated when the light power density was higher than 0.9 kW/m^2. This phenomenon can be attributed to the reduced effective trap states and therefore the light current cannot further increase. Figure 7b,c show the time-dependent responses of the photodetectors without and with Ag nanoparticles, respectively. Both devices were operated at the bias voltage of 2 V and the photocurrents were measured by turning on and off the solar simulator which had a power density of 1.03 kW/m^2. These measured results demonstrate that the photosensors behave well under continuous cycling illumination. To deeply study the response time of photosensors, the rising time was measured as the photocurrent was increased from 10% to 90%. The rising time for the device without and with Ag nanoparticles was 1.0 s and 1.5 s, respectively.

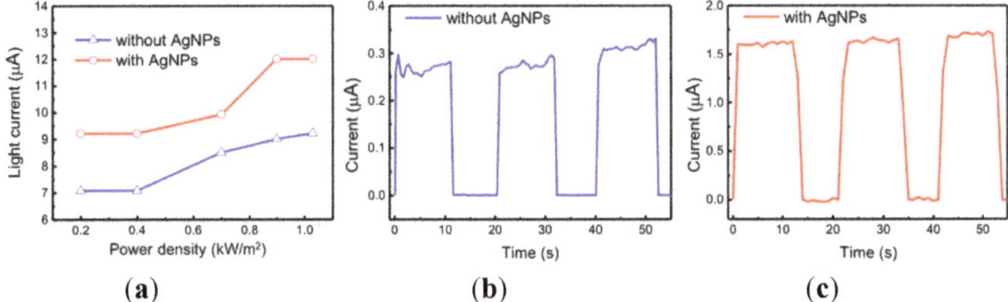

Figure 7. (**a**) Light current as a function of the input light power density under solar simulator illumination for photodetectors with and without Ag nanoparticles at the bias voltage of −2 V. Time-dependent current of photosensors (**b**) without and (**c**) with Ag nanoparticles recorded at the bias voltage 2 V under AM1.5G solar simulator illumination, turning on and off.

4. Conclusions

In summary, large-area rGO photosensors on SiO$_2$/Si substrates have been constructed via a cost-effective solution process. To further improve the performance of photosensors, hybrid materials composed of rGO and plasmonic Ag nanoparticles were formed and devices were operated under white-light illumination for outdoor environment applications. In this hybrid system, uniform and continuous rGO films acted as absorbing layer of photosensors and Ag nanoparticles promoted the photocurrent through the surface plasmon resonance effect. As a result, the Ag nanoparticle-rGO photosensors exhibited excellently improved photocurrent and 8-fold increase in the on/off ratio. Besides, the influence of the morphology of graphene on the surface plasmon resonance was discussed using FDTD simulated results. Our study paves a new strategy for the fabrication of an Ag nanoparticle-rGO hybrid system as well as for photosensors, which provides a possible way for the future development of the solution-process and large-area optoelectronic devices used in outdoor environments.

Author Contributions: Conceptualization, W.-C.T.; Data curation, X.-S.L.; Formal analysis, W.-C.T. and X.-S.L.; Resources, S.-L.C., M.-Y.L., W.-Y.U. and Y.-C.C.; Writing—original draft, W.-C.T. and X.-S.L.; Writing—review and editing, W.-C.T. and X.-S.L.

Funding: This research was funded by the Ministry of Science and Technology, Taiwan (grant number MOST 106-2221-E-033-053-MY2, MOST 107-2218-E-033-009-MY2).

Acknowledgments: W.-C.T. would like to thank Prof. Chiashain Chuang for many useful discussions.

Conflicts of Interest: The authors declare no conflict of interest.

References

1. Allen, M.J.; Tung, V.C.; Kaner, R.B. Honeycomb Carbon: A Review of Graphene. *Chem. Rev.* **2009**, *110*, 132–145. [CrossRef] [PubMed]
2. Neto, A.C.; Guinea, F.; Peres, N.M.; Novoselov, K.S.; Geim, A.K. The Electronic Properties of Graphene. *Rev. Mod. Phys.* **2009**, *81*, 109. [CrossRef]
3. Tu, W.-C.; Chen, J.-S.; Tsai, M.-L.; Wu, J.-R.; Li, G.-Y.; Lin, M.-Y.; Huang, C.-Y.; Uen, W.-Y. Improved Performance of All Solution-Processed Graphene Photodetectors Via Plasmonic Nanoparticles. *IEEE Photonics Technol. Lett.* **2017**, *29*, 423–426. [CrossRef]
4. Guo, X.; Wang, W.; Nan, H.; Yu, Y.; Jiang, J.; Zhao, W.; Li, J.; Zafar, Z.; Xiang, N.; Ni, Z. High-Performance Graphene Photodetector Using Interfacial Gating. *Optica* **2016**, *3*, 1066–1070. [CrossRef]
5. Guo, N.; Hu, W.; Jiang, T.; Gong, F.; Luo, W.; Qiu, W.; Wang, P.; Liu, L.; Wu, S.; Liao, L. High-Quality Infrared Imaging with Graphene Photodetectors at Room Temperature. *Nanoscale* **2016**, *8*, 16065–16072. [CrossRef] [PubMed]
6. Malko, D.; Neiss, C.; Vines, F.; Görling, A. Competition for Graphene: Graphynes with Direction-Dependent Dirac Cones. *Phys. Rev. Lett.* **2012**, *108*, 086804. [CrossRef] [PubMed]
7. Wu, Y.; Jenkins, K.A.; Valdes-Garcia, A.; Farmer, D.B.; Zhu, Y.; Bol, A.A.; Dimitrakopoulos, C.; Zhu, W.; Xia, F.; Avouris, P. State-of-the-Art Graphene High-Frequency Electronics. *Nano Lett.* **2012**, *12*, 3062–3067. [CrossRef] [PubMed]
8. Avouris, P. Graphene: Electronic and Photonic Properties and Devices. *Nano Lett.* **2010**, *10*, 4285–4294. [CrossRef]
9. Hill, E.W.; Vijayaragahvan, A.; Novoselov, K. Graphene Sensors. *IEEE Sens. J.* **2011**, *11*, 3161–3170. [CrossRef]
10. Goo Kang, C.; Kyung Lee, S.; Jin Yoo, T.; Park, W.; Jung, U.; Ahn, J.; Hun Lee, B. Highly Sensitive Wide Bandwidth Photodetectors Using Chemical Vapor Deposited Graphene. *Appl. Phys. Lett.* **2014**, *104*, 161902. [CrossRef]
11. Chitara, B.; Panchakarla, L.; Krupanidhi, S.; Rao, C. Infrared Photodetectors Based on Reduced Graphene Oxide and Graphene Nanoribbons. *Adv. Mater.* **2011**, *23*, 5419–5424. [CrossRef] [PubMed]
12. Yu, X.-X.; Yin, H.; Li, H.-X.; Zhao, H.; Li, C.; Zhu, M.-Q. A Novel High-Performance Self-Powered Uv-Vis-Nir Photodetector Based on a Cds Nanorod Array/Reduced Graphene Oxide Film Heterojunction and Its Piezo-Phototronic Regulation. *J. Mater. Chem. C* **2018**, *6*, 630–636. [CrossRef]
13. Gu, S.; Lou, Z.; Li, L.; Chen, Z.; Ma, X.; Shen, G. Fabrication of Flexible Reduced Graphene Oxide/Fe_2O_3 Hollow Nanospheres Based on-Chip Micro-Supercapacitors for Integrated Photodetecting Applications. *Nano Res.* **2016**, *9*, 424–434. [CrossRef]
14. Tu, W.-C.; Huang, C.Y.; Fang, C.W.; Lin, M.Y.; Lee, W.C.; Liu, X.S.; Uen, W.Y. Efficiency Enhancement of Pyramidal Si Solar Cells with Reduced Graphene Oxide Hybrid Electrodes. *J. Phys. D: Appl. Phys.* **2016**, *49*, 49LT02. [CrossRef]
15. Lee, W.-C.; Tsai, M.-L.; Chen, Y.-L.; Tu, W.-C. Fabrication and Analysis of Chemically-Derived Graphene/Pyramidal Si Heterojunction Solar Cells. *Sci. Rep.* **2017**, *7*, 46478. [CrossRef] [PubMed]
16. Chen, Z.; Li, X.; Wang, J.; Tao, L.; Long, M.; Liang, S.-J.; Ang, L.K.; Shu, C.; Tsang, H.K.; Xu, J.-B. Synergistic Effects of Plasmonics and Electron Trapping in Graphene Short-Wave Infrared Photodetectors with Ultrahigh Responsivity. *ACS Nano* **2017**, *11*, 430–437. [CrossRef] [PubMed]
17. Furchi, M.; Urich, A.; Pospischil, A.; Lilley, G.; Unterrainer, K.; Detz, H.; Klang, P.; Andrews, A.M.; Schrenk, W.; Strasser, G. Microcavity-Integrated Graphene Photodetector. *Nano Lett.* **2012**, *12*, 2773–2777. [CrossRef]
18. Ferreira, A.; Peres, N.; Ribeiro, R.; Stauber, T. Graphene-Based Photodetector with Two Cavities. *Phys. Rev. B* **2012**, *85*, 115438. [CrossRef]
19. Shiue, R.-J.; Gan, X.; Gao, Y.; Li, L.; Yao, X.; Szep, A.; Walker, D., Jr.; Hone, J.; Englund, D. Enhanced Photodetection in Graphene-Integrated Photonic Crystal Cavity. *Appl. Phys. Lett.* **2013**, *103*, 241109. [CrossRef]
20. Casalino, M.; Sassi, U.; Goykhman, I.; Eiden, A.; Lidorikis, E.; Milana, S.; De Fazio, D.; Tomarchio, F.; Iodice, M.; Coppola, G. Vertically Illuminated, Resonant Cavity Enhanced, Graphene–Silicon Schottky Photodetectors. *ACS Nano* **2017**, *11*, 10955–10963. [CrossRef]

21. Amendola, V.; Pilot, R.; Frasconi, M.; Marago, O.M.; Iati, M.A. Surface Plasmon Resonance in Gold Nanoparticles: A Review. *J. Phys. Condens. Matter* **2017**, *29*, 203002. [CrossRef] [PubMed]
22. Hutter, E.; Fendler, J.H. Exploitation of Localized Surface Plasmon Resonance. *Adv. Mater.* **2004**, *16*, 1685–1706. [CrossRef]
23. Ghosh, S.K.; Pal, T. Interparticle Coupling Effect on the Surface Plasmon Resonance of Gold Nanoparticles: From Theory to Applications. *Chem. Rev.* **2007**, *107*, 4797–4862. [CrossRef] [PubMed]
24. Agrawal, A.; Singh, A.; Yazdi, S.; Singh, A.; Ong, G.K.; Bustillo, K.; Johns, R.W.; Ringe, E.; Milliron, D.J. Resonant Coupling between Molecular Vibrations and Localized Surface Plasmon Resonance of Faceted Metal Oxide Nanocrystals. *Nano Lett.* **2017**, *17*, 2611–2620. [CrossRef] [PubMed]
25. Tu, W.-C.; Chang, Y.-T.; Yang, C.-H.; Yeh, D.-J.; Ho, C.-I.; Hsueh, C.-Y.; Lee, S.-C. Hydrogenated Amorphous Silicon Solar Cell on Glass Substrate Patterned by Hexagonal Nanocylinder Array. *Appl. Phys. Lett.* **2010**, *97*, 193109. [CrossRef]
26. Tu, W.-C.; Chang, Y.-T.; Wang, H.-P.; Yang, C.-H.; Huang, C.-T.; He, J.-H.; Lee, S.-C. Improved Light Scattering and Surface Plasmon Tuning in Amorphous Silicon Solar Cells by Double-Walled Carbon Nanotubes. *Sol. Energy Mater. Sol. Cells* **2012**, *101*, 200–203. [CrossRef]
27. Yan, B.; Li, X.; Bai, Z.; Lin, L.; Chen, G.; Song, X.; Xiong, D.; Li, D.; Sun, X. Superior Sodium Storage of Novel Vo 2 Nano-Microspheres Encapsulated into Crumpled Reduced Graphene Oxide. *J. Mater. Chem. A* **2017**, *5*, 4850–4860. [CrossRef]
28. Putri, L.K.; Ng, B.-J.; Ong, W.-J.; Lee, H.W.; Chang, W.S.; Chai, S.-P. Heteroatom Nitrogen-and Boron-Doping as a Facile Strategy to Improve Photocatalytic Activity of Standalone Reduced Graphene Oxide in Hydrogen Evolution. *ACS Appl. Mater. Interfaces* **2017**, *9*, 4558–4569. [CrossRef]
29. Wang, G.; Fu, Y.; Ma, X.; Pi, W.; Liu, D.; Wang, X. Reusable Reduced Graphene Oxide Based Double-Layer System Modified by Polyethylenimine for Solar Steam Generation. *Carbon* **2017**, *114*, 117–124. [CrossRef]
30. Kim, M.; Kang, P.; Leem, J.; Nam, S. A stretchable crumpled graphene photodetector with plasmonically enhanced photoresponsivity. *Nanoscale* **2017**, *9*, 4058–4065. [CrossRef]
31. Ferrari, A.C.; Meyer, J.; Scardaci, V.; Casiraghi, C.; Lazzeri, M.; Mauri, F.; Piscanec, S.; Jiang, D.; Novoselov, K.; Roth, S. Raman Spectrum of Graphene and Graphene Layers. *Phys. Rev. Lett.* **2006**, *97*, 187401. [CrossRef] [PubMed]
32. Malard, L.; Pimenta, M.; Dresselhaus, G.; Dresselhaus, M. Raman Spectroscopy in Graphene. *Phys. Rep.* **2009**, *473*, 51–87. [CrossRef]
33. Berciaud, S.; Cognet, L.; Tamarat, P.; Lounis, B. Observation of Intrinsic Size Effects in the Optical Response of Individual Gold Nanoparticles. *Nano Lett.* **2005**, *5*, 515–518. [CrossRef] [PubMed]
34. Catchpole, K.A.; Polman, A. Plasmonic Solar Cells. *Opt. Express* **2008**, *16*, 21793–21800. [CrossRef] [PubMed]

 *micromachines*

Article

Fabrication of Stable Carbon Nanotube Cold Cathode Electron Emitters with Post-Growth Electrical Aging

Jung Hyun Kim, Jung Su Kang and Kyu Chang Park *

Department of Information Display and Advanced Display Research Center, Kyung Hee University, Dongdaemun-gu, Seoul 02447, Korea; wjdgus791@khu.ac.kr (J.H.K.); kangjungsu@khu.ac.kr (J.S.K.)
* Correspondence: kyupark@khu.ac.kr; Tel.: +82-2-961-9447

Received: 8 November 2018; Accepted: 5 December 2018; Published: 7 December 2018

Abstract: We fabricated carbon nanotube (CNT) cold cathode emitters with enhanced and stable electron emission properties and long-time stability with electrical aging as a post-treatment. Our CNT field emitters showed improved electrical properties by electrical aging. We set the applied bias for effective electrical aging, with the bias voltage defined at the voltage where Joule heating appeared. At the initial stage of aging, the electron emission current started to increase and then was saturated within 3 h. We understood that 5 h aging time was enough at proper aging bias. If the aging bias is higher, excessive heating damages CNT emitters. With the electrical aging, we obtained improved electron emission current from 3 mA to 6 mA. The current of 6 mA was steadily driven for 9 h.

Keywords: carbon nanotube; field emitters; electrical aging; Joule heating; electron emission

1. Introduction

Carbon nanotubes (CNTs) have attracted much attention in vacuum nanoelectronic applications, especially in the field of cold cathode-based vacuum devices. CNTs have several excellent advantages as cold cathode emitters such as high aspect ratio, high thermal conductivity, and structural rigidity. Many post-treatment processes have been conducted to improve the electrical properties of CNTs [1–7]. Because of these various advantages, CNTs are suitable for electron sources. In this paper, we studied the enhancement of electrical properties of CNT emitters for their application as cold cathode emitters. [8,9] The stable and long-term lifetime of a CNT-based cold cathode electron beam (C-beam) is one of the most important requirements for commercialization. We developed the electrical aging techniques to fabricate high-performance CNT cold cathode emitters. It was confirmed that the thermal energy could change the crystallinity of CNTs through thermal annealing at 1000 °C [2,6]. In addition, the improved crystallinity could enhance the electrical characteristics, such as lower turn-on field and better stability of electron emission. In a high field emission current, Joule heating by an electron current through CNT emitters and non-uniform electron emission from each CNT emitter was a serious issue to be solved for lifetime enhancement. In previous studies, electrical aging removed the catalyst to induce tip opening and improved the emission current [3]. In this study, we observed the change of the electron emission characteristics by Joule heating. The crystallinity of CNT emitters is one of the most important factors to enhance electron emission. We used this thermal energy by resistive heating on CNT emitters for the structural modification of the CNT emitters. When Joule heating appears during field emission, the temperature of the CNT tip and body increase depending on the electron emission current. While growing CNTs, it is very difficult to control their structure effectively. As the aging time continues, at high currents, only a small amount of the CNT emitter from the non-uniform structure takes up the majority of the emission current, whereas surplus emission current incurs the destruction of these CNTs, which makes the current unstable. In addition, relatively weak CNT emitters are removed by resistive heating during electrical aging at high currents [10–20].

Thus, the aging allowed a relatively larger number, and more evenly distributed shorter multi-walled carbon nanotubes (MWCNTs) to become dominating emitters resulting in improving the field emission (FE) reproducibility [21,22]. According to numerous previous studies, other mechanisms involved in electrical aging are mainly the desorption of adsorbates caused by CNT heating [23,24], changes in geometric structure [25–31], and particle cleaning. These are the reasons why it is significant to carry out electrical aging to stabilize electron emission [32]. The heat generated by the resistance of the CNT emitter during electron emission could modify the structure of the CNT emitters. The Joule heat improved the crystallinity of the CNT emitters, resulting in stable electron emission. By adopting post-processing technology called electrical aging, we obtained a stable lifetime of 9 h at a 97 mA/cm^2 current density in a very simple and effective way.

2. Materials and Methods

2.1. Fabrication of Carbon Nanotube (CNT) Field Emitters

The CNT emitters were grown on a SiO$_2$ layer with an *n*-type silicon (Si) substrate, followed by a 30 nm Ni deposition with radio frequency magnetron sputtering. Then, through the photo-lithography process, the diameter of the CNT emitters were designed to be 3 μm and the distance between the emitters was set to 15 μm. Two hundred and twenty-five CNT emitters constituted one island, and a total of 128 (8 × 16) islands were patterned. They were grown by direct current plasma-enhanced chemical vapor deposition for 60 min. The growth temperature was 630 °C, the voltage of graphite susceptor (cathode) was −620 V, and the mesh voltage was + 320 V. Acetylene (C$_2$H$_2$) and ammonia (NH$_3$) gas were fed at rates of 20 sccm and 160 sccm, respectively, during the growth time [33,34]. After the growth, 1000 °C annealing was performed for 1 h in an argon gas atmosphere of 130 mTorr.

2.2. Electrical Aging Technology

Our electrical aging technique follows the schematic diagram of Figure 1. As shown in Figure 1a, the grown CNT emitters have non-uniform height and crystallinity. The uniformity of CNT height is difficult to control with deposition conditions. This structural issue causes a non-uniform field of individual CNT emitters. After various trials to set the aging condition, the starting point of electrical aging was defined as the bias where the Joule heat appeared in the overall emitter area. The Joule heat comes from current flow through the resistive nature of the CNT emitters. Figure 1b shows that we used the thermal energy by Joule heating as a structural modification of the emitters. Figure 1c shows that structurally unstable parts in the CNTs were burned out by the Joule heat and the height of the CNT emitters decreased, resulting in the homogenization of height. The heat also improved the crystallinity of the CNTs. As shown in Figure 1d, the homogenized structure and improved crystallinity formed a uniform field and enhance emission current, enabling stable electron emission characteristics at the higher current level.

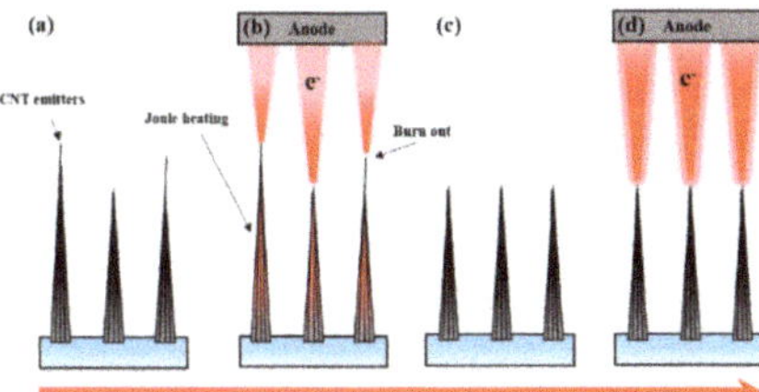

Figure 1. Schematic diagram of electrical aging technology. (**a**) Non-uniformity of height and crystallinity before aging; (**b**) electrical aging with Joule heating; (**c**) homogenization of height and crystallinity; (**d**) stable electron field emission by structural improvement.

This process was performed in a vacuum chamber. Figure 2 shows the schematic diagram and the digital photo image of the diode system used in the experiment. The base vacuum is 1×10^{-7} Torr and the distance between the cathode and the anode is 230 µm in the diode system. Figure 2c shows the Joule heating by high electron emission current from the observed CNT emitter. Joule heating images were observed with a digital single lens reflex camera (Canon, EOS-60D, Tokyo, Japan). The electron emission characteristics were measured with a direct current (DC) power supply (Spellman, SL 1200, Suffolk, NY, USA) and a multimeter (Agilent, 3441A, Santa Clara, CA, USA). We analyzed the crystallinity and morphology of the CNT emitters using Raman spectroscopy (Horiba Jobin-Yvon, LabRam ARAMIS, Kyoto, Japan) with a 514 nm excitation laser wavelength, scanning electron microscope (Hitachi, S-4700, Tokyo, Japan). All scanning electron microscope (SEM) images were taken with an accelerating voltage of 10 kV and a 45° tilt.

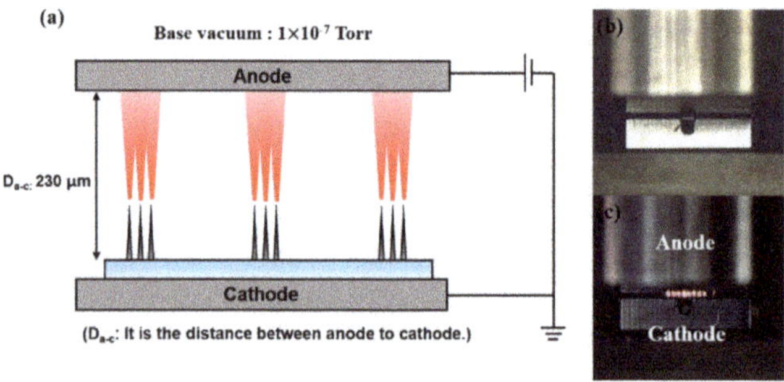

Figure 2. Diode system for measuring current-voltage (I-V) characteristics. (**a**) Schematic diagram of diode system; (**b**) digital single lens reflex camera image of the diode system; (**c**) observed Joule heating when the electron emission current is higher.

3. Results and Discussion

SEM images of the grown CNT field emitters are shown in Figure 3. As shown in Figure 3a, eight islands are patterned in 16 lines. Figure 3b shows one patterned island SEM image. One island has 15×15 array emitters. The CNT emitter is grown vertically on the Si substrate. Figure 3c shows that the single CNT emitter has a diameter of 3 µm and an average height of 40 µm, and unlike the non-patterned emitters in Figure 3d, the patterned CNT emitters have a cone shape and only one CNT at the tip.

Field emission characteristics over time during aging were associated with Joule heating. We categorized the emission currents in relation to Joule heating. The categories we classified are shown in Figure 4. At an emission current of 3 mA, overall Joule heating occurred and this bias was set as the starting point of electrical aging. The eight divided lights clearly show that the Joule heating appears in the emitter area. During 2 h of electrical aging, the emission current increased from 3 mA to 5.5 mA, and after 3 h the Joule heating disappeared completely. After aging was completed, stable electron emission was maintained for 5 h without visible light. The Joule heating began to reappear when the emission current increased, and the CNT emitter was damaged by the excessive Joule heating with increased electron emission current.

SEM images of the CNT emitter damaged by excessive Joule heating are shown in Figure 5. Figure 5a shows a number of broken emitters. Figure 5b shows the tips of one emitter that has been damaged. The bias set position of the electrical aging is very important. Excessive Joule heating causes damage to CNT emitters. However, at an aging current lower than the Joule heat light appeared, the aging effect is insignificant. In order to carry out electrical aging most effectively, we set the bias and current to those at which the Joule heating light appears in the entire emitter region as electrical aging conditions.

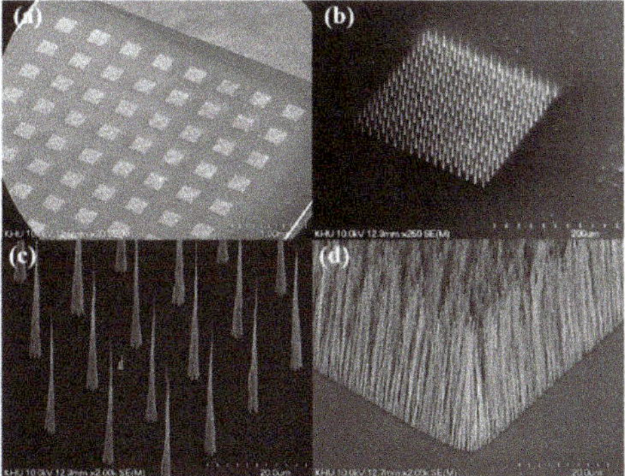

Figure 3. Scanning electron microscope (SEM) images of carbon nanotube (CNT) emitters grown by triode direct-current plasma-enhanced chemical vapor deposition (dc-PECVD). (**a**) Patterned CNT islands image; (**b**) magnified image of one island; (**c**) vertically aligned CNT emitters with cone shape in the island; (**d**) non-patterned CNT.

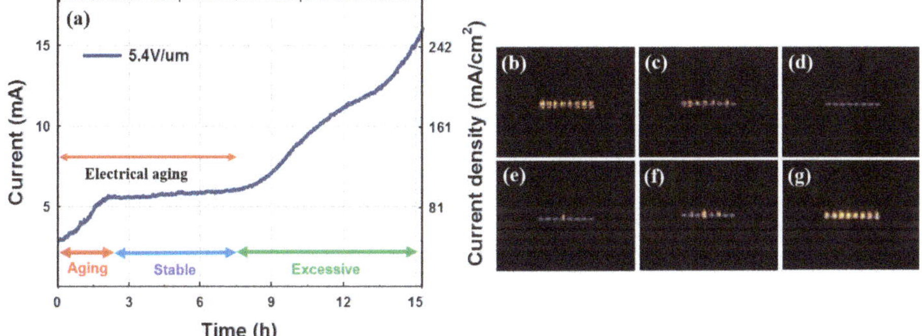

Figure 4. Definition of electrical aging. (**a**) Variation of electron emission current at a higher emission regime with time; (**b**) the section at which Joule heating occurs sufficiently; (**c**) the section at which Joule heating is reduced; (**d**) the section at which Joule heating disappears completely; (**e**) the section at which Joule heating reappears; (**f**) the section at which Joule heating increases; (**g**) excessive Joule heating that can cause damage to the CNT emitters.

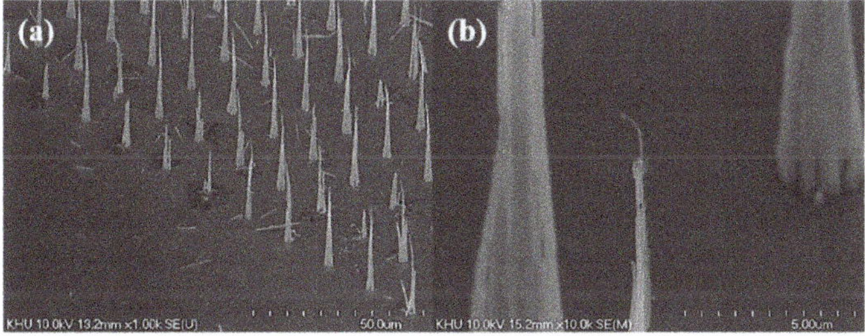

Figure 5. (**a**) Damaged CNT emitters by excessive Joule heating; (**b**) tip of a structurally modified emitter.

We were able to confirm the decrease in the turn-on field and the increase in the emission current density after the post-growth treatment. Current-voltage (I-V) characteristics for each treatment are shown in Figure 6 and summarized in Table 1. The electron emission characteristics could improve after annealing at 1000 °C. When electrical aging was completed, the current density increased to 97 mA/cm^2 at the same electric field, which is 20 times higher than the current density of as-grown CNTs. It is expected that the aging effect will cause the structural modification of the CNT emitter. So, we compared the structural properties and crystallinity of each treatment by SEM and Raman analysis.

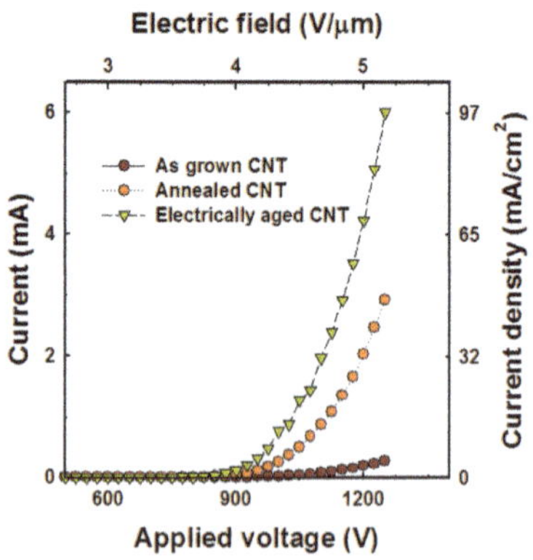

Figure 6. Field emission properties with treatment condition.

Table 1. Comparison of electrical properties of carbon nanotube (CNT) emitter with treatment condition. DC–direct current.

Treatment	Driving	Turn on Field (V/µm)	Current Density at 5.4 V/µm (mA/cm^2)
As-grown		3.6	4.5
Annealed	DC	3.3	48
Electrically aged		3	97

Figure 7 shows an SEM image to compare the structural features of the CNT emitters before and after electrical aging. In Figure 7a,b, the height reduction in the weak area of the CNT emitters were confirmed after electrical aging. A weak and unstable area of the CNT emitter burns out after aging. The uniform height contributes to a uniform electric field formation on CNT emitters. Uniform field formation allows individual nanotubes to emit a uniform amount of current and greatly contributes to improved lifetime.

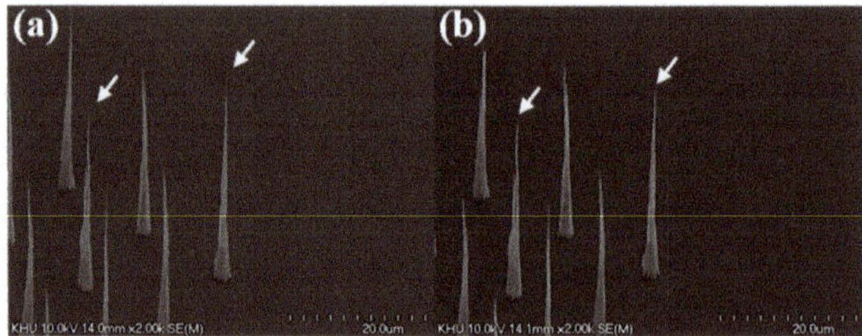

Figure 7. SEM image before and after aging. (**a**) CNT emitters before aging; (**b**) decreased CNT height (white arrow) after aging.

Figure 8 shows the Raman spectra of patterned and non-patterned as-grown CNT emitters. A D peak at 1350 cm^{-1}, and a G peak at around 1580 cm^{-1} appeared. As shown in Figure 8, the Raman spectrum revealed that non-patterned CNT emitters analyzed the Raman signals with several numbers of CNTs, but in the case of patterned CNT emitters, only one CNT emitter analyzed the Raman signals. The resolution of the Raman spectroscopy used was ~1 µm. The Raman signal of non-patterned CNTs showed an intensity ration of D and G band (I_D/I_G) ratio of 0.9 and a clear Raman signal because so many numbers of CNTs were measured. However, the patterned CNTs showed a large distribution of the I_D/I_G ratio. The variation of the G/D ratio is related to the crystallinity of each CNT. To compare the crystalline uniformity of CNT emitters, four CNT measurements were taken and compared to non-patterned CNT emitters with aging via Raman analysis.

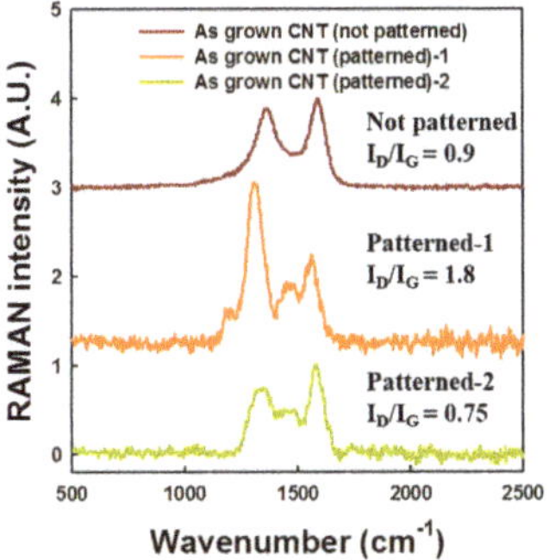

Figure 8. Normalized Raman spectra of patterned and non-patterned as-grown CNT samples. Raman spectra acquired with a 514 nm excitation laser.

Figure 9 shows the Raman spectra of patterned CNT emitters for each treatment. As shown in Figure 9a, four sample emitters were analyzed with Raman measurement. Two out of four samples show no Raman signal. It was found that as-grown samples had non-uniform crystallinity. Figure 9b shows that two Raman signals of 1000 °C annealed CNT were measured, similar to the as-grown patterned CNT emitters. However, the samples show a decrease in the variation of the I_D/I_G ratio. We confirmed the improvement of crystallinity by a thermal energy of 1000 °C annealing. However, the uniformity of the crystalline structure of the CNT emitters was insufficient by 1000 °C annealing. In Figure 9c, the Raman signal appeared for all four samples, and the I_D/I_G ratios were similar in all four signals. From the Raman analysis, it can be seen that the electrical aging contributes to the crystallinity enhancement of the CNT emitter and that a more uniform crystallinity appears with the electrical aging. This uniform structural characteristic would enable a long-term stable field emission.

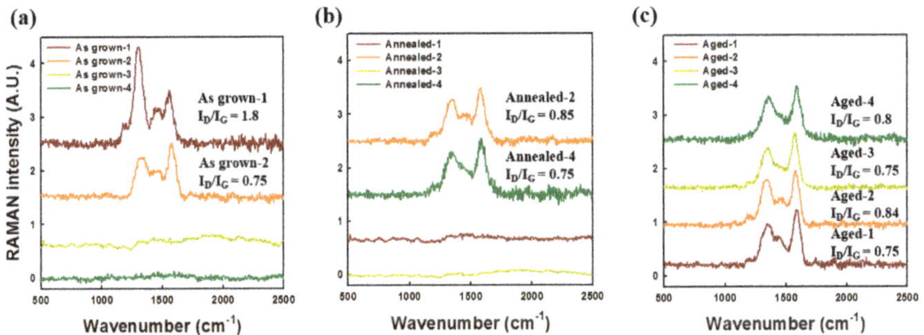

Figure 9. Normalized Raman spectra of patterned (**a**) as-grown, (**b**) annealed, (**c**) and electrically aged CNTs.

Figure 10 shows a comparison of the long-term stability of CNT emitters without and with electrical aging. The emission current was measured in constant voltage mode at 30-s steps. CNT emitters without electrical aging showed the degradation of the electron emission current from 6 mA to 5 mA after 9 h. However, for CNT emitters with electrical aging, the electron emission current did not degrade and a very uniform emission current was achieved. The fluctuation of the emission current was reduced by electrical aging to 0.3% ($\triangle$I = standard deviation/average current $\times$ 100). As a result, although the crystalline property of CNTs improved after annealing at 1000 °C, the crystallinity of each sample was not uniform and was insufficient for uniform long-time operation. However, the electrical aging induced uniform structural properties and crystallinity. The uniform and strong structure of electrically aged CNT emitter enabled stable operation even at high current emissions.

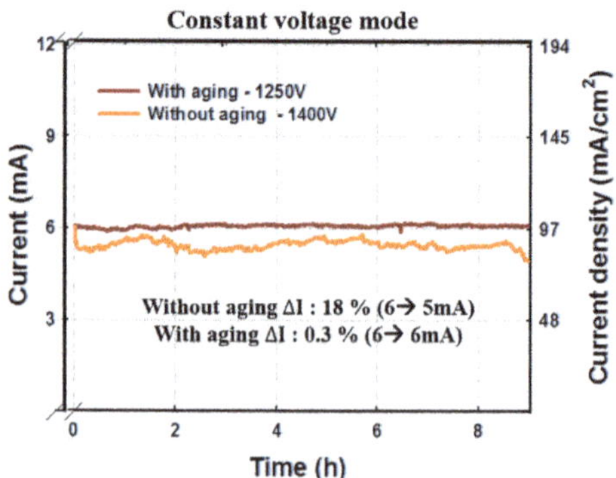

Figure 10. Comparison of stability with and without electric aging. Electrically aged emitters were shown to be more stable.

4. Conclusions

In this study, Joule heating generated at a high current emission on resistive CNT emitters was used to induce the self-annealing of the CNT emitters. With the thermal energy caused by Joule heating applied to the CNT emitters, the structure of individual CNT emitters changed and resulted in uniform emitters. In addition, owing to the elevated temperature during electron emission, we could enhance the crystallinity of the CNT emitters. With the electrical aging technique, structurally and electrically improved CNT emitters were made. We were able to obtain the stable long-term operation of CNT emitters with a current density of 97 mA/cm^2 for 9 h after electrical aging. If the Joule heating was excessive, the CNT emitters were damaged. To set the electrical aging current, just before the Joule heating light appeared was determined to be the best bias position of the current level. This electrical aging technique represents a simple and a better post-growth treatment process for the achievement of high-performance cold cathode CNT emitters that drive steadily, even at high current emission levels.

Author Contributions: K.C.P. conceived and designed the experiments and edited the paper; J.H.K. performed the experiments and wrote the paper; J.S.K. contributed to the fabrication of the CNT emitters.

Acknowledgments: This work was supported by the Radiation Technology R&D program through the National Research Foundation of Korea, funded by the Ministry of Science and ICT (NRF- 2017M2A2A4A01020658) and the BK21 Plus Program (Future-oriented innovative brain raising type, 21A20130000018) funded by the Ministry of Education (MOE, Korea) and the National Research Foundation of Korea (NRF).

Conflicts of Interest: The authors declare no conflict of interest. The funders had no role in the design of the study; in the collection, analyses, or interpretation of data; in the writing of the manuscript, or in the decision to publish the results.

References

1.   Kim, K.S.; Ryu, J.H.; Lee, C.S.; Jang, J.; Park, K.C. Enhanced and stable electron emission of carbon nanotube emitter arrays by post-growth hydrofluoric acid treatment. *J. Vac. Sci. Technol. B* **2007**, *20*, 120–124. [CrossRef]

2. Jung, S.I.; Jo, S.H.; Moon, H.S.; Kim, J.M.; Zang, D.-S.; Lee, C.J. Improved Crystallinity of Double-Walled Carbon Nanotubes after a High-Temperature Thermal Annealing and Their Enhanced Field Emission Properties. *J. Phys. Chem. C* **2007**, *111*, 4175–4179. [CrossRef]

3. Ryu, J.H.; Kim, K.S.; Lee, C.S.; Jang, J.; Park, K.C. Effect of electrical aging on field emission from carbon nanotube field emitter arrays. *J. Vac. Sci. Technol. B* **2008**, *26*, 856. [CrossRef]

4. Liu, C.; Kim, K.S.; Baek, J.; Cho, Y.; Han, S.; Kim, S.-W.; Min, N.; Choi, Y.; Kim, J.; Lee, C.J. Improved field emission properties of double-walled carbon nanotubes decorated with Ru nanoparticles. *Carbon* **2009**, *47*, 1158–1164. [CrossRef]

5. Wang, W.-P.; Wen, H.-C.; Jian, S.-R.; Cheng, H.-Z.; Jang, J.S.-C.; Juang, J.-Y.; Cheng, H.-C.; Chou, C.-P. Field emission characteristics of carbon nanotubes post-treated with high-density Ar plasma. *Appl. Surf. Sci.* **2010**, *256*, 2184–2188. [CrossRef]

6. Sun, Y.; Shin, D.H.; Yun, K.N.; Hwang, Y.M.; Song, Y.; Leti, G.; Jeon, S.; Kim, J.; Saito, Y.; Lee, C.J. Field emission behavior of carbon nanotube field emitters after high temperature thermal annealing. *AIP Adv.* **2014**, *4*, 077110. [CrossRef]

7. Lee, S.W.; Kang, J.S.; Lee, H.R.; Park, S.Y.; Jang, J.; Park, K.C. Enhanced and stable electron emission of carbon nanotube emitters with graphitization. *Vacuum* **2015**, *121*, 212–216. [CrossRef]

8. Kang, J.S.; Hong, J.H.; Chung, M.T.; Park, K.C. Highly stable carbon nanotube cathode for electron beam application. *J. Vac. Sci. Technol. B* **2016**, *34*, 02G104. [CrossRef]

9. Kang, J.S.; Park, K.C. Electron extraction electrode for a high-performance electron beam from carbon nanotube cold cathodes. *J. Vac. Sci. Technol. B* **2017**, *35*, 02C109. [CrossRef]

10. Park, J.H.; Jeon, S.Y.; Alegaonkar, P.S.; Yoo, J.B. Improvement of emission reliability of carbon nanotube emitters by electrical conditioning. *Thin Solid Films* **2008**, *516*, 3618–3621. [CrossRef]

11. Saito, Y.; Seko, K.; Kinoshita, J. Dynamic behavior of carbon nanotube field emitters observed by in situ transmission electron microscopy. *Diam. Relat. Mater.* **2005**, *14*, 1843–1847. [CrossRef]

12. Sveningsson, M.; Hansen, K.; Svensson, K.; Olsson, E.; Campbell, E.E.B. Quantifying temperature-enhanced electron field emission from individual carbon nanotubes. *Phys. Rev. B* **2005**, *72*. [CrossRef]

13. Machida, H.; Honda, S.; Fujii, S.; Himuro, K.; Kawai, H.; Ishida, K.; Oura, K.; Katayama, M. Effect of Electrical Aging on Field Electron Emission from Screen-Printed Carbon Nanotube Film. *Jpn. J. Appl. Phys.* **2007**, *46*, 867–869. [CrossRef]

14. Passacantando, M.; Bussolotti, F.; Santucci, S.; Di Bartolomeo, A.; Giubileo, F.; Iemmo, L.; Cucolo, A.M. Field emission from a selected multiwall carbon nanotube. *Nanotechnology* **2008**, *19*, 395701. [CrossRef] [PubMed]

15. Di, Y.; Xiao, M.; Zhang, X.; Wang, Q.; Li, C.; Lei, W.; Cui, Y. Large and stable emission current from synthesized carbon nanotube/fiber network. *J. Appl. Phys.* **2014**, *115*, 064305. [CrossRef]

16. Nilsson, L.; Groening, O.; Emmenegger, C.; Kuettel, O.; Schaller, E.; Schlapbach, L.; Kind, H.; Bonard, J.-M.; Kern, K. Scanning field emission from patterned carbon nanotube films. *Appl. Phys. Lett.* **2000**, *76*, 2071–2073. [CrossRef]

17. Liang, X.H.; Deng, S.Z.; Xu, N.S.; Chen, J.; Huang, N.Y.; She, J.C. On achieving better uniform carbon nanotube field emission by electrical treatment and the underlying mechanism. *Appl. Phys. Lett.* **2006**, *88*, 111501. [CrossRef]

18. Zou, Q.; Wang, M.Z.; Li, Y.G.; Zou, L.H. Ageing effect on the field emission reproducibility of multiwalled carbon nanotubes. *J. Exp. Nanosci.* **2011**, *6*, 270–280. [CrossRef]

19. Di Bartolomeo, A.; Scarfato, A.; Giubileo, F.; Bobba, F.; Biasiucci, M.; Cucolo, A.M.; Santucci, S.; Passacantando, M. A local field emission study of partially aligned carbon-nanotubes by atomic force microscope probe. *Carbon* **2007**, *45*, 2957–2971. [CrossRef]

20. Giubileo, F.; Di Bartolomeo, A.; Iemmo, L.; Luongo, G.; Urban, F. Field Emission from Carbon Nanostructures. *Appl. Sci.* **2018**, *8*, 526. [CrossRef]

21. Giubileo, F.; Di Bartolomeo, A.; Scarfato, A.; Iemmo, L.; Bobba, F.; Passacantando, M.; Santucci, S.; Cucolo, A.M. Local probing of the field emission stability of vertically aligned multi-walled carbon nanotubes. *Carbon* **2009**, *47*, 1074–1080. [CrossRef]

22. Lee, J.; Jung, Y.; Song, J.; Kim, J.S.; Lee, G.-W.; Jeong, H.J.; Jeong, Y. High-performance field emission from a carbon nanotube carpet. *Carbon* **2012**, *50*, 3889–3896. [CrossRef]

23. Dean, K.A.; Chalamala, B.R. Current saturation mechanisms in carbon nanotube field emitters. *Appl. Phys. Lett.* **2000**, *76*, 375–377. [CrossRef]

24. Hata, K.; Takakura, A.; Saito, Y. Field emission microscopy of adsorption and desorption of residual gas molecules on a carbon nanotube tip. *Surf. Sci.* **2001**, *490*, 296–300. [CrossRef]

25. Park, J.H.; Moon, J.S.; Nam, J.W.; Yoo, J.B.; Park, C.Y.; Kim, J.M.; Park, J.H.; Lee, C.J.; Choe, D.H. Field emission properties and stability of thermally treated photosensitive carbon nanotube paste with different inorganic binders. *Diamond Relat. Mater.* **2005**, *14*, 2113–2117. [CrossRef]

26. Cho, Y.; Song, H.; Choi, G.; Kim, D. A simple method to fabricate high-performance carbon nanotube field emitters. *J. Electroceram.* **2006**, *17*, 945–949. [CrossRef]

27. Guo, P.S.; Chen, T.; Chen, Y.; Zhang, Z.J.; Feng, T.; Wang, L.L.; Lin, L.F.; Sun, Z.; Zheng, Z.H. Fabrication of field emission display prototype utilizing printed carbon nanotubes/nanofibers emitters. *Solid-State Electron.* **2008**, *52*, 877–881. [CrossRef]

28. Wang, L.; Chen, Y.; Chen, T.; Que, W.; Sun, Z. Optimization of field emission properties of carbon nanotubes cathodes by electrophoretic deposition. *Mater. Lett.* **2007**, *61*, 1265–1269. [CrossRef]

29. Choi, Y.C.; Lee, N. Influence of length distributions of carbon nanotubes on their field emission uniformity in the paste-printed dot arrays. *Diam. Relat. Mater.* **2008**, *17*, 270–275. [CrossRef]

30. Park, J.H.; Alegaonkar, P.S.; Kim, D.Y.; Yoo, J.B. Electrical ageing of carbon nanotube composite cathode layers. *Diam. Relat. Mater.* **2008**, *17*, 980–985. [CrossRef]

31. Giubileo, F.; Di Bartolomeo, A.; Sarno, M.; Altavilla, C.; Santandrea, S.; Ciambelli, P.; Cucolo, A.M. Field emission properties of as-grown multiwalled carbon nanotube films. *Carbon* **2012**, *50*, 163–169. [CrossRef]

32. Dong, C.; Gupta, M.C. Influences of the surface reactions on the field emission from multiwall carbon nanotubes. *Appl. Phys. Lett.* **2003**, *83*, 159–161. [CrossRef]

33. Park, K.C.; Ryu, J.H.; Kim, K.S.; Yu, Y.Y.; Jang, J. Growth of carbon nanotubes with resist assisted patterning process. *J. Vac. Sci. Technol. B* **2007**, *25*, 1261–1264. [CrossRef]

34. Ryu, J.H.; Bae, N.Y.; Oh, H.M.; Zhou, O.; Jang, J.; Park, K.C. Stabilized electron emission from silicon coated carbon nanotubes for a high-performance electron source. *J. Vac. Sci. Technol. B* **2011**, *29*, 02B120. [CrossRef]

Article

Selective Detection of NO and NO$_2$ with CNTs-Based Ionization Sensor Array

Hui Song [1,*], **Kun Li** [2,*] **and Chang Wang** [3]

[1]   School of Software Engineering, Qufu Normal University, Qufu 273165, China
[2]   College of Engineering, Qufu Normal University, Rizhao 276825, China
[3]   Department of Microelectronics, School of Electronics and Information Engineering, Xi'an Jiaotong University, Xi'an 710049, China; wangc369@126.com
*   Correspondence: songh_0315@mail.qfnu.edu.cn (H.S.); lk_20160606@mail.qfnu.edu.cn (K.L.)

Received: 11 June 2018; Accepted: 6 July 2018; Published: 16 July 2018

Abstract: The accurate detection of NO$_x$ is an important issue, because nitrogen oxides are not only environmental pollutants, but also harm to human health. An array composed of two carbon nanotubes (CNTs)-based ionization sensors with different separations is proposed for NO and NO$_2$ selective detection. The experimental results indicate that the CNTs-based ionization sensor has an intrinsic, monotonically decreasing response to NO or NO$_2$. The sensor with 80 μm separations and 100 μm separations exhibited the highest sensitivity of −0.11 nA/ppm to 300 ppm NO and −0.49 nA /ppm to 70 ppm NO$_2$, respectively. Although the effect of the NO$_2$ concentration on the NO response is much stronger than that of NO on NO$_2$, the array of these two sensors still exhibits the ability to simultaneously detect the concentrations of NO and NO$_2$ in a gas mixture without component separation.

Keywords: ionization sensor array; NO$_x$; carbon nanotube (CNT); selectivity; non-self-sustaining discharge

---

1. Introduction

The oxides of nitrogen, collectively termed as NO$_x$ and mainly regarding NO + NO$_2$, are major atmospheric pollutants. Anthropogenic emissions of NO$_x$ are mostly from coal combustion and petroleum refining. Nitrogen oxides combine with water to form acid rain and with other pollutants to produce photochemical smog, which can greatly damage the environment. The accurate detection of NO$_x$ is an important issue, because the nitrogen oxides are not only environmental pollutants, but also harm to human health. Many kinds of NO$_x$ gas sensors have been investigated in recent years. In particular, nanostructured semiconducting metal oxides sensors [1–5] and carbon nanotubes (CNTs)-based sensors [6–12] have attracted much more concern. Due to the adsorption or desorption of gas molecules on the surface of metal oxides, the electrical conductivity of these sensors varies greatly with the changes in NO$_x$ concentration. However, the metal oxides gas sensors have the limitations of requiring high operating temperatures, and in addition, the response of those sensors largely depends on the morphology of the metal oxides' nanostructures [3].

A CNT, which has a large surface area to volume ratio and high electrical conductivity, can be an ideal candidate to operate at room temperature with high sensitivity [13]. One important use for CNTs is as a resistance sensor based on the change of electrical conductivity caused by gas adsorption and desorption. Almost all previously reported CNTs-based NO$_x$ gas sensors are based on this mechanism [6,14–17]. Another important application of CNTs is in a carbon nanotube field effect transistor (CNFET), which has characteristics that are affected by gas-induced changes of contact properties and by the doping level of the nanotube [18,19]. However, higher cross-sensitivity and lower recovery times at room temperature are the major problems of these sensors [8,13,15]. Thus, the improvement of selectivity and response times is the primary concern of researchers. With respect to selectivity, Yao et al. introduced moisture to selectively detect NO_2 and SO_2 [6]. The resistance of the gas sensor decreased independent of the moisture levels in the case of NO_2 and increased with SO_2 exposure at a high humidity level. But whether this approach can selectively detect any other gases has not been reported. With respect to response times, Dojin Kim et al. proposed a high-performance conduction model for a nanostructured sensor, and they revealed that the maximum response occurs at a diameter near the complete depletion condition [12]. The model has been experimentally proven for a metal oxides nanostructure, but whether it is valid for carbon nanotubes requires future confirmation.

In this paper, we propose a triple-electrode ionization sensor with a CNT film cathode (Figure 1a) for NO or NO_2 detection and an array consisting of two sensors with different electrode separations for simultaneously detecting a gas mixture of NO and NO_2. The discharge of the CNT-based triple-electrode ionization sensor has been demonstrated to be non-self-sustaining when the appropriate extracting voltage U_e is applied [20,21]. We measure the NO_X concentration with a discharge current, which is determined by the gas concentration at a given electrode separation and under ambient conditions based on the Townsend discharge theory [22]. We also demonstrated that the array, which is composed of those CNT-based ionization sensors with different electrode separations (insert of Figure 2), has the ability to directly discriminate between any different type of gas without component separation. Here, we used the array of two sensors with 80 μm and 100 μm separations to simultaneously detect and distinguish NO and NO_2. Gas sensing experiments were carried out at normal atmosphere and a relative humidity in the range of 21.1–21.9% RH. We also discuss the cross-sensitivity between the two gases.

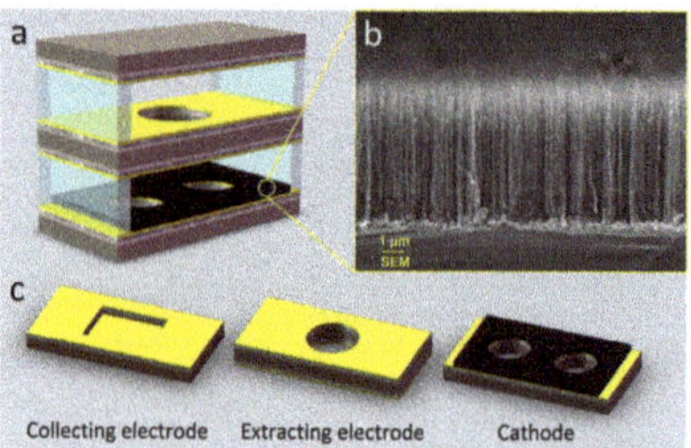

Figure 1. (**a**) Schematic diagram of the test system; (**b**) SEM image of the carbon nanotube (CNT) film; and (**c**) Schematic of the triple-electrode sensor structure.

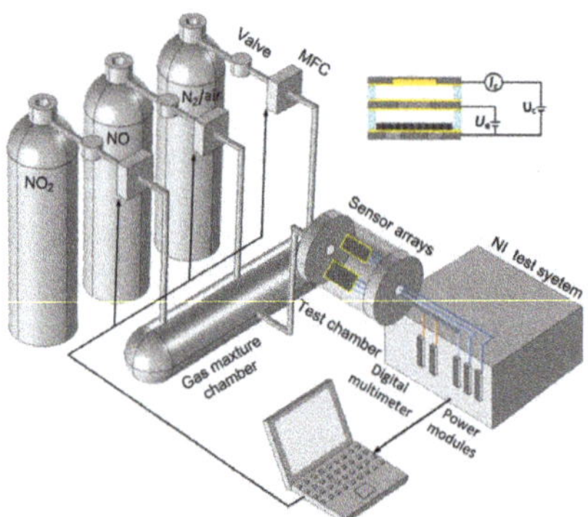

Figure 2. Schematic diagram of gas distribution and experimental testing system.

2. Experimental Detail

2.1. Synthesis of CNT Films and Three Electrodes

For the growth of CNT films on the cathode, a piece of substrate coated with iron phthalocyanine ($FeC_{32}N_8H_{16}$) is set in a tube furnace, which was preheated to 850 °C/890 °C with Ar/H_2 flow, sequentially. Then, a quartz boat loaded with the annealed cathode was pushed into furnace, and the growth process immediately started. At 920 °C, iron phthalocyanine decomposed for about 13 min until the dark green smoke disappeared. The substrates were naturally cooled to room temperature under Ar exposure in the furnace. Then, vertically aligned carbon nanotube (CNT) film was grown by thermal chemical vapor deposition (TCVD) method. As can be seen from the transmission electron microscopy (FEI, Quanta 250 FEG, Hillsboro, OR, USA) images of the samples, as shown in Figure 1b, the CNT film is homogeneous and dense, and the length of the suspended CNTs is ~5–6 μm.

Three silicon slices in sizes of 27 mm × 8 mm × 450 μm were processed through a mask, photolithography, dry etching, and cleaning, resulting in a cathode with two cooling holes of 4 mm diameter, the extracting electrode with one 3 mm radial round hole, and the collecting electrode with a square blind rectangle of 8×6 mm^2 in area and 200 μm in depth. Ti/Ni/Au films (50 nm/125 nm/400 nm) were sputtered in sequence on both sides of the extracting electrode, the inner side of the cathode, and the collecting electrode. Here, Ni film was used for preventing the inter-diffusion between the Ti and Au films. The three electrodes were rapidly annealed at 450 °C for about 50 s to enhance the bond strength between the substrate and the Ti/Ni/Au films (Figure 1c). Then, vertically aligned CNT film was transferred to the inner side of cathode with the wetting transfer method [23].

2.2. Fabrication of Sensors and Sensor Array

Polyester film, which is a good electrical insulating material due to its good insulation properties, exceptional thermal stability, and high impedance, was used to separate the cathode, the extracting electrode, and the collecting electrode. Polyester films with different thickness were cut to separate the electrodes. The gap distances between the cathode and the extracting electrode and between the extracting and the collecting electrodes were fixed as well (Figure 1a). Three electrodes with 100 μm separations were connected with three golden wires as the three pins of the sensor; thus, the NO_2 sensor was fabricated in the same way as the NO sensor device but with 80 μm separations. Hence, the sensor array composed of these two sensors with 80 μm and 100 μm separations was finished.

2.3. Experimental Testing System

The stable voltages applied to the sensor array were supplied by power modules (NI PXI-4132, NI, Austin, TX, USA). When the concentration of NO (NO_2) varied from 0 ppm to 1120 ppm (0–220 ppm), the collecting current was recorded by a precise digital multimeter (NI PXI-4071) controlled by a computer via the MXI interface of the test system (Figure 2).

The gas mixture was firstly prepared by mixing with dry air in the mixture chamber and then flowing it into the test chamber until atmospheric pressure was reached. The concentration of NO and NO_2 were controlled by three mass flow controllers (MFC, Line Tech M3030 V with 1% accuracy, Line Tech, Inchon, Korea) with full scales of 100 mL/min for NO, 50 mL/min for NO_2, and 1000 mL/min for dry air, respectively. Before each measurement, the test chamber was heated to 60 °C and pumped to a vacuum of ~5 kPa into which the well-mixed gas could flow. The gas sensing experiments were carried out at atmospheric pressure and a relative humidity of $20 \pm 2\%$ RH.

3. Results and Discussion

Because the applied extracting voltage U_e was lower than the breakdown voltage, our triple-electrode ionization gas sensor worked in a non-self-sustaining discharge state to reduce the damage to the CNTs caused by the electric breakdown. Thus, the non-self-sustaining discharge current could be used to measure the gas concentration with good stability, which had been demonstrated using the gold nanowires anode [24,25] and the CNTs-based cathode [26,27].

The electrons, emitted from the CNT-based cathode, collided with the gas molecules to produce more electrons and positive ions near the vicinity of the cathode, resulting in the occurrence of field ionization. A large number of positive ions migrated to the collecting electrode as collecting current I_c through diffusion and drift under the opposite electrical field. Based on the theory of Townsend discharge, the collecting current I_c, as part of the discharge current, was mainly determined by electrode separation and the first ionization coefficient α in the non-self-sustaining discharge at a given gas temperature as follows [28],

$$I_c = I_0 e^{\alpha d} \tag{1}$$

where I_0 is the initial current, d is the electrode separation between the CNT-based cathode and the extracting electrode, and $\alpha = APe^{-BP/E}$ is determined by the applied electric field E and partial pressure P of collision gases, because A and B are constants related to gas species and temperature. While E is proportional to the extracting voltage U_e at constant d, P is proportional to the gas concentration φ at constant gas temperature and volume. That is, the collecting current I_c is a function of the electrode separation d, the extracting voltage U_e, the gas species, and the gas concentration φ. Thus, the array composed of two sensors with different electrode separations d_1 and d_2 could simultaneously detect the NO and NO_2 concentration without component separation. The collecting currents I_{c1} and I_{c2} can be expressed as follow:

$$I_{c1} = f(\varphi_{NO}, \varphi_{NO_2}, d_1) \tag{2}$$

$$I_{c2} = f(\varphi_{NO}, \varphi_{NO_2}, d_2). \tag{3}$$

By measuring the collecting currents I_{c1} and I_{c2}, the gas concentrations φ_{NO} and φ_{NO_2} could be detected at a constant extracting voltage U_e and the given electrode separations. That is why we can use the sensor array to detect a gas mixture without any component separation.

The triple-electrode ionization sensors with 80 µm separations and 100 µm were are used to detect NO and NO_2, respectively (insert of Figure 3a,b). The collecting currents decreased monotonically with the increasing of the NO and NO_2 concentrations at a fixed U_c of 10 V and U_e of 150 V (Figure 3a,b). As the NO concentration increased from 0 to 1120 ppm, the collecting current decreased from 83.3 nA to 32.7 nA, and the sensor with 80 µm separations and 150 V extracting voltage exhibited the highest NO sensitivity of −0.11 nA/ppm to 300 ppm NO. For NO_2, the collecting current decreased from 77.3 nA to 8.7 nA with an increasing NO_2 concentration from 0 to 220 ppm, and the highest sensitivity was −0.49 nA/ppm to 70 ppm for the sensor with 100 µm separations at the 150 V extracting voltage. Here, the sensitivity S is defined as the ratio of the variation of the collecting current Δy and the variation of gas concentration Δx.

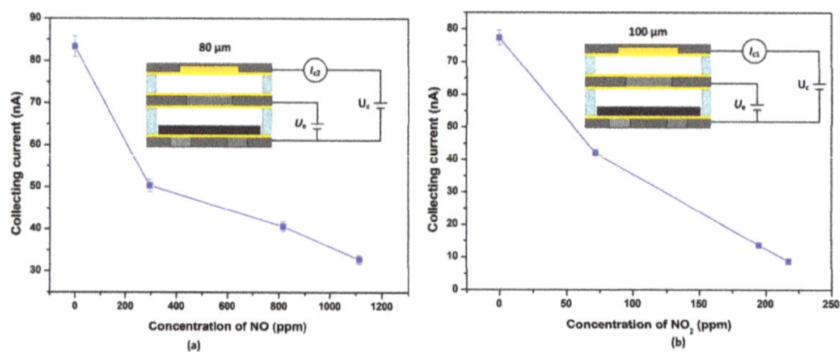

Figure 3. (a) NO response of the CNTs-based ionization sensor with 80 µm electrode separations and (b) NO$_2$ response of the sensor with 100 µm separations at U_e = 150 V and U_c = 10 V.

Cross-sensitivity is a key problem in the practical application of gas selective detecting. Figure 4a,b show the response of each sensor in the array to the mixture of NO and NO_2, respectively. The collecting current was reduced dramatically from 19.31 nA to 9.02 nA to 300 ppm NO with the NO_2 concentration increasing from 70 ppm to 220 ppm (Figure 4a). The average variation of the collecting current for each NO concentration was about 10 nA and as high as half of its response. In other words, the CNTs-based ionization sensor with 80 μm separations could detect the concentration of NO very well no matter how the NO_2 concentration in the gas mixture changed. While for NO_2, the minimum variation in the collecting current was also nearly 400 pA and occurred at 220 ppm when the concentration of NO in the gas mixture changed (Figure 4b). Accordingly, the ionization sensor with 100 μm separations still had the ability to detect NO_2 accurately in the gas mixture even though the effect of the NO concentration on the NO_2 response was weaker (insert in Figure 4b).

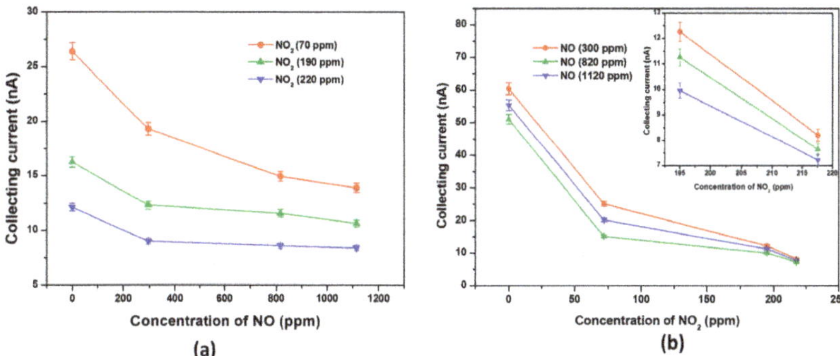

Figure 4. Simultaneous detection of NO and NO_2 in a gas mixture using a sensor array with 80 μm and 100 μm separations at 150 V U_e and 10 V U_c, respectively. The collecting currents decreased with increasing (**a**) NO and (**b**) NO_2 concentrations in dry air.

In addition, all the NO response curves show almost the same shape at different NO_2 concentrations (Figure 4a), and the same is true for NO_2 (Figure 4b). This indicates that each sensor had intrinsic gas sensing properties. Moreover, the array was thought to show a good repeatability without considering the variation of the collecting current caused by changes in the concentrations of interference components. This was attributed to the good stability of the sensor due to the long life of the CNTs.

4. Conclusions

The non-self-sustaining discharge of the triple-electrode CNTs-based sensor makes it possible to measure gas concentration with the non-self-sustaining discharge current with good stability. The array of these sensors shows a monotonically decreasing response to NO_X. The NO sensor with 80 μm separations and the NO_2 sensor with 100 μm separations exhibited the highest sensitivity −0.11 nA/ppm to 300 ppm NO and −0.49 nA/ppm to 70 ppm NO_2, respectively. The maximum variation in the collecting current of the NO sensor is about 10 nA, when the concentration of NO_2 in mixture is changed, while the minimum variation of the NO_2 sensor is also about 400 pA. Although NO_2 had a stronger effect on NO than that of NO on NO_2, the array of these two CNTs-based ionization sensors still has the ability to simultaneously detect the concentrations of NO_2 and NO in a gas mixture without component separation. In addition, the response shape of each sensor is almost the same. This indicates that the array has intrinsic gas sensing properties and good stability.

Author Contributions: H.S. wrote and revised the manuscript, K.L. organized and carried out experiments and C.W. is responsible for CNTs growth and sensors fabrication.

Funding: This research received no external funding.

Acknowledgments: Metal sputtering and the growth of CNTs were done in the Institute of Vacuum Microelectronics & microelectromechanical System, Xi'an Jiaotong University, Xi'an, China. The authors also thank Chang Wang for his help with the growth of CNTs and the senior lab manger, Kun Li, for his experimental guidance.

Conflicts of Interest: The authors declare no conflicts of interest.

References

1. Fine, G.F.; Cavanagh, L.M.; Afonja, A.; Binions, R. Metal oxide semi-conductor gas sensors in environmental monitoring. *Sensors* **2010**, *10*, 5469–5502. [CrossRef] [PubMed]

2. Rheaume, J.M.; Pisano, A.P. A review of recent progress in sensing of gas concentration by impedance change. *Ionics* **2011**, *17*, 99–108. [CrossRef]

3. Afzal, A.; Cioffi, N.; Sabbatini, L.; Torsi, L. NO_x sensors based on semiconducting metal oxide nanostructures: Progress and perspectives. *Sens. Actuators B Chem.* **2012**, *171*, 25–42. [CrossRef]

4. Privett, B.J.; Shin, J.H.; Schoenfisch, M.H. Electrochemical nitric oxide sensors for physiological measurements. *Chem. Soc. Rev.* **2010**, *39*, 1925–1935. [CrossRef] [PubMed]

5. Macam, E.R.; Blackburn, B.M.; Wachsman, E.D. The effect of La_2CuO_4 sensing electrode thickness on a potentiometric NO_x sensor response. *Sens. Actuator B-Chem.* **2011**, *157*, 353–360. [CrossRef]

6. Yao, F.; Dinh, L.D.; Lim, S.C.; Yang, S.B.; Hwang, H.R.; Yu, W.J.; Lee, I.H.; Gunes, F.; Lee, Y.H. Humidity-assisted selective reactivity between NO_2 and SO_2 gas on carbon nanotubes. *J. Mater. Chem.* **2011**, *21*, 4502–4508. [CrossRef]

7. Lu, R.J.; Shi, K.Y.; Zhou, W.; Wang, L.; Tian, C.G.; Pan, K.; Sun, L.; Fu, H.G. Highly dispersed Ni-decorated porous hollow carbon nanofibers: Fabrication, characterization, and NO_x gas sensors at room temperature. *J. Mater. Chem.* **2012**, *22*, 24814–24820. [CrossRef]

8. Iqbal, N.; Afzal, A.; Cioffi, N.; Sabbatini, L.; Torsi, L. NO_x sensing one- and two-dimensional carbon nanostructures and nanohybrids: Progress and perspectives. *Sens. Actuators B Chem.* **2013**, *181*, 9–21. [CrossRef]

9. Yun, J.H.; Kim, J.; Park, Y.C.; Song, J.W.; Shin, D.H.; Han, C.S. Highly sensitive carbon nanotube-embedding gas sensors operating at atmospheric pressure. *Nanotechnology* **2009**, *20*, 055503. [CrossRef] [PubMed]

10. Penza, M.; Rossi, R.; Alvisi, M.; Serra, E. Metal-modified and vertically aligned carbon nanotube sensors array for landfill gas monitoring applications. *Nanotechnology* **2010**, *21*, 105501. [CrossRef] [PubMed]

11. Mangu, R.; Rajaputra, S.; Singh, V.P. Mwcnt-polymer composites as highly sensitive and selective room temperature gas sensors. *Nanotechnology* **2011**, *22*, 215502. [CrossRef] [PubMed]

12. Vuong, N.M.; Jung, H.; Kim, D.; Kim, H.; Hong, S.K. Realization of an open space ensemble for nanowires: A strategy for the maximum response in resistive sensors. *J. Mater. Chem.* **2012**, *22*, 6716–6725. [CrossRef]

13. Llobet, E. Gas sensors using carbon nanomaterials: A review. *Sens. Actuators B Chem.* **2013**, *179*, 32–45. [CrossRef]

14. Lucci, M.; Reale, A.; Di Carlo, A.; Orlanducci, S.; Tamburri, E.; Terranova, M.L.; Davoli, I.; Di Natale, C.; D'Amico, A.; Paolesse, R. Optimization of a NO_x gas sensor based on single walled carbon nanotubes. *Sens. Actuator B-Chem.* **2006**, *118*, 226–231. [CrossRef]

15. Ueda, T.; Bhulyan, M.M.H.; Norimatsu, H.; Katsuki, S.; Ikegami, T.; Mitsugi, F. Development of carbon nanotube-based gas sensors for NO_x gas detection working at low temperature. *Physica E-Low-Dimens. Syst. Nanostruct.* **2008**, *40*, 2272–2277. [CrossRef]

16. Jang, D.M.; Jung, H.; Hoa, N.D.; Kim, D.; Hong, S.K.; Kim, H. Tin oxide-carbon nanotube composite for NO_x sensing. *J. Nanosci. Nanotechnol.* **2012**, *12*, 1425–1428. [CrossRef] [PubMed]

17. Yavari, F.; Castillo, E.; Gullapalli, H.; Ajayan, P.M.; Koratkar, N. High sensitivity detection of NO_2 and NH_3 in air using chemical vapor deposition grown graphene. *Appl. Phys. Lett.* **2012**, *100*, 203120. [CrossRef]

18. Boyd, A.; Dube, I.; Fedorov, G.; Paranjape, M.; Barbara, P. Gas sensing mechanism of carbon nanotubes: From single tubes to high-density networks. *Carbon* **2014**, *69*, 417–423. [CrossRef]

19. Dube, I.; Jiménez, D.; Fedorov, G.; Boyd, A.; Gayduchenko, I.; Paranjape, M.; Barbara, P. Understanding the electrical response and sensing mechanism of carbon-nanotube-based gas sensors. *Carbon* **2015**, *87*, 330–337. [CrossRef]

20. Shengbing, C.; Yong, Z.; Zhemin, D. Fabrication of gas sensor based on field ionization from SWCNTs with tripolar microelectrode. *J. Micromech. Microeng.* **2012**, *22*, 125017.

21. Zhang, Y.; Li, S.; Zhang, J.; Pan, Z.; Min, D.; Li, X.; Song, X.; Liu, J. High-performance gas sensors with temperature measurement. *Sci. Rep.* **2013**, *3*, 1267. [CrossRef] [PubMed]

22. Raizer, Y.P. *Gas Discharge Physics*; Springer: Berlin/Heidelberg, Germany, 1991.

23. Li, X.; Zhao, D.; Pang, K.; Pang, J.; Liu, W.; Liu, H.; Wang, X. Carbon nanotube cathode with capping carbon nanosheet. *Appl. Surf. Sci.* **2013**, *283*, 740–743. [CrossRef]

24. Sadeghian, R.B.; Kahrizi, M. A novel gas sensor based on tunneling-field-ionization on whisker-covered gold nanowires. *IEEE Sens. J.* **2008**, *8*, 161–169. [CrossRef]

25. Sadeghian, R.B.; Kahrizi, M. A novel miniature gas ionization sensor based on freestanding gold nanowires. *Sens. Actuators A Phys.* **2007**, *137*, 248–255. [CrossRef]

26. Nikfarjam, A.; Zad, A.I.; Razi, F.; Mortazavi, S.Z. Fabrication of gas ionization sensor using carbon nanotube arrays grown on porous silicon substrate. *Sens. Actuators A Phys.* **2010**, *162*, 24–28. [CrossRef]

27. Huang, J.; Wang, J.; Gu, C.; Yu, K.; Meng, F.; Liu, J. A novel highly sensitive gas ionization sensor for ammonia detection. *Sens. Actuators A Phys.* **2009**, *150*, 218–223. [CrossRef]

28. Raizer, Y.P.; Allen, J.E. *Gas Discharge Physics*; Springer: Berlin, Germany, 1997; Volume 2.

MDPI
St. Alban-Anlage 66
4052 Basel
Switzerland
Tel. +41 61 683 77 34
Fax +41 61 302 89 18
www.mdpi.com

Micromachines Editorial Office
E-mail: micromachines@mdpi.com
www.mdpi.com/journal/micromachines